普通高等教育"十一五"国家级规划教材　　高等学校Java课程系列教材

JSP 实用教程

（第4版）微课视频版

◎ 耿祥义　张跃平　主编

清华大学出版社
北京

内 容 简 介

JSP是一项动态Web技术标准,利用该Web技术可以建立安全、跨平台的先进的Web动态网站。本书叙述详细、通俗易懂、便于自学,不仅注重结合实例讲解难点和关键技术,而且特别注重培养在Web设计中正确使用MVC模式的能力。全书共分10章,内容包括JSP简介、JSP语法、Tag文件与Tag标记、内置对象、JSP与JavaBean、Java Servlet、MVC设计模式、数据库操作、文件操作以及"手机销售网"案例设计等。

本书所有知识点都结合具体实例进行介绍,力求详略得当,突出JSP在开发Web动态网站方面的强大功能,使读者快速掌握JSP的编程技巧。本书配有总计20小时的微课视频,扫描书中相应位置的二维码,可以在线学习。

本书可以作为计算机及相关专业的教材,也适合自学者及网站开发人员参考。

本书封面贴有清华大学出版社防伪标签,无标签者不得销售。
版权所有,侵权必究。 举报:010-62782989,beiqinquan@tup.tsinghua.edu.cn。

图书在版编目(CIP)数据

JSP实用教程:微课视频版/耿祥义,张跃平主编. —4版. —北京:清华大学出版社,2020.7(2024.12重印)
高等学校Java课程系列教材
ISBN 978-7-302-55980-1

Ⅰ. ①J… Ⅱ. ①耿… ②张… Ⅲ. ①JAVA语言-网页制作工具-高等学校-教材 Ⅳ. ①TP312.8 ②TP393.092.2

中国版本图书馆CIP数据核字(2020)第120370号

策划编辑:魏江江
责任编辑:王冰飞
封面设计:刘　键
责任校对:徐俊伟
责任印制:宋　林

出版发行:清华大学出版社
　　　网　　　址:https://www.tup.com.cn,https://www.wqxuetang.com
　　　地　　　址:北京清华大学学研大厦A座　　　邮　　编:100084
　　　社 总 机:010-83470000　　　邮　　购:010-62786544
　　　投稿与读者服务:010-62776969,c-service@tup.tsinghua.edu.cn
　　　质量反馈:010-62772015,zhiliang@tup.tsinghua.edu.cn
　　　课件下载:https://www.tup.com.cn,010-83470236
印 装 者:北京嘉实印刷有限公司
经　　销:全国新华书店
开　　本:185mm×260mm　　　印　张:22.5　　　字　数:576千字
版　　次:2003年5月第1版　2020年8月第4版　　　印　次:2024年12月第14次印刷
印　　数:199001~205000
定　　价:59.80元

产品编号:087694-01

前　言

党的二十大报告中指出：教育、科技、人才是全面建设社会主义现代化国家的基础性、战略性支撑。必须坚持科技是第一生产力、人才是第一资源、创新是第一动力，深入实施科教兴国战略、人才强国战略、创新驱动发展战略，这三大战略共同服务于创新型国家的建设。高等教育与经济社会发展紧密相连，对促进就业创业、助力经济社会发展、增进人民福祉具有重要意义。

本书是《JSP 实用教程》的第 4 版，继续保留原有的特点——注重教材的可读性和实用性，许多例题都经过精心的选择，既能帮助理解知识，又具有启发性。为了突出 MVC 在 Web 开发中的重要性，本版在内容结构上做了进一步的优化调整，使得在适合教学的基础上，更加突出实用性。本版中许多例子不仅注重实用性，而且特别注重培养读者在 Web 设计中使用 MVC 模式的能力。另外，本版的第 1~9 章配备了有针对性的上机实验内容，这些实验在内容上不仅注重趣味性，也更加注重实用性，对于巩固知识和扩展能力是非常有帮助的。这些实验注重实验步骤要求，指导学生按步骤完成实验，有利于提升学生的学习效果和 Web 设计能力。

全书共分 10 章。第 1 章介绍 JSP 重要性，对 Tomcat 服务器的安装与配置进行详细介绍。第 2 章讲解 JSP 页面的基本构成、常用的 JSP 标记。第 3 章讲解 Tag 文件与标记，特别重点强调怎样使用 Tag 文件实现代码复用。第 4 章讲解内置对象，特别强调这些内置对象在 JSP 应用开发中的重要性，结合实例使读者掌握内置对象的用法。第 5 章是 JSP 技术中很重要的内容，即怎样使用 Javabean 分离数据的显示和存储，这一章讲解了许多有一定应用价值的例子。第 6 章讲解 Servlet，对 Servlet 对象的运行原理给予细致的讲解，许多例子对于理解和掌握使用 Servlet 都是非常有帮助的。第 7 章讲解 MVC 开发模式，对 JSP 页面、bean 以及 Servlet 在 MVC 开发模式中的作用进行了重点介绍，并按照 MVC 模式给出了易于理解 MVC 设计模式的例子。第 8 章讲解数据库，也是 Web 应用开发的非常重要的一部分内容，采用 MySQL 数据库讲解主要知识点，讲解怎样使用数据库连接池技术实现数据库的操作，许多例子都是 Web 开发中经常使用的模块。第 9 章主要讲解怎样使用 Java 中的输入、输出流实现文件的读写操作，在实例上特别强调使用 MVC 模式实现文件的读写操作。第 10 章是"手机销售网"的案例设计，完全按照 MVC 模式开发设计，其目的是让读者掌握一般 Web 应用中常用基本模块的开发方法。

本书提供丰富的配套资源，包括教学大纲、教学课件、程序源码、电子教案、习题答案、上机实验、综合案例，本书还配有 20 小时的微课视频。

资源下载提示

课件等资源：扫描封底的"课件下载"二维码，在公众号"书圈"下载。

素材（源码）等资源：扫描目录上方的二维码下载。

视频等资源：扫描封底刮刮卡中的二维码，再扫码书里章节中的二维码，可以在线学习。

在线作业：扫描封底刮刮卡中的二维码，可以登录在线作业平台。

对于32学时教学计划(不含上机实验),可以采用第1～2章、第4～8章和第10章的内容;对于48学时或更多学时教学计划(不含上机实验),可以采用全部内容。

关注作者教学辅助微信公众号java-violin可获得本书相关资源。

希望本书能对读者学习JSP有所帮助,并请读者批评指正。

编　者

2020 年 4 月

本书介绍

源码下载

目录

第 1 章　JSP 简介

- 1.1　什么是 JSP 1
- 1.2　安装配置 JSP 运行环境 1
- 1.3　JSP 页面 6
  - 1.3.1　JSP 页面简介 6
  - 1.3.2　设置 Web 服务目录 7
- 1.4　JSP 运行原理 8
- 1.5　JSP 与 Java Servlet 的关系 10
- 1.6　HTML 与 JavaScript 11
  - 1.6.1　HTML 11
  - 1.6.2　JavaScript 12
- 1.7　上机实验 12
  - 1.7.1　输出英文字母表 13
  - 1.7.2　输出九九口诀表 14
  - 1.7.3　输出成绩单 15
- 1.8　小结 16
- 习题 1 17

第 2 章　JSP 语法

- 2.1　JSP 页面的基本结构 18
- 2.2　声明变量和定义方法 20
- 2.3　Java 程序片 22
- 2.4　Java 表达式 25
- 2.5　JSP 中的注释 26
- 2.6　JSP 指令标记 26
  - 2.6.1　page 指令标记 26
  - 2.6.2　include 指令标记 31
- 2.7　JSP 动作标记 33
  - 2.7.1　include 动作标记 33
  - 2.7.2　param 动作标记 34

2.7.3　forward 动作标记 ··· 35
　　　2.7.4　useBean 动作标记 ··· 37
2.8　上机实验 ·· 37
　　　2.8.1　实验 1　消费总和 ··· 37
　　　2.8.2　实验 2　日期时间 ··· 39
　　　2.8.3　实验 3　听英语 ·· 39
　　　2.8.4　实验 4　看电影 ·· 41
2.9　小结 ··· 44
习题 2 ·· 44

第 3 章　Tag 文件与 Tag 标记

3.1　Tag 文件 ··· 46
　　　3.1.1　Tag 文件的结构 ·· 46
　　　3.1.2　Tag 文件的保存 ·· 47
3.2　Tag 标记 ··· 48
　　　3.2.1　Tag 标记与 Tag 文件 ··· 48
　　　3.2.2　Tag 标记的使用 ·· 48
　　　3.2.3　Tag 标记的运行原理 ··· 49
3.3　Tag 文件中的常用指令 ··· 49
　　　3.3.1　tag 指令 ·· 49
　　　3.3.2　include 指令 ··· 50
　　　3.3.3　attribute 指令 ··· 50
　　　3.3.4　variable 指令 ··· 53
　　　3.3.5　taglib 指令 ·· 55
3.4　上机实验 ··· 55
　　　3.4.1　实验 1　解析单词 ··· 56
　　　3.4.2　实验 2　显示日历 ··· 57
习题 3 ·· 59

第 4 章　JSP 内置对象

4.1　request 对象 ··· 61
　　　4.1.1　获取用户提交的信息 ·· 62
　　　4.1.2　处理汉字信息 ·· 64
　　　4.1.3　常用方法举例 ·· 65
　　　4.1.4　处理 HTML 标记 ··· 67
　　　4.1.5　处理超链接 ·· 78
4.2　response 对象 ··· 79
　　　4.2.1　动态响应 contentType 属性 ······································· 79

		4.2.2 response 对象的 HTTP 文件头	81
		4.2.3 response 对象的重定向	82
	4.3	session 对象	83
		4.3.1 session 对象的 id	84
		4.3.2 session 对象与 URL 重写	85
		4.3.3 session 对象存储数据	86
		4.3.4 session 对象的生存期限	89
	4.4	application 对象	90
		4.4.1 application 对象的常用方法	90
		4.4.2 application 留言板	91
	4.5	out 对象	94
	4.6	上机实验	94
		4.6.1 实验 1 196 算法之谜	95
		4.6.2 实验 2 计算器	97
		4.6.3 实验 3 单词的频率	98
		4.6.4 实验 4 成绩与饼图	100
		4.6.5 实验 5 记忆测试	104
	4.7	小结	107
	习题 4		108

第 5 章　JSP 与 JavaBean

	5.1	编写和使用 JavaBean	110
		5.1.1 编写 JavaBean	110
		5.1.2 保存 bean 的字节码	111
		5.1.3 创建与使用 bean	111
	5.2	获取和修改 bean 的属性	115
		5.2.1 getProperty 动作标记	115
		5.2.2 setProperty 动作标记	116
	5.3	bean 的辅助类	119
	5.4	JSP 与 bean 结合的简单例子	121
		5.4.1 三角形 bean	121
		5.4.2 四则运算 bean	123
		5.4.3 浏览图像 bean	125
		5.4.4 日历 bean	126
		5.4.5 计数器 bean	129
	5.5	上机实验	131
		5.5.1 实验 1 小数表示为分数	131
		5.5.2 实验 2 记忆测试	133
		5.5.3 实验 3 成语接龙	137

5.6 小结 ········ 139
习题 5 ········ 140

第 6 章　Java Servlet 基础

6.1 servlet 的部署、创建与运行 ········ 142
　　6.1.1 源文件及字节码文件 ········ 142
　　6.1.2 编写部署文件 web.xml ········ 143
　　6.1.3 servlet 的创建与运行 ········ 145
　　6.1.4 向 servlet 传递参数的值 ········ 146
6.2 servlet 的工作原理 ········ 147
　　6.2.1 servlet 对象的生命周期 ········ 147
　　6.2.2 init 方法 ········ 148
　　6.2.3 service 方法 ········ 148
　　6.2.4 destroy 方法 ········ 148
6.3 通过 JSP 页面访问 servlet ········ 149
6.4 共享变量 ········ 154
6.5 doGet 和 doPost 方法 ········ 156
6.6 重定向与转发 ········ 159
6.7 使用 session ········ 164
6.8 上机实验 ········ 167
　　6.8.1 实验 1　绘制多边形数 ········ 167
　　6.8.2 实验 2　双色球福利彩票 ········ 170
　　6.8.3 实验 3　分析整数 ········ 174
6.9 小结 ········ 178
习题 6 ········ 179

第 7 章　MVC 模式

7.1 MVC 模式介绍 ········ 181
7.2 JSP 中的 MVC 模式 ········ 181
7.3 模型的生命周期与视图更新 ········ 182
　　7.3.1 request bean ········ 182
　　7.3.2 session bean ········ 183
　　7.3.3 application bean ········ 184
7.4 MVC 模式的简单实例 ········ 185
　　7.4.1 简单的计算器 ········ 185
　　7.4.2 表白墙 ········ 188
7.5 上机实验 ········ 195
　　7.5.1 实验 1　等差、等比级数和 ········ 195

7.5.2　实验 2　点餐 ………………………………………………………………… 199
7.6　小结 ………………………………………………………………………………… 204
习题 7 …………………………………………………………………………………… 204

第 8 章　JSP 中使用数据库

8.1　MySQL 数据库管理系统 ………………………………………………………… 207
　　8.1.1　下载、安装 MySQL ……………………………………………………… 207
　　8.1.2　启动 MySQL 数据库服务器 ……………………………………………… 208
　　8.1.3　MySQL 客户端管理工具 ………………………………………………… 210
8.2　连接 MySQL 数据库 ……………………………………………………………… 214
8.3　查询记录 …………………………………………………………………………… 218
　　8.3.1　结果集与查询 ……………………………………………………………… 218
　　8.3.2　随机查询 …………………………………………………………………… 219
　　8.3.3　条件查询 …………………………………………………………………… 222
8.4　更新、添加与删除记录 …………………………………………………………… 223
8.5　用结果集操作数据库中的表 ……………………………………………………… 225
　　8.5.1　更新记录 …………………………………………………………………… 225
　　8.5.2　插入记录 …………………………………………………………………… 226
8.6　预处理语句 ………………………………………………………………………… 228
　　8.6.1　预处理语句的优点 ………………………………………………………… 228
　　8.6.2　使用通配符 ………………………………………………………………… 230
8.7　事务 ………………………………………………………………………………… 232
8.8　分页显示记录 ……………………………………………………………………… 235
8.9　连接 SQL Server 与 Access ……………………………………………………… 242
　　8.9.1　连接 Microsoft SQL Server 数据库 ……………………………………… 242
　　8.9.2　连接 Microsoft Access 数据库 …………………………………………… 244
8.10　使用连接池 ………………………………………………………………………… 246
　　8.10.1　连接池简介 ……………………………………………………………… 246
　　8.10.2　建立连接池 ……………………………………………………………… 246
8.11　标准化考试训练 …………………………………………………………………… 250
　　8.11.1　功能概述 ………………………………………………………………… 250
　　8.11.2　数据库设计 ……………………………………………………………… 250
　　8.11.3　Web 应用设计 …………………………………………………………… 252
8.12　上机实验 …………………………………………………………………………… 260
　　8.12.1　实验 1　查询成绩 ……………………………………………………… 260
　　8.12.2　实验 2　管理学生成绩 ………………………………………………… 263
　　8.12.3　实验 3　小星星广告网 ………………………………………………… 266
8.13　小结 ………………………………………………………………………………… 284
习题 8 …………………………………………………………………………………… 284

# 第 9 章　JSP 中的文件操作

- 9.1 File 类 …………………………………………………………… 285
- 9.2 RandomAccessFile 类 …………………………………… 287
- 9.3 文件上传 ………………………………………………………… 290
- 9.4 文件下载 ………………………………………………………… 296
- 9.5 上机实验 ………………………………………………………… 298
    - 9.5.1 实验 1　查看 JSP 源文件 ……………………… 298
    - 9.5.2 实验 2　听学《新概念英语》 ………………… 300
- 9.6 小结 ……………………………………………………………… 304
- 习题 9 ………………………………………………………………… 304

第 10 章　手机销售网

- 10.1 系统模块构成 ………………………………………………… 305
- 10.2 Web 目录结构 ………………………………………………… 305
- 10.3 数据库设计与连接 …………………………………………… 306
    - 10.3.1 数据库设计 ……………………………………… 306
    - 10.3.2 数据库连接 ……………………………………… 309
- 10.4 Web 应用模块管理 …………………………………………… 309
    - 10.4.1 页面管理 ………………………………………… 309
    - 10.4.2 bean 与 servlet 管理 …………………………… 311
    - 10.4.3 web.xml(部署文件) …………………………… 311
    - 10.4.4 图像管理 ………………………………………… 313
- 10.5 会员注册 ……………………………………………………… 313
    - 10.5.1 视图(JSP 页面) ………………………………… 313
    - 10.5.2 模型(bean) ……………………………………… 314
    - 10.5.3 控制器(servlet) ………………………………… 315
- 10.6 会员登录 ……………………………………………………… 318
    - 10.6.1 视图(JSP 页面) ………………………………… 318
    - 10.6.2 模型(bean) ……………………………………… 319
    - 10.6.3 控制器(servlet) ………………………………… 319
- 10.7 浏览手机 ……………………………………………………… 322
    - 10.7.1 视图(JSP 页面) ………………………………… 322
    - 10.7.2 模型(bean) ……………………………………… 328
    - 10.7.3 控制器(servlet) ………………………………… 329
- 10.8 查看购物车 …………………………………………………… 333
    - 10.8.1 视图(JSP 页面) ………………………………… 333
    - 10.8.2 模型(bean) ……………………………………… 335

	10.8.3 控制器(servlet)	335
10.9	查询手机	341
	10.9.1 视图(JSP 页面)	341
	10.9.2 模型(bean)	342
	10.9.3 控制器(servlet)	342
10.10	查询订单	344
	10.10.1 视图(JSP 页面)	344
	10.10.2 模型(bean)	346
	10.10.3 控制器(servlet)	346
10.11	退出登录	346

第 1 章　JSP简介

本章导读

　　主要内容
　　　❖ 什么是 JSP
　　　❖ 安装配置 JSP 运行环境
　　　❖ JSP 页面
　　　❖ JSP 运行原理
　　　❖ JSP 与 Servlet 的关系
　　　❖ HTML 与 JavaScript
　　难点
　　　❖ JSP 运行原理
　　关键实践
　　　❖ 输出九九口诀表

1.1　什么是 JSP

视频讲解

　　JSP 是 Java Server Page 的缩写,是由 Sun 公司倡导,许多公司参与,于 1999 年推出的一种 Web 服务设计标准。JSP 是基于 Java Servlet 以及整个 Java 体系的 Web 开发技术,利用这一技术可以建立安全、跨平台的先进动态网站。JSP 以 Java 语言为基础,具有动态页面与静态页面分离、能够脱离硬件平台的束缚以及编译后运行等优点,已经成为开发动态网站的主流技术之一。需要强调的一点是:要想真正地掌握 JSP 技术,必须有较好的 Java 语言基础,以及基本的 HTML 语言方面的知识。

1.2　安装配置 JSP 运行环境

视频讲解

　　❶ 下载 Tomcat

　　网络应用中最常见的模式是 B/S 模式,即需要获取信息的用户使用浏览器向服务器发出请求,服务器对此作出响应,将有关信息发送给用户的浏览器。在 B/S 模式中,服务器上必须有所谓的 Web 应用程序,服务器通过运行这些 Web 应用程序来响应用户的请求。因此,基于 B/S 模式的网络程序的核心就是设计服务器端的 Web 应用程序。

　　一个服务器上可以有很多基于 JSP 的 Web 应用程序,以满足各种用户的需求。这些 Web 应用程序必须由一个软件来统一管理和运行,这样的软件被称作 JSP 引擎或 JSP 容器,而安装 JSP 引擎的计算机被称作一个支持 JSP 的 Web 服务器。支持 JSP 的 Web 服务器负责运行 JSP,并将运行结果返回给用户。有关 JSP 的运行原理将在 1.4 节讲解。

　　1999 年 10 月 Sun 公司将 Java Server Page 1.1 代码交给 Apache 组织,Apache 组织对

JSP进行了实用性研究,并将这个服务器项目称为Tomcat。从此,著名的Web服务器Apache开始支持JSP。这样,Jakarta—Tomcat就诞生了(Jakarta是JSP项目的最初名称)。目前,Tomcat能和大部分主流服务器一起高效率地工作。Tomcat是一个免费的开源JSP引擎,也称作Tomcat服务器。

读者可以登录http://tomcat.apache.org免费下载Tomcat。登录之后,首先在Download里选择要下载的Tomcat的版本号,这里我们选择的是Tomcat9,然后在Binary Distributions列出的下载选项中选择一项,对于Windows系统可以选择zip压缩格式的安装版本(这里选择这个版本,即apache-tomcat-9.0.26.zip,如图1.1所示),Linux系统可选择tar.gz压缩格式版本。

注：也可以扫描图1.2中的二维码,下载apache-tomcat-9.0.26.zip。

图1.1 下载Tomcat

图1.2 下载Tomcat的二维码

❷ 安装JDK

安装Tomcat之前必须首先安装JDK(版本不低于Java SE 8)。登录Oracle公司网站：

https://www.oracle.com/technetwork/java/javase/downloads/index.html

进入选择下载JDK的页面(如图1.3所示),这里下载的是jdk-13_windows-x64_bin.zip(Java SE 13,网站显示的最新版本)。如果使用其他的操作系统,可以下载相应的JDK。

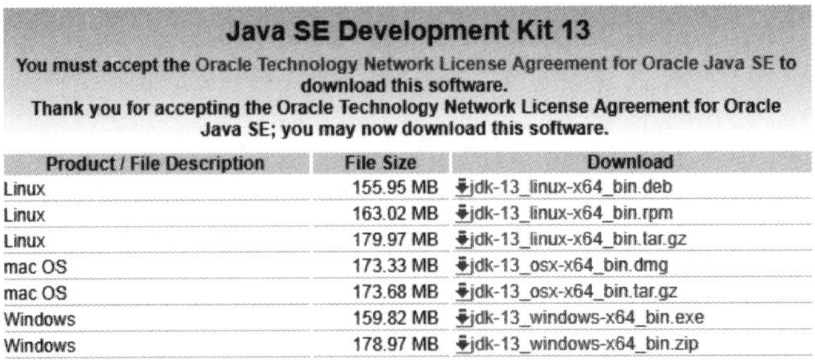

图1.3 下载JDK

注：也可以扫描图1.4中的二维码，下载jdk-13_windows-x64_bin.zip。

将下载的jdk-13_windows-x64_bin.zip解压到D:\磁盘，形成如图1.5所示的目录结构，其中D:\jdk-13为默认的安装目录，用户可以将这个目录重命名为自己喜欢的名字，这里使用默认的目录：D:\jdk-13。

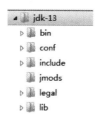

图1.4　下载JDK的二维码　　　　图1.5　JDK的安装目录

JDK本身包含了Java运行环境（Java Runtime Environment，JRE），该环境由Java虚拟机（Java Virtual Machine，JVM）、类库以及一些核心文件组成。JDK 9以及之后的版本将Java虚拟机（Java Virtual Machine，JVM）、类库以及一些核心文件分别存放在JDK根目录的\bin子目录中和\lib子目录中（不再有单独的JRE目录）。

❸ 设置系统环境变量

1）系统环境变量JAVA_HOME

右击"我的电脑"/"计算机"，在弹出的快捷菜单中选择"属性"命令，弹出"系统特性"对话框，再单击该对话框中的"高级属性设置"，然后单击"环境变量"按钮，添加系统环境变量JAVA_HOME（不分大小写），让该系统环境变量的值是安装JDK后的根目录，例如D:\jdk-13，如图1.6所示。Tomcat通过当前机器设置的系统变量JAVA_HOME的值来寻找所需要的JDK。

图1.6　设置系统变量JAVA_HOME

2）系统环境变量path

JDK提供的Java编译器（javac.exe）和Java解释器（java.exe）位于JDK根目录的\bin文件夹中，为了能在任何目录中使用编译器和解释器，应在系统中设置path。系统环境变量path在安装操作系统后就已经有了，所以不需要再添加path，只需要为其增加新的取值（path可以有多个值）。对于Windows 7/Windows XP，右击"计算机"/"我的电脑"，在弹出的快捷菜单中选择"属性"命令，弹出"系统"对话框，再单击该对话框中的"高级系统设置"/"高级选项"，然后单击"环境变量"按钮，弹出"环境变量设置"对话框，在该对话框中的"系统变量（S）"中找到path，单击"编辑（I）"按钮，弹出"编辑系统变量"对话框（如图1.6）。在"编辑系统变量"对话框中为path添加的新值是%JAVA_HOME%\bin（因为Java编译器（javac.exe）和Java解释器（java.exe）位于\bin中）。由于已经设置了系统环境变量JAVA_HOME的值是D:\jdk-13，因此可以用%JAVA_HOME%代替D:\jdk-13（%系统环境变量%是该系统环境变量的全部取值)。对于Windows 7系统,在编辑系统环境变量path的界面里,path的两个值之间必须用分号分隔(如图1.7所示)。对于Windows 10系统,在编辑系统环境变量path的界面里,单击"新建"按钮,就可以为path增加新的值,path每个值独占一行,因此不需要用分号分隔。

❹ 安装与启动Tomcat服务器

将安装了Tomcat的计算机称作一个Tomcat服务器。

1) 安装Tomcat服务器

将下载的apache-tomcat-9.0.26.zip解压到磁盘某个分区,比如解压到D:\,会产生名字是apache-tomcat-9.0.26的目录,出现如图1.8所示的目录结构。

图1.7 编辑path

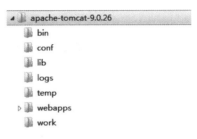

图1.8 安装Tomcat服务器

2) 启动Tomcat服务器

执行Tomcat安装根目录中bin文件夹中的startup.bat,启动Tomcat服务器。上述操作会占用一个MS-DOS窗口(如图1.9所示的界面),如果关闭当前MS-DOS窗口,将关闭Tomcat服务器。使用startup.bat启动Tomcat服务器,Tomcat服务器将使用JAVA_HOME环境变量设置的JDK。

图1.9 启动Tomcat服务器

注:对于Tomcat9版本,启动Tomcat服务器,控制台会出现乱码现象(不影响功能),如果不希望出现乱码,可以修改D:\apache-tomcat-9.0.26\conf下的logging.properties文件,用记事本打开该文件,找到其中的:

java.util.logging.ConsoleHandler.encoding = UTF - 8

将其修改为:

java.util.logging.ConsoleHandler.encoding = GB2312

或

```
java.util.logging.ConsoleHandler.encoding = GBK
```

保存修改后的 logging.properties 文件,然后重新运行 startup.bat。

3) 测试 Tomcat 服务器。

在浏览器的地址栏中输入 http://localhost:8080 或 http://127.0.0.1:8080,会出现如图 1.10 所示的 Tomcat 服务器的测试页面。

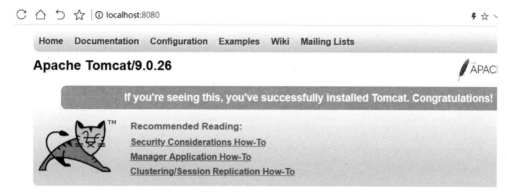

图 1.10 测试 Tomcat 服务器

注:Tomcat 服务器默认占用 8080 端口,如果 Tomcat 所使用的端口已经被占用,Tomcat 服务器将无法启动。有关端口的配置稍后讲解。

4) 配置端口

8080 是 Tomcat 服务器默认占用的端口,可以通过修改 Tomcat 服务器安装目录中 conf 文件下的主配置文件 server.xml 来更改端口。用记事本打开 server.xml 文件,找到出现:

```
<Connector port = "8080" protocol = "HTTP/1.1"
          connectionTimeout = "20000"
          redirectPort = "8443" />
```

的部分,将其中的 port="8080" 更改为新的端口(比如将 8080 更改为 9090 等),并重新启动 Tomcat 服务器即可。如果 Tomcat 服务器所在的计算机没有启动占用 80 端口的其他网络程序,也可以将 Tomcat 服务器的端口设置成 80,这样用户在访问 Tomcat 服务器时可以省略端口,例如:

```
http://127.0.0.1/
```

5) 系统环境变量 CATALINA_HOME

使用 startup.bat 启动 Tomcat 服务器时,如果发现当前操作系统没有设置系统环境变量 CATALINA_HOME 的值,在启动 Tomcat 服务器的过程中会设置 CATALINA_HOME 的值,但不保证所有的操作系统都允许 startup.bat 这样做。因此,官方推荐在正式启动 Tomcat 服务器之前,也设置 CATALINA_HOME 值,即添加系统环境变量 CATALINA_HOME(不分大小写),让该系统环境变量的值是 Tomcat 服务器的根目录,比如 D:\apache-tomcat-9.0.26。

1.3　JSP 页面

视频讲解

▶ 1.3.1　JSP 页面简介

将在第 2 章详细讲解 JSP 页面的结构及有关语法，本节简单了解 JSP 页面即可。简单地说，一个 JSP 页面中可以有普通的 HTML 标记和 JSP 规定的 JSP 标记，以及通过标记符号"<%""%>"之间加入的 Java 程序片。例 1_1 是一个简单的 JSP 页面。

例 1_1

example1_1.jsp（效果如图 1.11 所示）

```
<%@ page contentType = "text/html" %>
<%@ page pageEncoding = "utf-8" %>
<HTML><body bgcolor = pink>
<h1>这是一个简单的 JSP 页面</h1>
<% int i, sum = 0;
    for(i = 1;i <= 100;i++){
        sum = sum + i;
    }
%>
<p style = "font-family:宋体;font-size:36;color:blue">
1 到 100 的连续和是:<% = sum %>
</p></body></HTML>
```

这是一个简单的JSP页面

1到100的连续和是:5050

图 1.11　简单的页面

（1）JSP 页面的编码。

指定 JSP 页面的编码。将 JSP 页面的编码设置成 UTF-8 是为了 JSP 页面能适用于更多的平台，因此 JSP 页面中使用了如下 page 指令（有关细节在 2.6.1 节讨论）：

```
<%@ page pageEncoding = "utf-8" %>
```

指定了 JSP 文件的编码。

（2）JSP 页面的保存。

JSP 页面按文本文件保存，扩展名是.jsp。比如，用文本编辑器"记事本"编辑 JSP 页面，在保存 JSP 页面时，将"保存类型（T）"选择为"所有文件（*.*）"，将"编码（E）"选择为"UTF-8"（因为 JSP 页面指定的编码是 UTF-8），如图 1.12 所示。

在保存 JSP 页面时，文件的名字遵守标识符规定：名字可以由字母、下画线、美元符号和数字组成，并且第一个字符不能是数字。需要注意的是，JSP 技术基于 Java 语言，名字区分大小写，比如，Example.jsp 和 example.jsp 是不相同的 JSP 文件。

尽管有一些编辑器是专门用来编写 JSP 页面的，但对于 JSP 的学习阶段，建议使用纯文

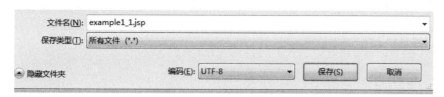

图 1.12 JSP 页面的保存

本编辑器来编辑 JSP 页面(本书用 Windows 的"记事本"作为编辑器)。有时为了明显地区分普通的 HTML 标记和 Java 程序片以及 JSP 标记,可以用大写字母书写普通的 HTML 标记(这不是必需的,HTML 标记不区分大小写)。

▶ 1.3.2 设置 Web 服务目录

必须将编写好的 JSP 页面文件保存到 Tomcat 服务器的某个 Web 服务目录中,只有这样,远程的用户才可以通过浏览器访问该 Tomcat 服务器上的 JSP 页面。人们常说的一个网站,实际上就是一个 Web 服务目录。

❶ 根目录

如果 Tomcat 服务器的安装目录是 D:\apache-tomcat-9.0.26,那么 Tomcat 的 Web 服务目录的根目录是 D:\ apache-tomcat-9.0.26\webapps\root。

用户如果准备访问根目录中的 JSP 页面,那么访问 JSP 页面的 URL 格式是:

```
http://Tomcat 服务器的 IP 地址(或域名):端口/JSP 页面的名字
```

必须省略 Web 根目录的名字 root。假设根目录中存放的 JSP 页面的名字是 example1_1.jsp,那么用户可以在浏览器地址栏输入:

```
http://192.168.1.100:8080/example1_1.jsp
```

来访问 example1_1.jsp 页面。也许你没有为 Tomcat 服务器所在的机器设置过一个有效的 IP 地址,那么为了调试 JSP 页面,你可以打开 Tomcat 服务器所在机器上的浏览器,在浏览器的地址栏中输入:

```
http://127.0.0.1:8080/example1_1.jsp
```

❷ webapps 下的 Web 服务目录

Tomcat 服务器安装目录(比如 D:\apache-tomcat-9.0.26)下的 webapps 目录下的任何一个子目录都可以作为一个 Web 服务目录。如果将 JSP 页面文件保存到 webapps 下的 Web 服务目录中,那么访问 JSP 页面的 URL 格式是:

```
http://Tomcat 服务器的 IP 地址(或域名):端口/Web 服务目录/JSP 页面的名字
```

比如,在 webapps 下新建子目录 ch1,那么 ch1 目录就成为一个 Web 服务目录。如果将 JSP 页面文件 example1_1.jsp 保存在 ch1 目录中,那么访问 JSP 页面 example1_1.jsp 的 URL 格式是:

```
http://127.0.0.1:8080/ch1/example1_1.jsp
```

在浏览器地址栏中输入 http://127.0.0.1:8080/ch1/example1_1.jsp 后的效果如图 1.11 所示。

注：在 webapps 下新建 Web 服务目录，不必重新启动 Tomcat 服务器。

❸ 新建 Web 服务目录

可以将 Tomcat 服务器所在计算机的某个目录（非 webapps 下的子目录）设置成一个 Web 服务目录，并为该 Web 服务目录指定虚拟目录，即隐藏 Web 服务目录的实际位置，用户只能通过虚拟目录访问 Web 服务目录中的 JSP 页面。

可以通过修改 Tomcat 服务器安装目录下 conf 文件夹中的 server.xml 文件来设置新的 Web 服务目录。假设要将 D:\Book\zh 以及 C:\wang 作为 Web 服务目录，并让用户分别使用 apple 和 cloud 虚拟目录访问 Web 服务目录 D:\Book\zh 和 C:\wang 下的 JSP 页面，首先用记事本打开 conf 文件夹中的主配置文件 server.xml，找到出现 </Host> 的部分（接近 server.xml 文件尾部处），然后在 </Host> 的前面加入：

```
< Context path = "/apple" docBase = "D:\Book\zh" debug = "0" reloadable = "true"/>
< Context path = "/cloud" docBase = "C:\wang" debug = "0" reloadable = "true"/>
```

注：XML 文件是区分大小写的，不可以将 <Context> 写成 <context>。

主配置文件 server.xml 修改后，必须重新启动 Tomcat 服务器。重启后，用户就可以通过虚拟目录 apple 或 cloud 访问放到 D:\Book\zh 或 C:\wang 中的 JSP 页面。比如，将 example1_1.jsp 保存到 D:\Book\zh 或 C:\wang 中，在浏览器地址栏中输入：

```
http://127.0.0.1:8080/apple/example1_1.jsp
```

或

```
http://127.0.0.1:8080/cloud/example1_1.jsp
```

访问 example1_1.jsp 页面。

❹ 相对目录

Web 服务目录下的目录称为相对 Web 服务目录。比如，我们可以在 Web 服务目录 D:\Book\zh 下再建立一个子目录 image，将 example1_1.jsp 文件保存到 image 中。那么可以在浏览器的地址栏中输入：

```
http://127.0.0.1:8080/apple/image/example1_1.jsp
```

来访问 example1_1.jsp。

1.4　JSP 运行原理

视频讲解

当服务器上的一个 JSP 页面被第一次请求执行时，Tomcat 服务器根据 JSP 页面生成一个 Java 文件，并编译这个 Java 文件生成字节码文件，然后执行字节码文件响应用户的请求。而当这个 JSP 页面再次被请求执行时，Tomcat 服务器将直接执行字节码文件来响应用户，而 JSP 页面的首次执行往往由服务器管理者来执行。字节码文件的主要工作是：

（1）把 JSP 页面中的 HTML 标记符号（页面的静态部分）交给客户端浏览器负责显示。

（2）负责处理 JSP 标记，并将有关的处理结果（用字符串形式）发送到客户端浏览器。

（3）执行"<%"和"%>"之间的 Java 程序片（JSP 页面中的动态部分），并把执行结果（用字符串形式）交给客户端浏览器显示。

（4）当多个用户请求一个 JSP 页面时，Tomcat 服务器为每个用户启动一个线程，该线程负责执行常驻内存的字节码文件来响应相应用户的请求。这些线程由 Tomcat 服务器来管理，将 CPU 的使用权在各个线程之间快速切换，以保证每个线程都有机会执行字节码文件（如图 1.13 所示），与传统的 CGI 为每个用户启动一个进程相比较，效率要高得多。

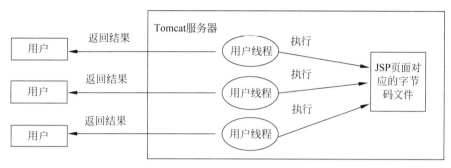

图 1.13　JSP 的运行原理

注：在 Web 设计中，"用户"（"客户"）一词通常指用户（客户）端计算机上驻留的浏览器。

注：如果对 JSP 页面进行了修改、保存，那么 Tomcat 服务器会生成新的字节码文件。

如果 JSP 页面保存在 ch1 目录中，那么，可以在 Tomcat 服务器下的 D:\apache-tomcat-9.0.26\work\Catalina\localhost\ch1\org\apache\jsp 目录中，找到 Tomcat 服务器生成的 JSP 页面对应的 Java 文件，以及编译 Java 文件得到的字节码文件。

需要特别注意的是，当 JSP 页面指定的编码是 UTF-8 时：

<%@ page pageEncoding = "utf-8" %>

保存 JSP 页面应当将"编码"选择为"UTF-8"，原因是，Tomcat 服务器根据 JSP 页面生成 Java 文件时，Tomcat 服务器的解析器是按照 UTF-8 编码来解析 JSP 页面中的数据（JSP 页面本质上是一个 XML 文档）产生对应 Java 文件。如果保存 JSP 页面时，不小心将"编码"选择为其他编码（不是 UTF-8），那么 Tomcat 服务器的解析器产生的对应 Java 文件中的某些字符串就可能有乱码现象，而这些字符串又发送到了客户端浏览器，所以用户浏览器显示信息就出现了乱码现象（比如需要客户浏览器显示的非 ASCII 字符就可能呈现"乱码"状态）。

以下是例 1_1 中 example1_1.jsp 对应的 Java 文件（example1_005f1_jsp.java）的片段，我们把 Tomcat 服务器负责执行和客户端浏览器负责显示的信息分别给出了注释。

example1_005f1_jsp.java

```
package org.apache.jsp;
import javax.servlet. * ;
import javax.servlet.http. * ;
import javax.servlet.jsp. * ;
public final class example1_005f1_jsp extends HttpJspBase
```

```
    implements JspSourceDependent, JspSourceImports {
        …                                                  //省略部分
        out = pageContext.getOut();
        _jspx_out = out;
        out.write("  \r\n");                                //客户端浏览器负责显示
        out.write("<HTML><BODY BGCOLOR = cyan>\r\n");       //客户端浏览器负责显示
        out.write("<h1>这是一个简单的JSP页面</h1>\r\n");      //客户端浏览器负责显示
        out.write("  ");                                    //客户端浏览器负责显示
        int i, sum = 0;                                     //Tomcat 服务器负责执行
        for(i = 1;i<= 100;i++){                             //Tomcat 服务器负责执行
            sum = sum + i;
        }
        out.write("\r\n");                                  //客户端浏览器负责显示
        out.write
        ("<p style = \"font - family:宋体;font - size:36;color:blue\">\r\n");
        out.write("1 到 100 的连续和是:");                   //客户端浏览器负责显示
        out.print(sum );                                    //客户端浏览器负责显示
        out.write(" \r\n");                                 //客户端浏览器负责显示
        out.write("</p></body></HTML>    \r\n");            //客户端浏览器负责显示
        …                                                   //省略部分
    }
```

下面是客户端浏览器查看到的 example1_1.jsp 的源代码(在浏览器的页面上右击,查看源文件):

```
<HTML><body bgcolor = pink>
<h1>这是一个简单的JSP页面</h1>
<p style = "font - family:宋体;font - size:36;color:blue">
1 到 100 的连续和是:5050
</p></body></HTML>
```

1.5 JSP 与 Java Servlet 的关系

视频讲解

Java Servlet 就是编写在服务器端创建对象的 Java 类,习惯上称之为 Servlet 类,Servlet 类的对象习惯上称之为一个 servlet(第 6 章讲述)。在 JSP 技术出现之前,Web 应用开发人员就是自己编写 Servlet 类,并负责编译生成字节码文件,复制这个字节码文件到服务器的特定目录中,以便服务器使用这个 Servlet 类的字节码,创建一个 servlet 来响应用户的请求。Java Servlet 的最大缺点是不能有效地管理页面的逻辑部分和页面的输出部分,导致 Servlet 类的代码非常混乱,单独用 Java Servlet 来管理网站变成一件很困难的事情。为了克服 Java Servlet 的缺点,Sun 公司用 Java Servlet 作为基础,推出了 Java Server Page。JSP 技术就是以 Java Servlet 为基础,提供了 Java Servlet 的几乎所有好处,当用户请求一个 JSP 页面时,Tomcat 服务器自动生成 Java 文件,编译 Java 文件,并用编译得到的字节码文件在服务器端创建一个 servlet。但是 JSP 页面不是 Java Servlet 技术的全部,它只是 Java Servlet 技术的一个成功应用。JSP 页面技术屏蔽了 servlet 创建的过程,使得 Web 程序设计者只要关心 JSP 页面本身的结构、设计好各种标记,比如使用

HTML 标记设计页面的视图，使用 JavaBean 标记有效地分离页面的视图和数据存储等。

　　有些 Web 应用可能只需要 JSP＋JavaBean 就能设计得很好(第 5 章讲解 JavaBean)，但是，对于某些 Web 应用，就可能需要 JSP＋JavaBean＋Servlet 来完成(第 7 章讲解 MVC 模式)，即需要服务器再创建一些 servlet 对象，配合 JSP 页面来完成整个 Web 应用程序的工作。

1.6　HTML 与 JavaScript

视频讲解

▶ 1.6.1　HTML

　　HTML(Hyper Text Markup Language,超文本标记语言)是用来编写 Web 页面(俗称的网页)的语言。HTML 不体现数据的组织结构，只是描述数据的显示格式或提交方式。目前的 HTML 大约有一百多个标记(这些标记由浏览器负责解释执行)，每个标记(不区分大小写)都用于体现怎样显示数据或怎样提交数据。本书假设读者有一定 HTML 的基本知识，但对于非常重要的 HTML 标记，会结合具体应用在第 4.1.4 节介绍。对于非常简单、易于理解的 HTML 标记，本书直接使用。如果读者对 HTML 语言比较陌生，建议补充这方面的知识。HTML 语言是非常容易掌握的一门语言，理由是，只要知道标记的含义，即可会使用。读者想了解某个 HTML 标记的使用细节，可以在网络上随时查阅有关知识，比如查询 HTML Table 标记的用法等。

　　下列 HTML 文件 show.html 将数据分别用黑体 1、黑体 2 和黑体 3 来显示，用浏览器打开 show.html 的效果如图 1.14 所示。

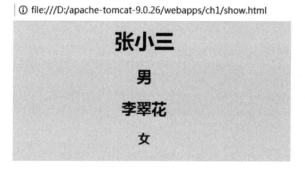

图 1.14　用浏览器打开 HTML 文件

show.html

```
<html><body bgcolor=yellow>
<center>
<H1>张小三
    <H2>男 </H2>
</H1>
<H2>李翠花
    <H3>女 </H3>
</H2>
</center>
</body></html>
```

1.6.2 JavaScript

JavaScript(简称JS)是一种解释型的脚本语言(和Java语言没有关系),由浏览器负责解释执行。JavaScript常被用来美化网页的效果、添加一些动态的显示效果(如滚动的文字)。可以在JSP页面里使用script标记插入JavaScript的代码,这些JavaScript的代码由客户的浏览器负责解释执行。本书重点学习JSP的内容(侧重服务器端功能),且默认读者初步了解HTML语言。因为掌握JSP并不需要JavaScript,因此,不要求读者掌握或熟悉JavaScript(如果读者熟悉JavaScript,在掌握JSP之后,在JSP页面中加入JavaScript内容就是非常简单的一件事情)。

下面的例1_2,在JSP页面中使用script标记插入JavaScript的代码计算了1~100的连续和,并显示了客户端浏览器的时间(即浏览器所驻留的计算机的时间),注意,这些JavaScript的代码完全由浏览器负责执行。

例 1_2

example1_2.jsp(效果如图1.15所示)

```
<%@ page contentType="text/html" %>
<%@ page pageEncoding="utf-8" %>
<HTML><body>
<script>    <!-- JavaScript(JS)标记 -->
var sum = 0;
var i = 1;
  for(i = 1;i <= 100;i++) {
     sum = sum + i;
  }
document.write("<h1>1~100 连续和是:" + sum + "<br></h1>");
var userTime = new Date();
var hour = userTime.getHours();
var minute = userTime.getMinutes();
var second = userTime.getSeconds();
var millisecond = userTime.getMilliseconds();
document.write("<h2>浏览器时间:" +
          hour + ":" + minute + ":" + second + ":" + millisecond + "<br></h2>");
</script>
</body></HTML>
```

1~100连续和是:5050

浏览器时间:22:22:17:60

图1.15 JSP页面里嵌入JS脚本

1.7 上机实验

提供了详细的实验步骤要求,按步骤完成,提升学习效果,积累经验,不断提高Web设计能力。

视频讲解

1.7.1 输出英文字母表

❶ 实验目的

掌握怎样在 Tomcat 服务器的 webapps 中建立新的 Web 服务目录,怎样访问 Web 服务目录下的 JSP 页面。

❷ 实验要求

(1) 在 Tomcat 服务器的 webapps 目录下(比如,D:\apache-tomcat-9.0.26\webapps)新建一个名字是 ch1_practice_one 的目录,即新建一个 Web 服务目录 ch1_practice_one。

(2) 编写 JSP 页面。用文本编辑器编写一个简单的 JSP 页面 letter.jsp(见后面的参考代码),该 JSP 页面可以显示大小写英文字母表。将 JSP 页面 letter.jsp 保存到 Web 服务目录 ch1_practice_one 中。

(3) letter.jsp 的背景是一幅图像。JSP 页面中的<body>标记使用 background 属性设置页面的背景图像是 java.jpg。在 ch1_practice_one 下新建文件夹 image,将 java.jpg 文件保存在 image 中。

(4) 用浏览器访问 JSP 页面 letter.jsp。

❸ 参考代码

学生可按照实验要求,参考本代码编写自己的实验代码。

letter.jsp(效果如图 1.16 所示)

```jsp
<%@ page contentType="text/html" %>
<%@ page pageEncoding="utf-8" %>
<HTML>
<BODY background="image/java.jpg">
<p style="font-family:黑体;font-size:36">
<br> 英文字母表:</br>
</p>
<p style="font-family:宋体;font-size:25;color:blue">
<% char upperCase;                    //存放大写字母
   char lowerCase;
   for(upperCase = 'A';upperCase <= 'Z';upperCase++) {
      lowerCase = (char)(upperCase + 32);
      out.print(upperCase + "(" + lowerCase + ")" + " ");
      if(upperCase == 'M')
         out.print("<br>");
   }
%>
</p>
</BODY></HTML>
```

图 1.16 输出英文字母表

1.7.2 输出九九口诀表

❶ 实验目的

掌握怎样在 Tomcat 服务器之外建立新的 Web 服务目录,怎样访问 Web 服务目录下的 JSP 页面。

❷ 实验要求

(1) 设置 Web 服务目录。在硬盘分区 C:\下新建一个名字是 ch1_practice_two 的目录。打开 Tomcat 服务器中 conf 目录里的 server.xml 文件(比如,D:\apache-tomcat-9.0.26\conf),找到出现</Host>的部分(server.xml 文件尾部),然后在</Host>的前面加入:

```
<Context path = "/number" docBase = "c:/ch1_practice_two" debug = "0" reloadable = "true" />
```

即,将 ch1_practice_two 设置为 Web 服务目录,并为该 Web 服务目录指定名字为 number 的虚拟目录。

(2) 编写 JSP 页面。用文本编辑器编写 JSP 页面 outputNumber.jsp(见后面的参考代码),该 JSP 页面可以输出乘法口诀表。将 JSP 页面 outputNumber.jsp 保存到 Web 服务目录 ch1_practice_two 中。

(3) 用浏览器访问 JSP 页面 outputNumber.jsp。

❸ 参考代码

学生可按照实验要求,参考本代码编写自己的实验代码。

outputNumber.jsp(效果如图 1.17 所示)

```
<%@ page contentType = "text/html" %>
<%@ page pageEncoding = "utf-8" %>
<HTML><body>
<h3>乘法表</h3>
<p style = "font-family:宋体;font-size:15;color:green">
    <%
        for(int j = 1;j <= 9;j++){
            for(int i = 1;i <= j;i++) {
                int n = i * j;
                out.print(i + "×" + j + " = " + n + " ");
            }
            out.print("<br>");
        }
    %>
</p></body></HTML>
```

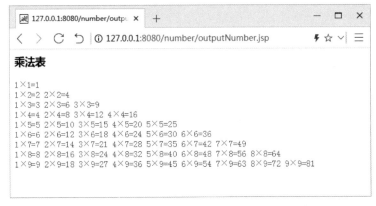

图 1.17 输出乘法口诀表

1.7.3 输出成绩单

❶ 实验目的

复习 HTML 中的<table>标记。

❷ 实验要求

(1) 在 Tomcat 服务器的 webapps 目录下(比如,D:\apache-tomcat-9.0.26\webapps)新建一个名字是 ch1_practice_three 的目录,即新建一个 Web 服务目录 ch1_practice_three。

(2) 编写 JSP 页面。用文本编辑器编写 JSP 页面 outputReport.jsp(见后面的参考代码),该 JSP 页面可以输出成绩单。将 JSP 页面 outputReport.jsp 保存到 Web 服务目录 ch1_practice_three 中。

(3) 用浏览器访问 JSP 页面 outputReport.jsp。

❸ 参考代码

学生可按照实验要求,参考本代码编写自己的实验代码。

outputReport.jsp(效果如图 1.18 所示)

```jsp
<%@ page contentType = "text/html" %>
<%@ page pageEncoding = "utf-8" %>
<HTML><body>
<p style = "font-family:黑体;font-size:22">
    数学、英语和语文成绩单。<br>单科满分是 150 分。
</p>
<% int math = 98;
    int english = 90;
    int chinese = 110;
    int sum = math + english + chinese;
%>
<p style = "font-family:宋体;font-size:20">
<table border = 2>
<tr>
  <td>姓名</td><td>数学成绩</td><td>英语成绩</td><td>语文成绩</td>
  <td>总成绩</td>
</tr>
<tr>
  <td>张三</td><td><% = math %></td><td><% = english %></td>
  <td><% = chinese %></td>
  <td><% = sum %></td>
</tr>
<% math = 115;
    english = 70;
    chinese = 120;
    sum = math + english + chinese;
%>
```

```
<tr>
 <td>李四</td><td><% = math %></td><td><% = english %></td>
 <td><% =chinese %></td>
 <td><% = sum %></td>
<tr>
<% math=88;
   english=100;
   chinese=98;
   sum = math+english+chinese;
%>
<tr>
  <td>王五</td><td><% = math %></td><td><% = english %></td>
  <td><% =chinese %></td>
  <td><% = sum %></td>
</tr>
</table>
</p>
</body></HTML>
```

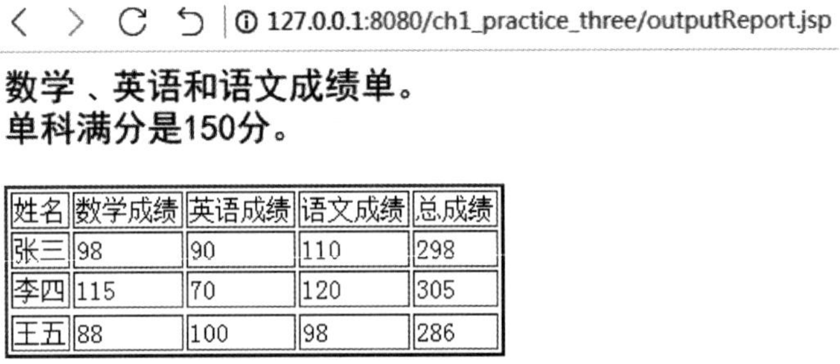

姓名	数学成绩	英语成绩	语文成绩	总成绩
张三	98	90	110	298
李四	115	70	120	305
王五	88	100	98	286

图 1.18 输出成绩单

1.8 小结

- JSP 技术不仅是开发 Web 应用的先进技术,而且是进一步学习相关技术的基础。
- JSP 引擎是支持 JSP 程序的 Web 容器,负责运行 JSP,并将有关结果发送到客户端。目前流行的 JSP 引擎之一是 Tomcat。
- 安装 Tomcat 服务器,首先要安装 JDK,并需要设置 JAVA_HOME 环境变量。
- JSP 页面必须保存在 Web 服务目录中。Tomcat 服务器的 webapps 下的目录都可以作为 Web 服务目录。如果想让 webapps 以外的其他的目录作为 Web 服务目录,必须修改 Tomcat 服务器下 conf 文件夹中的 server. xml 文件,并重新启动 Tomcat 服务器。
- 当服务器上的一个 JSP 页面被第一次请求执行时,Tomcat 服务器首先将 JSP 页面文件转译成一个 Java 文件,再将这个 Java 文件编译生成字节码文件,然后通过执行字节码文件响应客户的请求。
- 当多个客户请求一个 JSP 页面时,Tomcat 服务器为每个客户启动一个线程,该线程负责执行常驻内存的字节码文件来响应相应客户的请求。这些线程由 Tomcat 服务器来管理，将 CPU 的使用权在各个线程之间快速切换，以保证每个线程都有机会执行字节码文件。

习题 1

1. 安装 Tomcat 的计算机需要事先安装 JDK 吗？
2. 怎样启动和关闭 Tomcat 服务器？
3. Boy.jsp 和 boy.jsp 是不是相同的 JSP 文件名字？
4. 请在 D:\ 下建立一个名字为 water 的目录，并将该目录设置成一个 Web 服务目录，然后编写一个简单 JSP 页面，保存到该目录中，让用户使用虚拟目录 fish 来访问该 JSP 页面。
5. 假设 Dalian 是一个 Web 服务目录，其虚拟目录为 moon。A.jsp 保存在 Dalian 的子目录 sea 中，假设 Tomcat 服务器的端口为 8080，则正确访问 A.jsp 的方式是：

   A. http://127.0.0.1:8080:/A.jsp
   B. http://127.0.0.1:8080:/Dalian/A.jsp
   C. http://127.0.0.1:8080:/moon/A.jsp
   D. http://127.0.0.1:8080:/moon/sea/A.jsp

6. 如果想修改 Tomcat 服务器的端口号，应当修改哪个文件？能否将端口号修改为 80？
7. 在 Tomcat 服务器的 webapps 目录下新建一个名字是 letter 的 Web 服务目录。编写 JSP 页面 letter.jsp，保存在 letter 的 Web 服务目录中，该 JSP 页面可以显示希腊字母表。

第 2 章　JSP 语法

本章导读

　　主要内容
- JSP 页面的基本结构
- 声明变量和定义方法
- Java 程序片
- Java 表达式
- JSP 指令标记
- JSP 动作标记

　　难点
- Java 程序片
- JSP 动作标记

　　关键实践
- 消费总和
- 听英语

本章在 webapps 目录下新建一个 Web 服务目录 ch2,除非特别约定,本章例子中的 JSP 页面均保存在 ch2 目录中,另外,在 ch2 目录下建立一个名字是 image 的目录,用于存放 JSP 页面需要的图像文件。

2.1　JSP 页面的基本结构

视频讲解

在传统的 HTML 页面文件中加入 Java 程序片和 JSP 标记就构成了一个 JSP 页面。一个 JSP 页面可由 5 种元素组合而成:
- 普通的 HTML 标记和 JavaScript 标记。
- JSP 标记,如指令标记、动作标记。
- 变量和方法的声明。
- Java 程序片。
- Java 表达式。

当 Tomcat 服务器上的一个 JSP 页面被第一次请求执行时,Tomcat 服务器首先将 JSP 页面文件转译成一个 Java 文件,再将这个 Java 文件编译生成字节码文件,然后通过执行字节码文件响应用户的请求。当多个用户请求一个 JSP 页面时,Tomcat 服务器为每个用户启动一个线程,该线程负责执行常驻内存的字节码文件来响应相应用户的请求。这些线程由 Tomcat 服务器来管理,将 CPU 的使用权在各个线程之间快速切换,以保证每个线程都有机会执行字节码文件。这个字节码文件的任务就是:
- 把 JSP 页面中普通的 HTML 标记和 JavaScript 标记交给用户端浏览器执行显示。

- JSP 标记、方法的定义、Java 程序片由服务器负责处理和执行,将需要显示的结果发送给用户端浏览器。
- Java 表达式由服务器负责计算,并将结果转化为字符串,然后交给用户端浏览器负责显示。

> 注:由于本书重点讲述 JSP 的内容(侧重服务器端功能),因此默认读者初步了解 HTML 语言,但不要求掌握 JavaScript,即俗称的 JS 脚本语言(如果读者熟悉 JS,在掌握 JSP 之后,在 JSP 页面中加入 JS 内容是非常简单的事情)。如果读者想了解 HTML 语言的知识,可以在网络上随时查阅有关知识点,比如查询 HTML 标记的用法等。

在下面的例 2_1 中,example2_1.jsp 页面包含了 5 种元素,其中使用 HTML 的 img 标记显示一幅图像(需要将一幅图像 time.jpg 保存在 Web 服务目录 ch2 的 image 目录中),使用 JavaScript 显示客户端浏览器的时间,使用 Java 程序片显示 Tomcat 服务器端的时间,使用 Java 表达式显示一些变量的值。对于普通的 HTML 标记和 JavaScript 标记,Tomcat 服务器要发送到用户端浏览器执行、显示。客户端浏览器访问例 2_1 的 example2_1.jsp 页面,可以发现客户端浏览器时间和 Tomcat 服务器端的时间差(这里相差了 20 毫秒,见图 2.1,相差的时间值依赖于网络传输的速度以及浏览器的速度)。

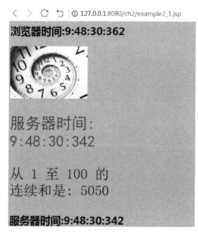

图 2.1 含有 5 种元素的 JSP 页面

例 2_1

example2_1.jsp(效果如图 2.1 所示)

```jsp
<%@ page contentType = "text/html" %>   <!-- jsp 指令标记 -->
<%@ page pageEncoding = "utf-8" %>      <!-- jsp 指令标记 -->
<%@ page import = "java.time.LocalTime" %><!-- jsp 指令标记 -->
<%!    public int continueSum(int start,int end){        //定义方法
         int sum = 0;
         for(int i = start;i <= end;i++)
             sum = sum + i;
         return sum;
     }
%>
<HTML><body bgcolor = pink>
```

```
<script>    <!-- JavaScript(JS)标记 -->
   var userTime = new Date();
   var hour = userTime.getHours();
   var minute = userTime.getMinutes();
   var second = userTime.getSeconds();
   var millisecond = userTime.getMilliseconds();
   document.write("<h2>浏览器时间:" +
           hour + ":" + minute + ":" + second + ":" + millisecond + "<br></h2>");
</script>
<img src = "image/time.jpg"  width = 180 height = 120 />   <!-- html 标记 -->
<p style = "font-family:黑体;font-size:36;color:red"> <!-- html 标记 -->
<%   //Java 程序片:
      LocalTime timeServer = LocalTime.now();
      int hour = timeServer.getHour();
      int minute = timeServer.getMinute();
      int second = timeServer.getSecond();
      int nano =  timeServer.getNano() ;              //纳秒
      int millisecond = nano/1000000;
      out.print("服务器时间:<br>" +
            hour + ":" + minute + ":" + second + ":" + millisecond);
      int start = 1;
      int end = 100;
      int sum = continueSum(start,end);
%>
</p>
<p style = "font-family:宋体;font-size:33;color:blue">
   从
   <%= start %>     <!-- Java 表达式 -->
   至
   <%= end %>       <!-- Java 表达式 -->
   的<br>连续和是:
   <%= sum %>       <!-- Java 表达式 -->
</p>
<script>    <!-- JavaScript(JS)标记 -->
document.write("<h2>服务器时间:" +
<%= hour %> + ":" + <%= minute %> + ":" + <%= second %> + ":" + <%= millisecond %> + "</h2>");
</script>
</body></HTML>
```

2.2 声明变量和定义方法

视频讲解

在"<%!"和"%>"标记符号之间声明变量和定义方法。"<%!"和"%>"标记符号的内容习惯上放在 JSP 页面指令之后,<HTML>之前,也可以写在<HTML>与</HTML>之间。

❶ 声明变量

"<%!"和"%>"之间声明的变量在整个 JSP 页面内都有效,与"<%!""%>"标记符在 JSP 页面中所在的书写位置无关,但习惯上把"<%!""%>"标记符写在 JSP 页面的前面。Tomcat 服务器将 JSP 页面转译成 Java 文件时,将"<%!""%>"标记符之间声明的变量作为类的成员变量,这些变量占有的内存空间直到 Tomcat 服务器关闭才释放。当多个用户请求一个 JSP 页面时,Tomcat 服务器为每个用户启动一个线程,这些线程由 Tomcat 服务器来管理,这些线程共享 JSP 页面的成员变量,因此任何一个用户对 JSP 页面成员变量操作的结果,都会影响到其他用户。

例 2_2 利用成员变量被所有用户共享这一性质,实现了一个简单的计数器。

例 2_2

example2_2.jsp(效果如图 2.2 所示)

```jsp
<%@ page contentType="text/html" %>
<%@ page pageEncoding="utf-8" %>
<HTML><body bgcolor=yellow>
<%! int i=0;
%>
<% i++;
%>
<p style="font-family:宋体;font-size:36">
您是第<%= i %>个访问本站的用户。
</p>
</body></HTML>
```

图 2.2 简单的计数器

❷ 定义方法和类

可以在"<%!"和"%>"标记符号之间定义方法,可以在 Java 程序片中调用该方法。方法内声明的变量只在该方法内有效,当方法被调用时,方法内声明的变量被分配内存,方法被调用完毕即可释放这些变量所占的内存。

可以在"<%!"和"%>"标记符号之间定义类,可以在 Java 程序片中使用该类创建对象。

例 2_3 在"<%!"和"%>"之间定义了两个方法 multi(int x,int y)和 div(int x,int y),然后在程序片中调用这两个方法。在"<%!"和"%>"之间定义了一个 Circle 类,然后在程序片中使用该类创建对象,计算了圆的面积。

例 2_3

example2_3.jsp(效果如图 2.3 所示)

```jsp
<%@ page contentType="text/html" %>
<%@ page pageEncoding="utf-8" %>
<HTML><body bgcolor=#ffccff>
<p style="font-family:宋体;font-size:36;color:blue">
<%! double multi(double x,double y){          //定义方法
      return x*y;
    }
    double div(double x,double y){             //定义方法
      return x/y;
    }
    class Circle {                             //定义类
```

```
            double r;
            double getArea(){
                return 3.1415926 * r * r;
            }
        }
%>
<% double x = 8.79;
    double y = 20.8;
    out.print("调用 multi 方法计算" + x + "与" + y + "的积:<br>");
    out.print(multi(x,y));
    out.print("<br>调用 div 方法计算" + y + "除以" + x + "的商,<br>");
    String s = String.format("小数点保留 3 位: %10.3f",div(y,x));
    out.println(s);
    Circle circle = new Circle();                    //用 Circle 类创建对象
    circle.r = 3.6;
    out.print("<br>半径是" + circle.r + "的圆面积:" + circle.getArea());
%>
</p></body></HTML>
```

图 2.3 定义方法和类

2.3 Java 程序片

可以在"<%"和"%>"之间插入 Java 程序片。一个 JSP 页面可以有许多程序片,这些程序片将被 Tomcat 服务器按顺序执行。在程序片中声明的变量称作 JSP 页面的局部变量。局部变量的有效范围与其声明的位置有关,即局部变量在 JSP 页面后继的所有程序片以及表达式部分内都有效。Tomcat 服务器将 JSP 页面转译成 Java 文件时,将各个程序片的这些变量作为类中某个方法的变量,即局部变量。

Java 程序片可以写在<HTML>之前,也可以写在<HTML>和</HTML>之间或</HTML>之后。

当多个用户请求一个 JSP 页面时,Tomcat 服务器为每个用户启动一个线程,该线程负责执行字节码文件响应用户的请求。Tomcat 服务器使用多线程来处理程序片,特点如下:

- 操作 JSP 页面的成员变量。Java 程序片中操作的成员变量是各个线程(用户)共享的变量,任何一个线程对 JSP 页面成员变量操作的结果,都会影响到其他线程。
- 调用 JSP 页面的方法。Java 程序片中可以出现方法调用语句,所调用的方法必须是 JSP 页面曾定义的方法。
- 操作局部变量。当一个线程享用 CPU 资源时,Tomcat 服务器让该线程执行 Java 程序片,这时,Java 程序片中的局部变量被分配内存空间,当轮到另一个线程享用 CPU 资源时,Tomcat 服务器让该线程再次执行 Java 程序片,那么,Java 程序片中的局部变量会再次分配内存空间。也就是说,Java 程序片已经被执行了两次,分别运行在不同的线程中,即运行在不同的时间片内。运行在不同线程中的 Java 程序片的局部变量互不干扰,即一个用户改变 Java 程序片中的局部变量的值不会影响其他用户的 Java 程序片中的局部变量。当一个线程将 Java 程序片执行完毕,运行在该线程中的 Java 程序片的局部变量释放所占的内存。

Java 程序片执行特点如图 2.4 所示。

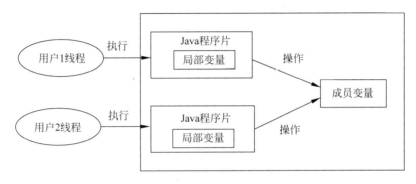

图 2.4 程序片的执行

根据 Java 程序片的上述特点,对于某些特殊情形必须给予特别注意。比如,如果一个用户在执行 Java 程序片时调用 JSP 页面的方法操作成员变量,可能不希望其他用户也调用该方法操作成员变量,以免对其产生不利的影响(成员变量被所有的用户共享),那么就应该将操作成员变量的方法用 synchronized 关键字修饰。当一个线程在执行 Java 程序片期间调用 synchronized 方法时,其他线程在其 Java 程序片中调用这个 synchronized 方法时就必须等待,直到正在调用 synchronized 方法的线程调用执行完该方法。在下面的例 2_4 中,通过 synchronized 方法操作一个成员变量来实现一个简单的计数器。

例 2_4

example2_4.jsp

```jsp
<%@ page contentType = "text/html" %>
<%@ page pageEncoding = "utf-8" %>
<HTML><body>
<p style = "font-family:宋体;font-size:36;color:blue">
<%! int count = 0;                          //被用户共享的 count
    synchronized void setCount() {          //synchronized 修饰的方法
       count++;
    }
%>
<% setCount();
   out.println("您是第" + count + "个访问本站的客户");
%>
</p></body></HTML>
```

一个 JSP 页面中的 Java 程序片会按其在页面中的顺序被执行,而且某个 Java 程序片中声明的局部变量在其后继的所有 Java 程序片以及表达式部分内都有效。利用 Java 程序片的这个性质,有时候可以将一个 Java 程序片分割成几个 Java 程序片,然后在这些 Java 程序片之间再插入其他标记元素。在程序片中插入 HTML 中的标记的技巧对于灵活显示数据是非常重要的。通常的格式是:

```
HTML 中的标记
<%
    Java 程序片
%>
HTML 中的标记
<%
    Java 程序片
%>
HTML 中的标记
```

下面的例 2_5 获得一个 7～19 的随机数,如果获得的数小于或等于 13 就显示一幅小学生的图像,否则显示一幅中学生的图像。显示图像需要在程序片之间插入用于显示图像的

```
<imagesrc=图像的URL>文字说明</image>
```

HTML 标记(将两幅名字分别为 xiao.jpg 和 zhong.jpg 的图像保存到 Web 服务目录 ch2 的 image 目录中,见本章开始的约定)。

例 2_5

example2_5.jsp(效果如图 2.5(a)或(b)所示)

```
<%@ page contentType="text/html" %>
<%@ page pageEncoding="utf-8" %>
<HTML><body bgcolor=cyan>
<%  //Math.random()是[0,1)之间的随机数
    int number = 7 + (int)(Math.random() * 13);
    if(number <= 13) {
%>
        <center><h2>显示小学生图片</h2><!-- 插入其他标记 -->
        <image src='image/xiao.jpg' width=180 height=178>小学生</image>
<%    }
    else {
%>
        <center><h2>显示中学生图片</h2>
        <image src='image/zhong.jpg' width=180 height=178>中学生</image>
<%    }
%>
</body></HTML>
```

(a) 随机数不大于13　　　　　　　(b) 随机数大于13

图 2.5　显示图像

2.4 Java 表达式

视频讲解

可以在“<%=”和“%>”之间插入一个可求值的表达式(注意：不可插入语句，“<%=”是一个完整的符号，“<%”和“=”之间不要有空格)。表达式的值由服务器负责计算，并将计算结果用字符串形式发送到用户端显示。

Java 表达式可以写在< HTML >之前，也可以写在< HTML >和</HTML >之间或</HTML >之后。

需要注意的是，在 JSP 页面中，表达式的值被表示成一个字符串的形式，即 Tomcat 服务器将表达式的结果转换成字符串，然后发送给用户的浏览器。因此，在编写 JSP 页面时，要把 Java 表达式按普通的文本来使用。

下面的例 2_6 计算几个表达式的值。

例 2_6

example2_6.jsp(效果如图 2.6 所示)

```
<%@ page contentType = "text/html" %>
<%@ page pageEncoding = "utf-8" %>
<HTML><body bgcolor = pink >
<%   int x = 12, y = 9;
%>
<p style = "font-family:宋体;font-size:36">
  计算表达式 x+y+x%y,即<% = x %>+<% = y %>+<% = x %>%<% = y %>的值:
  <% = x+y+x%y %>
<br>计算表达式 x>y 即<% = x %>><% = y %>的值:
  <% = x>y %>
<br>计算表达式 sin(<% = Math.PI %>/2)的值:
    <% = Math.sin(Math.PI/2) %>
<br>
<%   if(x-y>=0) {
%>
        如果<% = x %>大于<% = y %>,计算<% = x %> - <% = y %>即
        <% = x-y %>的平方根:
        <% = Math.sqrt(x-y) %>
<%   }
%>
</p></body></HTML>
```

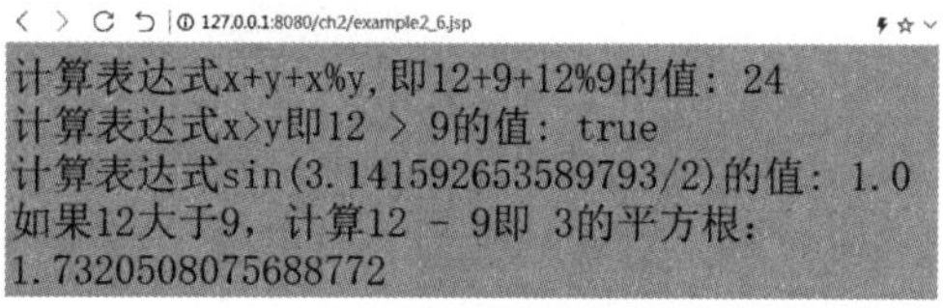

127.0.0.1:8080/ch2/example2_6.jsp
计算表达式x+y+x%y,即12+9+12%9的值：24
计算表达式x>y即12 > 9的值：true
计算表达式sin(3.141592653589793/2)的值：1.0
如果12大于9，计算12 - 9即 3的平方根：
1.7320508075688772

图 2.6 计算表达式的值

2.5 JSP 中的注释

视频讲解

注释可以增强 JSP 页面的可读性，使 JSP 页面易于维护。JSP 页面中的注释可分为两种。

(1) HTML 注释：在标记符号"<!--"和"-->"之间加入注释内容：

```
<!-- 注释内容 -->
```

JSP 引擎把 HTML 注释交给用户，因此用户通过浏览器查看 JSP 页面的源文件时，能够看到 HTML 注释。

(2) JSP 注释：在标记符号"<%--"和"--%>"之间加入注释内容：

```
<%-- 注释内容 --%>
```

Tomcat 服务器忽略 JSP 注释，即在编译 JSP 页面时忽略 JSP 注释。

下面的例 2_7 使用了 HTML 和 JSP 注释。

例 2_7

example2_7.jsp

```jsp
<%@ page contentType="text/html" %>
<%@ page pageEncoding="utf-8" %>
<HTML><body>
<%-- 下面是 Java 程序片 --%>
<% String str = "C:\\jspfile\\example2_7.jsp";
   int index = str.lastIndexOf("\\");
   str = str.substring(index+1);
%>
<!-- 以下字体的颜色为蓝色 -->
<p style="font-family:黑体;font-size:20;color:blue">
抽取字符串<%= str %>中的 JSP 文件名字
<!-- 以下字体的颜色为红色 -->
<font size="6" color="red"><br>
<%-- 下面是 Java 表达式 --%>
<%= str %>
</font>
</p>
</body></HTML>
```

2.6 JSP 指令标记

视频讲解

▶ 2.6.1 page 指令标记

page 指令用来定义整个 JSP 页面的一些属性和这些属性的值，属性值用单引号或双引号括起来。可以使用多个 page 指令分别为每个属性指定值，如：

```
<%@ page 属性1="属性1的值" %>
```

```
<%@ page 属性2 = "属性2的值"%>
...
<%@ page 属性n = "属性n的值"%>
```

也可以用一个 page 指令指定多个属性的值,如:

```
<%@ page 属性1 = "属性1的值" 属性2 = "属性2的值"  ... %>
```

例如,前面各个例子中的 2 个 page 指令:

```
<%@ page contentType = "text/html" %>
<%@ page pageEncoding = "utf-8" %>
```

也可以合并成 1 个 page 指令:

```
<%@ page contentType = "text/html" pageEncoding = "utf-8" %>
```

page 指令的作用对整个 JSP 页面有效,与其书写的位置无关。习惯上把 page 指令写在 JSP 页面的最前面。

page 指令标记可以为 contentType、import、language、session、buffer、autoFlush、isThreadSafe、pageEncoding、inform 等属性指定值。以下将分别讲述这些属性的设置与作用。

❶ contentType 属性

我们已经知道,当用户请求一个 JSP 页面时,Tomcat 服务器负责解释执行 JSP 页面,并将某些信息发送到用户的浏览器,以便用户浏览这些信息。Tomcat 服务器同时负责通知用户的浏览器使用怎样的方法来处理所接收到的信息。这就要求 JSP 页面必须设置响应的 MIME(Multipurpose Internet Mail Extention)类型,即设置 contentType 属性的值。contentType 属性值确定 JSP 页面响应的 MIME 类型。属性值的一般形式是:

```
"MIME 类型"
```

比如,如果我们希望用户的浏览器启用 HTML 解析器来解析执行所接收到的信息(即所谓的网页形式),就可以如下设置 contentType 属性的值:

```
<%@ page contentType = "text/html" %>
```

如果希望用户的浏览器启用本地的 MS-Word 应用程序来解析执行收到的信息,就可以如下设置 contentType 属性的值:

```
<%@ page contentType = "application/msword" %>
```

如果不使用 page 指令为 contentType 指定一个值,那么 contentType 属性的默认值是 "text/html"。JSP 页面使用 page 指令只能为 contentType 指定一个值,不允许两次使用 page 指令给 contentType 属性指定不同的属性值,下列用法是错误的:

```
<%@ page contentType = "text/html" %>
<%@ page contentType = "application/msword" %>
```

用 page 指令为 contentType 指定一个值的同时,也可以为 contentType 的附加属性 charset 指定一个值(默认值是 iso-8859-1),例如:

```
<%@ page contentType = "text/html;charset = gb2312" %>
```

contentType 的附加属性 charset 的值是通知用户浏览器用怎样的编码解析收到的字符，当 JSP 页面用 page 指令指定设置 charset 的值是 GB2312 时，浏览器会将编码切换成 GB2312。但是，如果 JSP 页面用 page 指定了 JSP 的页面本身的编码，例如：<%@ page pageEncoding = "UTF-8" %>，那么 charset 的值和 JSP 的页面编码保持一致，即 UTF-8（目前的浏览器都支持 UTF-8 编码，所以一般不需要再指定 charset 的值，使其和 JSP 的页面编码保持一致即可）。

可以使用 page 指令为 contentType 属性指定的值有 text/html、text/plain、image/gif、image/x-xbitmap、image/jpeg、image/pjpeg、application/x-shockwave-flash、application/vnd.ms-powerpoint、application/vnd.ms-excel、application/msword 等。

如果用户的浏览器不支持某种 MIME 类型，那么用户的浏览器就无法用相应的手段处理所接收到的信息。比如，使用 page 指令设置 contentType 属性的值是 "application/msword"，如果用户浏览器所驻留的计算机没有安装 MS-Word 应用程序，那么浏览器就无法处理所接收到的信息。

注：如果想深入了解 MIME 类型，可以在流行的搜索引擎中搜索 MIME 关键字。

下面的例 2_8 中 example2_8.jsp 页面使用 page 指令设置 contentType 属性的值是 "image/jpeg"，当用户请求 example2_8.jsp 页面时，用户的浏览器将启用图形解码器来解析执行收到的信息。

例 2_8

example2_8.jsp（效果如图 2.7 所示）

```
<%@ page contentType = "image/jpeg" %>
<%@ page import = "java.awt.*" %>
<%@ page import = "java.io.OutputStream" %>
<%@ page import = "java.awt.image.BufferedImage" %>
<%@ page import = "java.awt.geom.*" %>
<%@ page import = "javax.imageio.ImageIO" %>
<%  int width = 320, height = 300;
    BufferedImage image =
    new BufferedImage(width,height,BufferedImage.TYPE_INT_RGB);
    Graphics g = image.getGraphics();
    g.setColor(Color.lightGray);
    g.fillRect(0, 0, width, height);
    Graphics2D g_2d = (Graphics2D)g;
    Ellipse2D ellipse = new Ellipse2D.Double (70,90,120,60);
    g_2d.setColor(Color.blue);
    AffineTransform trans = new  AffineTransform();
    for(int i = 1;i <= 24;i++) {
       trans.rotate(15.0 * Math.PI/180,160,130);
       g_2d.setTransform(trans);
       g_2d.draw(ellipse);
    }
    g_2d.setColor(Color.red);
    Arc2D arc = new Arc2D.Double (200,220,100,40,0,270,Arc2D.PIE);
    g_2d.fill(arc);
```

```
arc.setArc(5,5,100,40,0,-270,Arc2D.PIE);
g_2d.fill(arc);
g_2d.setColor(Color.black);
g_2d.setFont(new Font("",Font.BOLD,18));
g_2d.drawString("耿祥义, Graphic Drawer",10,280);
g_2d.dispose();
OutputStream outClient = response.getOutputStream();    //指向用户端的输出流
boolean boo = ImageIO.write(image,"jpeg",outClient);
%>
```

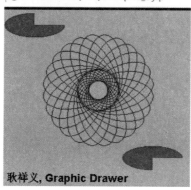

图 2.7　contentType 为 image/jpeg

❷ pageEncoding 属性

pageEncoding 属性的默认值是 UTF-8。需要注意的是,和 contentType 的附加属性 charset 的值的意义不同,pageEncoding 属性值是定义 JSP 页面使用的编码,即是告诉 Tomcat 服务器的解析器用怎样的编码解析 JSP 页面中的字符,比如当 JSP 页面指定的编码是 UTF-8 时:

```
<%@ page pageEncoding = "utf-8" %>
```

保存 JSP 页面应当将"编码"选择为"UTF-8",原因是,Tomcat 服务器根据 JSP 页面产生成 Java 文件时,Tomcat 服务器的解析器是按照 UTF-8 编码来解析 JSP 页面中的字符数据 (JSP 页面本质上是一个 XML 文档)产生对应 Java 文件。如果保存 JSP 页面时,不小心将"编码"选择为其他编码(不是 UTF-8),那么 Tomcat 服务器的解析器产生的对应 Java 文件中的某些字符串就可能有乱码现象,而这些字符串又发送到了客户端浏览器,所以用户浏览器显示信息就出现了乱码现象(比如需要客户浏览器显示的非 ASCII 字符就可能呈现"乱码"状态)。

另外,JSP 页面使用 page 指令只能为 pageEncoding 指定一个值,不允许两次使用 page 指令给 pageEncoding 属性指定不同的或相同的属性值。

❸ language 属性

language 属性定义 JSP 页面使用的脚本语言,该属性的值目前只能取"java"。

为 language 属性指定值的格式是:

```
<%@ page language = "java" %>
```

language 属性的默认值是"java",即如果在 JSP 页面中没有使用 page 指令指定该属性的值的,那么,JSP 页面默认有如下的 page 指令:

```
<%@ page language = "java" %>
```

❹ import 属性

该属性的作用是为 JSP 页面引入 Java 运行环境提供的包中的类,这样就可以在 JSP 页面的程序片部分、变量及方法定义部分以及表达式部分使用包中的类。可以为该属性指定多个值,该属性的值可以是某包中的所有类或一个具体的类,例如:

```
<%@ page import = "java.io.*", "java.time.LocalDate" %>
```

JSP 页面默认 import 属性已经有如下的值:

"java.lang.*" "javax.servlet.*" ""javax.servlet.jsp.*" ""javax.servlet.http.*"

当为 import 指定多个属性值时,比如:

```
<%@ page import = "java.util.*" %>
<%@ page import = "java.io.*" %>
```

那么,JSP 引擎把 JSP 页面转译成的 Java 文件中会有如下的 import 语句:

```
import java.util.*;
import java.io.*;
```

❺ session 属性

session 属性用于设置是否需要使用内置的 session 对象。session 的属性值可以是 true 或 false。session 属性默认的属性值是 true。

❻ buffer 属性

内置输出流对象 out 负责将服务器的某些信息或运行结果发送到用户端显示。buffer 属性用来指定 out 设置的缓冲区的大小或不使用缓冲区。例如:

```
<%@ page buffer = "24kb" %>
```

buffer 属性的默认值是 8kb。buffer 属性可以取值"none",即设置 out 不使用缓冲区。

❼ autoFlush 属性

autoFlush 属性指定 out 的缓冲区被填满时,缓冲区是否自动刷新。

autoFlush 可以取值 true 或 false。autoFlush 属性的默认值是 true。当 autoFlush 属性取值 false 时,如果 out 的缓冲区填满,就会出现缓存溢出异常。当 buffer 的值是"none"时,autoFlush 的值就不能设置成 false。

❽ isThreadSafe 属性

isThreadSafe 属性用来设置访问 JSP 页面是否是线程安全的。isThreadSafe 的属性值可取 true 或 false。当 isThreadSafe 属性值设置为 true 时,JSP 页面能同时响应多个用户的请求;当 isThreadSafe 属性值设置成 false 时,JSP 页面同一时刻只能响应一个用户的请求,其他用户须排队等待。isThreadSafe 属性的默认值是 true。

当 isThreadSafe 属性值为 true 时,CPU 的使用权在各个线程间快速切换。也就是说,即使一个用户的线程没有执行完毕,CPU 的使用权也可能要切换给其他的线程,如此轮流,直到各个线程执行完毕;当 JSP 使用 page 指令将 isThreadSafe 属性值设置成 false 时,该 JSP 页面同一时刻只能响应一个用户的请求,其他用户须排队等待。也就是说,CUP 要保证一个线程将 JSP 页面执行完毕才会把 CPU 使用权切换给其他线程。

❾ info 属性

info 属性的属性值是一个字符串,其目的是为 JSP 页面准备一个常用但可能要经常修改的字符串。例如,

```
<%@ page info = "we are students" %>
```

可以在 JSP 页面中使用方法:

```
getServletInfo();
```

获取 info 属性的属性值。

注:当 JSP 页面被转译成 Java 文件时,转译成的类是 Servlet 的一个子类,所以在 JSP 页面中可以使用 Servlet 类的方法:getServletInfo()。

例 2_9 使用 getServletInfo()方法获取 info 的属性值,效果如图 2.8 所示(需要将一幅图像 tsinghua.jpg 存放到 Web 服务目录 ch2 的 image 目录中)。

图 2.8 获取 info 属性的值

例 2_9
example2_9.jsp(效果如图 2.8 所示)

```
<%@ page info = "清华大学图像 tsinghua.jpg" %>
<% String s = getServletInfo();
   String str[] = s.split("图像");
%>
<HTML><center>
<body background = "image/<% = str[1] %>">
<p style = "font-family:宋体;font-size:36;color:blue">
<br><% = str[0] %>出版社是中国著名出版社
<br><% = str[0] %>是全国著名的高等学府
</p></body></center><HTML>
```

▶ 2.6.2 include 指令标记

如果需要在 JSP 页面内某处整体嵌入一个文件,就可以考虑使用 include 指令标记,其语法格式如下:

```
<%@ include file = "文件的 URL" %>
```

include 指令标记的作用是在 JSP 页面出现该指令的位置处，静态嵌入一个文件，该文件的编码必须和当前 JSP 页面一致，比如二者都是 UTF-8 编码。被嵌入的文件必须是可以访问或可以使用的。如果该文件和当前 JSP 页面在同一 Web 服务目录中，那么"文件的 URL"就是文件的名字；如果该文件在 JSP 页面所在的 Web 服务目录的一个子目录中，比如 fileDir 子目录中，那么"文件的 URL"就是"fileDir/文件的名字"。所谓静态嵌入，就是当前 JSP 页面和嵌入的文件合并成一个新的 JSP 页面，然后 Tomcat 服务器再将这个新的 JSP 页面转译成 Java 文件。因此，嵌入文件后，必须保证新合并成的 JSP 页面符合 JSP 语法规则，即能够成为一个 JSP 页面文件。比如，被嵌入的文件是一个 JSP 页面，该 JSP 页面使用 page 指令为 contentType 属性设置了值：

```
<%@ page contentType = "application/msword" %>
```

那么，合并后的 JSP 页面就两次使用 page 指令为 contentType 属性设置了不同的属性值，导致出现语法错误。因为 JSP 页面中的 page 指令只能为 contentType 指定一个值。

Tomcat 5.0 版本以后的服务器每次都要检查 include 指令标记嵌入的文件是否被修改过，因此，JSP 页面成功静态嵌入一个文件后，如果对嵌入的文件进行了修改，那么 Tomcat 服务器会重新编译 JSP 页面，即将当前的 JSP 页面和修改后的文件合并成一个 JSP 页面，然后 Tomcat 服务器再将这个新的 JSP 页面转译成 Java 类文件。

使用 include 指令可以实现代码的复用。比如，每个 JSP 页面上都可能都需要一个导航条，以便用户在各个 JSP 页面之间方便地切换，那么每个 JSP 页面都可以使用 include 指令在页面的适当位置整体嵌入一个相同的文件。

需要特别注意的是，允许被嵌入的文件使用 page 指令指定 contentType 属性的值，但指定的值必须和嵌入该文件的 JSP 页面中的 page 指令指定的 contentType 属性的值相同。

例 2_10 中两个 JSP 页面使用 include 指令标记嵌入同一个文本文件：ok.txt，该文本文件的内容是关于这两个 JSP 页面之间（北京大学与清华大学之间）的超链接。例 2_10 中的 ok.txt 文件保存（必须用 UTF-8 编码保存）在当前 JSP 页面所在的 Web 服务目录 ch2 的子目录 myfile 中，需要将两幅能分别代表北京大学和清华大学的图像 beida.jpg 和 tsinghua.jpg 保存在 Web 服务目录 ch2 的子目录 image 中，将格式为 .wav 或 .mp3 的音频文件保存在 Web 服务目录 ch2 的子目录 sound 中。运行效果如图 2.9(a) 和图 2.9(b) 所示。

(a) 北京大学　　　　　　　(b) 清华大学

图 2.9　用 include 指令嵌入文件

ok.txt

```
<%@ page contentType = "text/html" %>
<center>
```

```
< A href = "example2_10_a.jsp">北京大学</A>
< A href = "example2_10_b.jsp">清华大学</A>
```

例 2_10

example2_10_a.jsp(效果如图 2.9(a)所示)

```
<%@ page contentType = "text/html" %>
<%@ page pageEncoding = "utf-8" %>
<%@ include file = "myfile/ok.txt" %>
<HTML><center><body background = "image/beida.jpg">
<bgsound src = "sound/beida.mp3" loop = 1 >
<h1 >这里是北京大学 </h1>
</body></HTML>
```

example2_10_b.jsp(效果如图 2.9(b)所示)

```
<%@ page contentType = "text/html" %>
<%@ page pageEncoding = "utf-8" %>
<%@ include file = "myfile/ok.txt" %>
<HTML><center><body background = "image/tsinghua.jpg">
<bgsound src = "sound/tsinghua.mp3" loop = "1" >
<h1 >这里是清华大学 </h1>
</body></HTML>
```

2.7 JSP 动作标记

动作标记是一种特殊的标记,它影响 JSP 运行时的功能。

视频讲解

▶ 2.7.1 include 动作标记

include 动作标记语法格式为:

```
< jsp:include page = "文件的 URL"/>
```

或

```
< jsp:include page = "文件的 URL">
    param 子标记
</jsp:include >
```

需要注意的是,当 include 动作标记不需要 param 子标记时,必须使用第一种形式。

include 动作标记告诉 JSP 页面动态包含一个文件,即 JSP 页面运行时才将文件加入。与静态嵌入文件的 include 指令标记不同,当 Tomcat 服务器根据 JSP 页面产生成 Java 文件时,不把 JSP 页面中动作指令 include 所包含的文件与原 JSP 页面合并为一个新的 JSP 页面,而是告诉 Java 解释器,这个文件在 JSP 运行(Java 文件的字节码文件被加载执行)时才包含进来。如果包含的文件是普通的文本文件,就将文件的内容发送到用户端,由用户端的浏览器负责显示;如果包含的文件是 JSP 文件,Tomcat 服务器就执行这个文件,然后将执行的结果发送到用户端,并由用户端的浏览器负责显示这些结果。

尽管 include 动作标记和 include 指令标记的作用都是处理所需要的文件，但是处理方式和处理时间是不同的。include 指令标记是在编译阶段就处理所需要的文件，被处理的文件在逻辑和语法上依赖于当前的 JSP 页面，其优点是页面的执行速度快；而 include 动作标记是在 JSP 页面运行时才处理文件，被处理的文件在逻辑和语法上独立于当前 JSP 页面，其优点是可以使用 param 子标记更灵活地处理所需要的文件（见后面的 param 动作标记），缺点是执行速度要慢一些。

注：书写 include 动作标记<jsp:include page···/>时要注意："jsp"":""include"三者之间不要有空格。

▶ 2.7.2 param 动作标记

param 标记以"名字-值"对的形式为其他标记提供附加信息，param 标记不能独立使用，须作为 jsp:include、jsp:forward 标记的子标记来使用。

param 动作标记语法格式是：

```
<jsp:param name = "参数" value = "参数的值" />
```

当该标记与 jsp:include 动作标记一起使用时，可以将 param 标记中参数的值传递到 include 动作标记要加载的文件中去，被加载的 JSP 文件可以使用 Tomcat 服务器提供的 request 内置对象获取 include 动作标记的 param 子标记中 name 给出的参数的值，如图 2.10 所示。因此 include 动作标记通过使用 param 子标记来处理加载的文件，比 include 指令标记更为灵活。

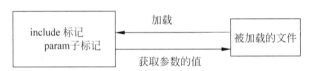

图 2.10 使用 include 动作标记加载文件

下面例 2_11 中，example2_11.jsp 使用 include 动作标记加载 JSP 文件 triangle.jsp，triangle.jsp 页面保存在当前 Web 服务目录 ch2 的子目录 myfile 中。triangle.jsp 页面可以计算并显示三角形的面积，当 triangle.jsp 被加载时获取 example2_11.jsp 页面 include 动作标记的 param 子标记提供的三角形三边的长度。效果如图 2.11 所示。

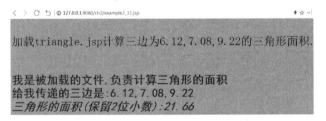

图 2.11 用 param 子标记向加载的文件传值

例 2_11

example2_11.jsp（效果如图 2.11 所示）

```
<%@ page contentType = "text/html" %>
```

第2章 JSP语法

```jsp
<%@ page pageEncoding="utf-8" %>
<HTML><body bgcolor=cyan>
<% double a=6.12,b=7.08,c=9.22;
%>
<p style="font-family:宋体;font-size:36">
<br>加载 triangle.jsp 计算三边为<%=a%>,<%=b%>,<%=c%>的三角形面积.
    <jsp:include page="myfile/triangle.jsp">
        <jsp:param name="sideA" value="<%=a%>"/>
        <jsp:param name="sideB" value="<%=b%>"/>
        <jsp:param name="sideC" value="<%=c%>"/>
    </jsp:include>
</p></body></HTML>
```

triangle.jsp

```jsp
<%@ page contentType="text/html" %>
<%@ page pageEncoding="utf-8" %>
<%! public String getArea(double a,double b,double c) {
        if(a+b>c&&a+c>b&&c+b>a) {
            double p=(a+b+c)/2.0;
            double area=Math.sqrt(p*(p-a)*(p-b)*(p-c));
            String result=String.format("%.2f",area);    //保留两位小数
            return result;
        }
        else {
            return(""+a+","+b+","+c+"不能构成一个三角形,无法计算面积");
        }
    }
%>
<% String sideA=request.getParameter("sideA");           //获取参数 sideA 的值
   String sideB=request.getParameter("sideB");
   String sideC=request.getParameter("sideC");
   double a=Double.parseDouble(sideA);
   double b=Double.parseDouble(sideB);
   double c=Double.parseDouble(sideC);
%>
<p style="font-family:黑体;font-size:36;color:blue">
<br><b>我是被加载的文件,负责计算三角形的面积<br>
    给我传递的三边是:<%=sideA%>,<%=sideB%>,<%=sideC%></b>
<br><b><i>三角形的面积(保留2位小数):<%= getArea(a,b,c) %></i></b></i>
</p>
```

▶ 2.7.3 forward 动作标记

forward 动作标记的语法格式是：

```jsp
<jsp:forward page="要转向的页面" />
```

或

```jsp
<jsp:forward page="要转向的页面">
    param 子标记
</jsp:forward>
```

该指令的作用是：从该指令处停止当前页面的执行，而转向执行 page 属性指定的 JSP 页面。需要注意的是，当 forward 动作标记不需要 param 子标记时，必须使用第一种形式。

forward 标记可以使用 param 动作标记作为子标记，向转向的页面传送信息。forward 动作标记指定的要转向的 JSP 文件可以使用 Tomcat 服务器提供的 request 内置对象获取 param 子标记中 name 指定的属性值。

需要注意的是，当前页面使用 forward 动作标记转向后，尽管用户看到了转向后的页面的效果，但浏览器地址栏中显示的仍然是转向前的 JSP 页面的 URL 地址，因此，如果刷新浏览器的显示，将再次执行当前浏览器地址栏中显示的 JSP 页面。

例 2_12 中的页面均保存在 Web 服务目录 ch2 中。example2_12.jsp 使用 forward 标记转向 example2_12_a.jsp 或 example2_12_b.jsp 页面。在 example2_12.jsp 页面随机产生一个 1~10 之间的随机数，若该数大于 5 就转向页面 example2_12_a.jsp，否则转向页面 example2_12_b.jsp。example2_12.jsp 使用 param 子标记将随机数传递给要转向的页面。example2_12_a.jsp 和 example2_12_b.jsp 所使用的图像文件 pic_a.jpg 和 pic_b.jpg 保存在当前 Web 服务目录的 image 子目录中。example2_12_a.jsp 页面和 example2_12_b.jsp 页面的效果如图 2.12(a)和图 2.12(b)所示。

(a) example2_12.a.jsp效果　　　　　(b) example2_12.b.jsp效果

图 2.12　forward 动作标记

例 2_12

example2_12.jsp

```jsp
<%@ page contentType = "text/html" %>
<%@ page pageEncoding = "utf-8" %>
<HTML><body>
<h1>产生一个 1~10 之间的随机数
<%  double i = (int)(Math.random() * 10) + 1;
    if(i <= 5) {
%>      <jsp:forward page = "example2_12_a.jsp" >
            <jsp:param name = "number" value = "<% = i %>" />
        </jsp:forward>
<%  }
    else {
%>      <jsp:forward page = "example2_12_b.jsp" >
            <jsp:param name = "number" value = "<% = i %>" />
        </jsp:forward>
```

```
<%  }
%>
</body></HTML>
```

example2_12_a.jsp（效果如图 2.12(a)所示）

```
<%@ page contentType = "text/html" %>
<%@ page pageEncoding = "utf-8" %>
<HTML><body bgcolor = cyan>
<p style = "font-family:宋体;font-size:36">
<% String s = request.getParameter("number");
   out.println("传递过来的值是" + s);
%>
<br><img src = image/pic_a.jpg width = 300 height = 280/>
</p></body></HTML>
```

example2_12_b.jsp（效果如图 2.12（b）所示）

```
<%@ page contentType = "text/html" %>
<%@ page pageEncoding = "utf-8" %>
<HTML><body bgcolor = yellow>
<p style = "font-family:宋体;font-size:36">
<% String s = request.getParameter("number");
   out.println("传递过来的值是" + s);
%>
<br><img src = image/pic_b.jpg width = 300 height = 280 />
</p></body></HTML>
```

▶ 2.7.4 useBean 动作标记

将在第 5 章详细讨论 useBean 标记。useBean 标记用来创建并使用一个 JavaBean,是非常重要的一个动作标记。Sun 公司倡导的是 JavaBean 负责存储数据,JSP 页面显示 JavaBean 中的数据,而 servlet 负责管理 JavaBean 中的数据(见第 7 章)。

2.8 上机实验

提供了详细的实验步骤要求,按步骤完成,提升学习效果,积累经验,不断提高 Web 设计能力。

视频讲解

▶ 2.8.1 实验 1 消费总和

❶ 实验目的

掌握 JSP 页面的基本结构。

❷ 实验要求

(1) 在 JSP 页面 showPriceSum.jsp 中定义一个方法 public double getPriceSum(String input),该方法可以返回参数 input 含有的各个数字的代数和。比如 String str = "麻辣豆腐:20.6 元,红烧肉:68.9 元",那么 getPriceSum(str)返回的值是 89.5。

(2) Java 程序片中用 String 对象 str 封装表示菜单的字符序列。

（3）使用 Java 程序片显示 str 以及 getPriceSum(mess)返回的值。

（4）在 Tomcat 服务器的 webapps 目录下（比如，D:\apache-tomcat-9.0.26\webapps）新建名字是 ch2_practice_one 的 Web 服务目录。把 showPriceSum.jsp 保存到 ch2_practice_one 目录中。

（5）用浏览器访问 JSP 页面 showPriceSum.jsp。

❸ 参考代码

参考代码运行效果如图 2.13 所示。

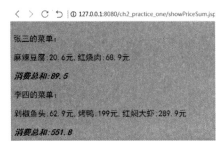

图 2.13　显示消费总和

showPriceSum.jsp

```jsp
<%@ page contentType="text/html" %>
<%@ page pageEncoding="utf-8" %>
<%@ page import="java.util.regex.Pattern" %>
<%@ page import="java.util.regex.Matcher" %>
<%!
public double getPriceSum(String input){            //定义方法
    Pattern pattern;                                 //模式对象
    Matcher matcher;                                 //匹配对象
    String regex = "-?[0-9][0-9]*[.]?[0-9]*";       //匹配数字的正则表达式
    pattern = Pattern.compile(regex);                //初始化模式对象
    matcher = pattern.matcher(input);                //初始化匹配对象,用于检索 input
    double sum = 0;
    while(matcher.find()) {
        String str = matcher.group();
        sum += Double.parseDouble(str);
    }
    return sum;
}
%>
<HTML><body bgcolor=cyan>
<p style="font-family:黑体;font-size:20">
<br>张三的菜单:</br>
<% String str = "麻辣豆腐:20.6元,红烧肉:68.9元";
%>
<br><%= str %></br>
<br><b><i>消费总和:<%= getPriceSum(str) %></i></b></br>
<br>李四的菜单:</br>
<% str = "剁椒鱼头:62.9元,烤鸭:199元,红焖大虾:289.9元";
%>
<br><%= str %></br>
<br><b><i>消费总和:<%= getPriceSum(str) %></i></b></br>
</p>
</body></HTML>
```

2.8.2 实验 2 日期时间

❶ 实验目的

掌握怎样在 JSP 页面中使用 include 指令标记在 JSP 页面中静态插入一个文件的内容。

❷ 实验要求

（1）页面 time.jsp 负责显示该页面被访问时的日期时间。

（2）编写 JSP 页面 userTime.jsp，要求 useTime.jsp 使用 include 指令标记在当前页面中嵌入 time.jsp。

（3）在 Tomcat 服务器的 webapps 目录下（比如，D:\apache-tomcat-9.0.26\webapps）新建名字是 ch2_practice_two 的 Web 服务目录。time.jsp 和 useTime.jsp 保存到 ch2_practice_two 目录中。

（4）用浏览器访问 JSP 页面 userTime.jsp。

❸ 参考代码

参考代码运行效果如图 2.14 所示。

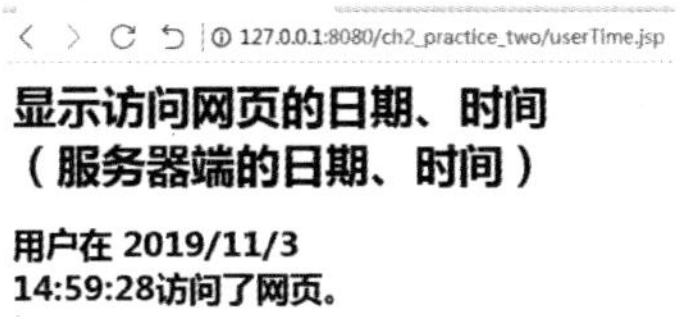

图 2.14　显示日期时间

time.jsp

```
<%@ page import = "java.time.LocalDate" %>
<%@ page import = "java.time.LocalTime" %>
<%
    LocalDate date = LocalDate.now();
    LocalTime time = LocalTime.now();
%>
<h2>
用户在 <%= date.getYear() %>/<%= date.getMonthValue() %>/
<%= date.getDayOfMonth() %><br>
<%= time.getHour() %>:<%= time.getMinute() %>:<%= time.getSecond() %>访问了网页。
</h2>
```

userTime.jsp

```
<%@ page contentType = "text/html" %>
<%@ page pageEncoding = "utf-8" %>
<HTML><body>
<h1>显示访问网页的日期、时间<br>(服务器端的日期、时间)</h1>
<%@ include file = "time.jsp" %>
</body></HTML>
```

2.8.3 实验 3 听英语

❶ 实验目的

掌握怎样在 JSP 页面中使用 include 动作标记在 JSP 页面中动态加载一个文本文件和音

频文件。

❷ 实验要求

（1）在 Tomcat 服务器的 webapps 目录下（比如，D:\apache-tomcat-9.0.26\webapps）新建名字是 ch2_practice_three 的 Web 服务目录，在 ch2_practice_three 目录下新建一个名字是 english 的目录。

（2）编写 JSP 页面 listenEnglish.jsp，要求 listenEnglish.jsp 使用两个 include 动作标记分别加载一个文本文件 english.txt 和一个能播放音频文件 english.mp3 的 JSP 文件 audio.jsp。要求 english.txt 内容是一篇英文课文（编码是 UTF-8），english.mp3 是 english.txt 的英文朗读。listenEnglish.jsp 和 audio.jsp 保存在 ch2_practice_three 目录中，english.txt 和 english.mp3 保存在 english 目录中。

（3）用浏览器访问 JSP 页面 listenEnglish.jsp。

❸ 参考代码

参考代码运行效果如图 2.15 所示。

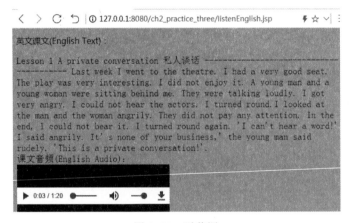

图 2.15　听英语

listenEnglish.jsp

```
<%@ page contentType = "text/html" %>
<%@ page pageEncoding = "utf-8" %>
<HTML><body bgcolor = cyan>
<br>英文课文(English Text)：</br>
<p style = "font-family:宋体;font-size:18;color:black">
<jsp:include page = "english/english.txt" />
<br>课文音频(English Audio)：</br>
  <jsp:include page = "audio.jsp" />
</p></body></HTML>
```

audio.jsp

```
<%@ page contentType = "text/html" %>
<%@ page pageEncoding = "utf-8" %>
<HTML><body bgcolor = pink>
<embed src = "english/english.mp3" autostart = false>
课文音频
</embed>
</body></HTML>
```

▶ 2.8.4 实验4 看电影

❶ 实验目的

通过模拟用 10 元购买票价是 2 元的电影票一张,掌握怎样在 JSP 页面中使用 forward 动作标记转向其他页面。

❷ 实验要求

(1) 在 Tomcat 服务器的 webapps 目录下(比如,D:\apache-tomcat-9.0.26\webapps)新建名字是 ch2_practice_four 的 Web 服务目录。在 ch2_practice_four 目录下新建一个名字是 movie 的目录,将名字是 movie.mp4 的视频保存在该目录中。

(2) 编写 watchMovie.jsp、change5.jsp、change2.jsp 和 change1.jsp 页面,并保存到 ch2_practice_four 目录中。其他要求如下:

- watchMovie.jsp 负责用 10 元购买票价是 2 元的电影票一张,需要找赎(找零)8 元。
- watchMovie.jsp 使用 forward 动作标记转向 change5.jsp 页面,并向其传递正整数 8。
- change5.jsp 负责用 5 元找赎(找零),完成找零 1 张 5 元,将剩余的 3 元交给 change2.jsp,即使用 forward 动作标记转向 change2.jsp 页面,并向其传递正整数 3。
- change2.jsp 负责用 2 元找赎(找零),完成找零 1 张 2 元,将剩余的 1 元交给 change1.jsp,即使用 forward 动作标记转向 change1.jsp 页面,并向其传递正整数 1。
- change1.jsp 负责用 1 元找赎(找零),完成找零任务。然后提供播放视频的 HTML 标记。

(3) 用浏览器访问 JSP 页面 watchMovie.jsp。

❸ 参考代码

参考代码运行效果如图 2.16 所示。

图 2.16 看电影

watchMovie.jsp

```
<%@ page contentType = "text/html" %>
<%@ page pageEncoding = "utf-8" %>
<HTML><body bgcolor = yellow>
<h1>用一张 10 元购买票价是 2 元的电影票一张。
<%   int backMoney = 0;
```

```jsp
         backMoney = 10 - 2;
%>
<jsp:forward page = "change5.jsp">
    <jsp:param name = "number" value = "<% = backMoney %>" />
    <jsp:param name = "mess" value = "" />
</jsp:forward>
</body></HTML>
```

change5.jsp

```jsp
<%@ page contentType = "text/html" %>
<%@ page pageEncoding = "utf-8" %>
<HTML><body bgcolor = cyan>
<% request.setCharacterEncoding("gb2312");
   String backMoneyStr = request.getParameter("number");
   String mess = request.getParameter("mess");
   int backMoney = 0;
   int count = 0;
   int coin = 5;
   backMoney = Integer.parseInt(backMoneyStr);
   while(true){
      count ++;
      if(count * coin > backMoney)
         break;
   }
   backMoney = backMoney - (count - 1) * coin;
   mess = mess + "<br>找赎" + (count - 1) + "张面值" + coin + "元的钱币";
   if(backMoney > 0) {
%>     <jsp:forward page = "change2.jsp">
         <jsp:param name = "number" value = "<% = backMoney %>" />
         <jsp:param name = "mess" value = "<% = mess %>" />
       </jsp:forward>
<%  }
    else {
       out.print("<br>" + mess);
%>     <br><embed src = "movie/movie.mp4" width = 300 height = 270
             autostart = false>
       看电影
       </embed>
<%  }
%>
</body></HTML>
```

change2.jsp

```jsp
<%@ page contentType = "text/html" %>
<%@ page pageEncoding = "utf-8" %>
<HTML><body bgcolor = cyan>
<% request.setCharacterEncoding("gb2312");
   String backMoneyStr = request.getParameter("number");
   String mess = request.getParameter("mess");
   int backMoney = 0;
```

```jsp
            int count = 0;
            int coin = 2;
            backMoney = Integer.parseInt(backMoneyStr);
            while(true){
                count ++;
                if(count * coin > backMoney)
                    break;
            }
            backMoney  = backMoney - (count - 1) * coin;
            mess = mess + "<br>找赎" + (count - 1) + "张面值" + coin + "元的钱币";
            if(backMoney > 0) {
%>
                <jsp:forward page = "change1.jsp" >
                    <jsp:param name = "number" value = "<% = backMoney %>" />
                    <jsp:param name = "mess" value = "<% = mess %>" />
                </jsp:forward>
<%          }
            else {
                //out.print("<br>" + mess);
%>          <br><embed src = "movie/movie.mp4" width = 300 height = 270
                autostart = false >
                看电影
                </embed>
<%      }
%>
</body></HTML>
```

change1.jsp

```jsp
<%@ page contentType = "text/html" %>
<%@ page pageEncoding = "utf-8" %>
<HTML><body bgcolor = cyan>
<% request.setCharacterEncoding("gb2312");
    String backMoneyStr = request.getParameter("number");
    String mess = request.getParameter("mess");
    int backMoney = 0;
    int count = 0;
    int coin = 1;
    backMoney = Integer.parseInt(backMoneyStr);
    while(true){
        count ++;
        if(count * coin > backMoney)
            break;
    }
    backMoney  = backMoney - (count - 1) * coin;
    mess = mess + "<br>找赎" + (count - 1) + "张面值" + coin + "元的钱币";
    out.print("<br>" + mess);
%>
<br><embed src = "movie/movie.mp4" width = 300 height = 270   autostart = false>
        看电影
</embed>
</body></HTML>
```

2.9 小结

- 一个 JSP 页面可由普通的 HTML 标记、JSP 标记、成员变量和方法的声明、Java 程序片和 Java 表达式组成。JSP 引擎把 JSP 页面中的 HTML 标记交给用户的浏览器执行显示,负责处理 JSP 标记、变量和方法,同时负责运行 Java 程序片、计算 Java 表达式,并将需要显示的结果发送给用户的浏览器。
- JSP 页面中的成员变量是被所有用户共享的变量。Java 程序片可以操作成员变量,任何一个用户对 JSP 页面成员变量操作的结果,都会影响到其他用户。
- 如果多个用户访问一个 JSP 页面,那么该页面中的 Java 程序片就会被执行多次,分别运行在不同的线程中,即运行在不同的时间片内。运行在不同线程中的 Java 程序片的局部变量互不干扰,即一个用户改变 Java 程序片中的局部变量的值不会影响其他用户的 Java 程序片中的局部变量。
- page 指令用来定义整个 JSP 页面的一些属性和这些属性的值。比较常用的两个属性是 contentType 和 import。page 指令只能为 contentType 指定一个值,但可以为 import 属性指定多个值。
- include 指令标记是在编译阶段就处理所需要的文件,被处理的文件在逻辑和语法上依赖于当前 JSP 页面,其优点是页面的执行速度快;而 include 动作标记是在 JSP 页面运行时才处理文件,被处理的文件在逻辑和语法上独立于当前 JSP 页面,其优点是可以使用 param 子标记更灵活地处理所需要的文件。

习题 2

1. "<%!"和"%>"之间声明的变量与"<%"和"%>"之间声明的变量有何不同?
2. 如果有两个用户访问一个 JSP 页面,该页面中的 Java 程序片将被执行几次?
3. 是否允许 JSP 页面同时含有如下两条 page 指令:

```
<%@ page contentType = "text/html;charset = gb2312" %>
<%@ page contentType = "application/msword" %>
```

4. 是否允许 JSP 页面同时含有如下两条 page 指令:

```
<%@ page import = "java.util.*" %>
<%@ page import = "java.sql.*" %>
```

5. 假设有两个用户访问下列 JSP 页面 hello.jsp,请问第一个访问和第二个访问 hello.jsp 页面的用户看到的页面的效果有何不同?

hello.jsp

```
<%@ page contentType = "text/html" %>
<%@ page pageEncoding = "utf-8" %>
<%@ page isThreadSafe = "false" %>
<HTML><BODY>
```

```
<%!   int sum = 1;
      void add(int m){
          sum = sum + m;
      }
%>
<% int n = 100;
   add(n);
%>
<% = sum %>
</BODY></HTML>
```

6. 请编写一个简单的 JSP 页面,显示希腊字母表。

7. 请简单叙述 include 指令标记和 include 动作标记的不同。

8. 编写三个 JSP 页面:main.jsp、circle.jsp 和 ladder.jsp,将三个 JSP 页面保存在同一 Web 服务目录中。main.jsp 使用 include 动作标记加载 circle.jsp 和 ladder.jsp 页面。circle.jsp 页面可以计算并显示圆的面积,ladder.jsp 页面可以计算并显示梯形的面积。当 circle.jsp 和 ladder.jsp 被加载时获取 main.jsp 页面 include 动作标记的 param 子标记提供的圆的半径以及梯形的上底、下底和高的值。

第 3 章 Tag文件与Tag标记

本章导读

  主要内容
- Tag 文件的结构
- Tag 标记
- Tag 文件中的常用指令

  难点
- Tag 文件中的 attribute 指令
- Tag 文件中的 variable 指令

  关键实践
- 解析单词

  一个 Web 应用中的许多 JSP 页面可能需要使用某些相同的信息,如都需要使用相同的导航栏、标题等。如果能将许多页面都需要的共同的信息形成一种特殊文件,而且各个 JSP 页面都可以使用这种特殊的文件,那么这样的特殊文件就是可复用的代码。代码复用是软件设计的一个重要方面、是衡量软件可维护性的重要指标之一。

  第 2 章学习了 include 指令标记和 include 动作标记,使用这两个标记可以实现代码的复用。但是,在某些情况下,使用 include 指令标记和 include 动作标记有一定的缺点,比如,如果 include 指令标记或动作标记要处理的文件是一个 JSP 文件,那么用户可以在浏览器的地址栏中直接输入该 JSP 文件所在 Web 服务目录访问这个 JSP 文件,这可能不是 Web 应用所希望发生的,因为该 JSP 文件也许仅仅是个导航条,仅仅供其他 JSP 文件使用 include 指令标记或动作标记来嵌入或动态加载的,而不是让用户直接访问的。另外,include 指令标记和 include 动作标记允许所要处理的文件存放在 Web 服务目录中的任意子目录中,不仅显得杂乱无章,而且使得 include 标记和所处理文件的所在目录的结构形成了耦合,不利于 Web 应用的维护。

  本章我们将学习一种特殊的文本文件:Tag 文件。Tag 文件和 JSP 文件很类似,可以被 JSP 页面动态加载调用,实现代码的复用(但用户不能通过该 Tag 文件所在 Web 服务目录直接访问 Tag 文件)。

  本章在 webapps 目录下新建一个 Web 服务目录 ch3,除非特别约定,例子中的 JSP 页面均保存在 ch3 目录中。

3.1 Tag 文件

视频讲解

▶ 3.1.1 Tag 文件的结构

  Tag 文件是扩展名为.tag 的文本文件,其结构和 JSP 文件类似。一个 Tag 文件中可以有普通的 HTML 标记符、某些特殊的指令标记(见 3.4 节)、成员变量声明和方法的定义、Java

程序片和 Java 表达式。以下是一个简单的 Tag 文件 oddNumberSum.tag,负责计算 100 内的全部奇数的代数和。

oddNumberSum.tag

```
<%@ tag pageEncoding="utf-8"%>
<p style="font-family:宋体;font-size:36">
1~100 内的奇数之和：
<%   int sum = 0,i = 1;
        for(i = 1;i<=100;i++){
            if(i%2 == 1)
                sum = sum + i;
        }
        out.println(sum);
%>
</p>
```

▶ 3.1.2　Tag 文件的保存

❶ Tag 文件所在目录

Tag 文件可以实现代码的复用,即 Tag 文件可以被许多 JSP 页面使用。为了能让一个 Web 应用中的 JSP 页面使用某一个 Tag 文件,必须把这个 Tag 文件存放到 Tomcat 服务器指定的目录中,也就是说,如果某个 Web 服务目录下的 JSP 页面准备调用一个 Tag 文件,那么必须在该 Web 服务目录下,建立如下的目录结构：

```
Web 服务目录\WEB-INF\tags
```

例如：

```
ch3\WEB-INF\tags
```

其中的 WEB-INF(字母大写)和 tags 都是固定的目录名称,而 tags 下的子目录的名称可由用户给定。

一个 Tag 文件必须保存到 tags 目录或其下的子目录中。这里把 3.1.1 节中的 oddNumberSum.tag 保存到 ch3\WEB-INF\tags 目录中。

❷ Tag 文件的编码

保存 Tag 文件时按照 Tag 文件指定的编码保存,例如 Tag 文件使用 tag 指令(见稍后的 3.4 节)：

```
<%@ tag pageEncoding="utf-8"%>
```

指定的编码是 UTF-8,因此需要按照 UTF-8 编码保存 Tag 文件。例如,用文本编辑器"记事本"编辑 Tag 文件,在保存该 Tag 文件时,将"保存类型(T)"选择为"所有文件(*.*)",将"编码(E)"选择为"UTF-8"。

3.2 Tag 标记

3.2.1 Tag 标记与 Tag 文件

视频讲解

某个 Web 服务目录下的 Tag 文件只能由该 Web 服务目录中的 JSP 页面调用,JSP 页面必须通过 Tag 标记来调用一个 Tag 文件。

Tag 标记的名字和 Tag 文件的名字一致,也就是说,当我们编写了一个 Tag 文件并保存到特定目录中后(见 3.1.2 节),也就给出了一个 Tag 标记,该标记的格式为:

```
<Tag 文件的名字 />
```

或

```
<Tag 文件的名字> 其他内容(称为标体内容)</Tag 文件的名字>
```

一个 Tag 文件对应着一个 Tag 标记,把全体 Tag 标记称之为一个自定义标记库或简称为标记库。

3.2.2 Tag 标记的使用

一个 JSP 页面通过使用 Tag 标记来调用一个 Tag 文件。Web 服务目录下的一个 JSP 页面在使用 Tag 标记来调用一个 Tag 文件之前,必须首先使用 taglib 指令标记引入该 Web 服务目录下的标记库,只有这样,JSP 页面才可以使用 Tag 标记调用相应的 Tag 文件。

taglib 指令的格式如下:

```
<%@ taglib tagdir = "标记库的位置" prefix = "前缀">
```

例如:

```
<%@ taglib tagdir = "/WEB-INF/tags" prefix = "computer" %>
```

引入标记库后,JSP 页面就可以使用带前缀的 Tag 标记调用相应的 Tag 文件,其中的前缀由<taglib>指令中的 prefix 属性指定。例如 JSP 如下使用 Tag 标记调用相应的 Tag 文件:

```
<computer:oddNumberSum />
```

taglib 指令中的 prefix 给出的前缀由用户自定义,其好处是,通过前缀可以有效地区分不同标记库中具有相同名字的标记文件。

注:JSP 页面使用 Tag 标记时,冒号:的左右不要有空格。

例 3_1 中的 JSP 页面使用 Tag 标记调用 oddNumberSum.tag(该 Tag 文件见 3.1.1 节)计算 100 之内的奇数和。

例 3_1

example3_1.jsp(效果如图 3.1 所示)

```
<%@ page contentType = "text/html" %>
```

```
<%@ page pageEncoding = "utf-8" %>
<%@ taglib tagdir = "/WEB-INF/tags" prefix = "computer" %>
<HTML><body bgcolor = cyan>
<h1>调用 Tag 文件计算 100 内奇数和：</h1>
<computer:oddNumberSum /><%-- 使用 Tag 标记 --%>
</body></HTML>
```

图 3.1　使用 Tag 标记

如果把 3.1.1 节中的 oddNumberSum.tag 保存在 ch3\WEB-INF\tags\example1 目录中,那么只要将上述 JSP 页面的 taglib 指令修改为：

```
<%@ taglib tagdir = "/WEB-INF/tags/example1" prefix = "computer" %>
```

即可。

3.2.3　Tag 标记的运行原理

Tomcat 服务器处理 JSP 页面中的 Tag 标记的原理如下：
- 如果该 Tag 标记对应的 Tag 文件是首次被 JSP 页面调用,那么 Tomcat 服务器会将 Tag 文件转译成一个 Java 文件,并编译这个 Java 文件生成字节码文件,然后执行这个字节码文件(这和执行 JSP 页面的原理类似)。
- 如果该 Tag 文件已经被转编译为字节码文件,Tomcat 服务器将直接执行这个字节码文件。
- 如果对 Tag 文件进行了修改,那么 Tomcat 服务器会重新将 Tag 文件转译成一个 Java 文件,并编译这个 Java 文件生成字节码文件,然后执行这个字节码文件。

3.3　Tag 文件中的常用指令

视频讲解

与 JSP 文件类似,Tag 文件中也有一些常用指令,这些指令将影响 Tag 文件的行为。Tag 文件中经常使用的指令有 tag、taglib、include、attribute、variable。

以下将分别讲述上述指令在 Tag 文件中的作用和用法。

3.3.1　tag 指令

Tag 文件中的 tag 指令类似于 JSP 文件中的 page 指令。Tag 文件通过使用 tag 指令可以指定某些属性的值,以便从总体上影响 Tag 文件的处理和表示。tag 指令的语法如下：

```
<%@ tag 属性1 = "属性值" 属性2 = "属性值" … 属性n = "属性值" %>
```

在一个 Tag 文件中可以使用多个 tag 指令,因此我们经常使用多个 tag 指令为属性指定需要的值:

```
<%@ tag 属性1 = "属性值" %>
<%@ tag 属性2 = "属性值" %>
...
<%@ tag 属性n = "属性值" %>
```

❶ language 属性

language 属性的值指定 Tag 文件使用的脚本语言,目前只能取值 Java,其默认值就是 Java,因此在编写 Tag 文件时,没有必要使用 tag 指令指定 language 属性的值。

❷ import 属性

import 属性的作用是为 Tag 文件引入包中的类,这样就可以在 Tag 文件的程序片部分、变量及方法定义部分、表达式部分使用包中的类。import 属性可以取多个值,import 属性默认已经有如下值:"java.lang.*" "javax.servlet.*" "javax.servlet.jsp.*" "javax.servlet.http.*"。

❸ pageEncoding

该属性指定 Tag 文件的字符编码,其默认值是 ISO-8859-1。目前,为了避免显示信息出现乱码现象,Tag 文件需要将该属性值设置为 UTF-8。

▶ 3.3.2 include 指令

在 Tag 文件中也有和 JSP 文件类似的 include 指令标记,其使用方法和作用与 JSP 文件中的 include 指令标记类似。

▶ 3.3.3 attribute 指令

Tag 文件充当着可复用代码的角色,如果一个 Tag 文件允许使用它的 JSP 页面向该 Tag 文件传递数据,就使得 Tag 文件的功能更为强大。在 Tag 文件中通过使用 attribute 指令让使用它的 JSP 页面向该 Tag 文件传递需要的数据。attribute 指令的格式如下:

```
<%@ attribute name = "对象名字" required = "true"|"false" type = "对象的类型" %>
```

例如 Tag 文件 myTag.tag 中有如下 attribute 指令:

```
<%@ attribute name = "result" required = "true" type = "java.lang.Double" %>
```

那么就相当于 Tag 文件中有了一个名字是 result 的对象,但 Tag 文件不需要创建该对象 result,而是等待 JSP 页面将一个 Double 型的对象的引用传递给 result。

attribute 指令中的 name 属性是必需的,该属性的值是一个对象的名字。JSP 页面在调用 Tag 文件时,可向 name 属性指定的对象传递一个引用。需要特别注意的是,type 在指定对象类型时,必须使用包名,比如,不可以将 java.lang.Double 简写为 Double。如果 attribute 指令中没有使用 type 指定对象的类型,那对象的类型默认是 java.lang.String 类型。

JSP 页面使用 Tag 标记向所调用的 Tag 文件中 name 指定的对象传递一个引用,方式如下:

```
<前缀:Tag 文件名字 对象名字 = "对象的引用" />
```

比如,JSP 页面使用 Tag 标记(假设标记的前缀为 computer)调用 myTag.tag:

```
< computer:myTag result = "new Double(3.1415926)" />
```

就向 myTag.tag 中 attribute 指令给出的对象 result 对象传递了一个 Double 对象的引用。

attribute 指令中的 required 属性也是可选的,如果省略 required 属性,那么 required 的默认值是 false。当指定 required 的值是 true 时,调用该 Tag 文件的 JSP 页面必须向该 Tag 文件中 attribute 指令中的 name 属性给出的对象传递一个引用,当指定 required 的值是 false 时,调用该 Tag 文件的 JSP 可以向该 Tag 文件中 attribute 指令中的 name 属性给出的对象传递或不传递对象的引用。

注:在 Tag 文件中不可以再定义和 attribute 指令中的 name 属性给出的对象具有相同名字的变量,否则将隐藏 attribute 指令中给出的对象,使其失效。

在下面的例 3_2 中,triangle.tag 存放在 ch3\WEB-INF\tags\example2 目录中,该 Tag 文件负责计算、显示三角形的面积。example3_2.jsp 页面保存在 ch3 中,使用 Tag 标记调用 triangle.tag 文件,并且向 triangle.tag 传递三角形三边的长度。

例 3_2

example3_2.jsp(效果如图 3.2 所示)

```jsp
<%@ page contentType = "text/html" %>
<%@ page pageEncoding = "utf-8" %>
<%@ taglib tagdir = "/WEB-INF/tags/example2" prefix = "getTriangleArea" %>
<HTML><body bgcolor = yellow>
<p style = "font-family:宋体;font-size:36;color:blue">
<%-- 使用 Tag 标记: --%>
<getTriangleArea:triangle sideA = "15" sideB = "16" sideC = "20"/>
</p>
</body></HTML>
```

triangle.tag

```jsp
<%@ tag pageEncoding = "utf-8" %>
<%@ attribute name = "sideA" required = "true" %>
<%@ attribute name = "sideB" required = "true" %>
<%@ attribute name = "sideC" required = "true" %>
<%! public String getArea(double a,double b,double c) {
       if(a + b > c&&a + c > b&&c + b > a) {
           double p = (a + b + c)/2.0;
           double area = Math.sqrt(p * (p - a) * (p - b) * (p - c));
           String result = String.format("%.2f",area);
           return "<br>三角形面积(小数点保留2位):" + result;
       }
       else
           return("<br>" + a + "," + b + "," + c + "不能构成一个三角形,无法计算面积");
    }
%>
<% out.println("<BR>三边: " + sideA + "," + sideB + "," + sideC);
   double a = Double.parseDouble(sideA);
```

```
        double b = Double.parseDouble(sideB);
        double c = Double.parseDouble(sideC);
        out.println(getArea(a,b,c));
%>
```

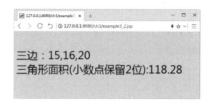

图 3.2　调用 Tag 文件计算面积

下面的例 3_3 中，JSP 页面 example3_3.jsp 只负责将一组随机数据存放到链表(java.util.LinkedList 类型对象)中，然后将链表传递给 sort.tag，sort.tag 负责按低到高顺序显示链表中的数据。sort.tag 存放在 ch3\WEB-INF\tags\example3 目录中，example3_3.jsp 保存在 ch3 目录中。

例 3_3

example3_3.jsp（效果如图 3.3 所示）

```
<%@ page contentType = "text/html" %>
<%@ page pageEncoding = "utf-8" %>
<%@ page import = "java.util.LinkedList" %>
<%@ page import = "java.util.Random" %>
<%@ taglib tagdir = "/WEB-INF/tags/example3" prefix = "sortNumber" %>
<HTML><body bgcolor = #CCCCCC>
<% LinkedList<Double> listNumber = new LinkedList<Double>();
   Random random = new Random();
   for(int i = 0;i < 3;i++) {
       double d = random.nextDouble();        //[0,1]之间的随机数
       listNumber.add(d);
   }
%>
<p style = "font-family:宋体;font-size:36;color:blue">
排序数据
<sortNumber:sort list = "<% = listNumber %>"/>   <%-- 使用Tag标记 --%>
</body></HTML>
```

sort.tag

```
<%@ attribute name = "list" required = "true" type = "java.util.LinkedList" %>
<%@ tag import = "java.util.Collections" %>
<%@ tag import = "java.util.Iterator" %>
<%  Collections.sort(list);                    //排序链表
    Iterator<Double> ite = list.iterator();    //得到迭代器
    while(ite.hasNext()) {                     //遍历链表
        out.print("<br>" + ite.next());
    }
%>
```

图 3.3 调用 Tag 文件排序数据

3.3.4 variable 指令

Tag 文件通过使用 attribute 指令，可以使得调用该 Tag 文件的 JSP 页面动态地向其传递数据。在某些 Web 应用中，JSP 页面不仅希望向 Tag 文件传递数据，而且希望 Tag 文件能返回数据给 JSP 页面。比如，许多 JSP 页面可能都需要调用某个 Tag 文件对某些数据进行基本的处理，但不希望 Tag 文件做进一步的特殊处理以及显示数据，因为各个 JSP 页面对数据的进一步处理或显示格式的要求是不同的。因此，JSP 页面希望 Tag 文件将数据的基本处理结果存放在某些对象中，将这些对象返回给当前 JSP 页面即可。

Tag 文件通过使用 variable 指令可以将 Tag 文件中的对象返回给调用该 Tag 文件的 JSP 页面。

❶ variable 指令的格式

variable 指令的格式如下：

```
<%@ variable name-given = "对象名" variable-class = "对象类型" scope = "有效范围" %>
```

variable 指令中属性 name-given 的值就是 Tag 文件返回给 JSP 页面的对象。该对象的名字必须符合标识符规定，即名字可以由字母、下画线、美元符号和数字组成，并且第一个字符不能是数字字符。variable 指令中属性 variable-class 的值是返回的对象的类型，对象的类型必须带有包名，比如 java.lang.Double、java.time.LocalDate 等类型。如果 variable 指令中没有使用 variable-class 给出对象的类型，那么对象的类型是 java.lang.String 类型。

variable 指令中 scope 属性的值指定对象的有效范围，scope 的值可以取 AT_BEGIN、NESTED 和 AT_END。当 scope 的值是 AT_BEGIN 时，JSP 页面一旦开始使用 Tag 标记，就得到了 variable 指令返回给 JSP 页面的对象，JSP 页面就可以在 Tag 标记的标记体中或 Tag 标记结束后的各个部分中使用 variable 指令返回给 JSP 页面的对象。当 scope 的值是 NESTED 时，JSP 页面只可以在 Tag 标记的标记体中使用 variable 指令返回给 JSP 页面的对象。当 scope 的值是 AT_END 时，JSP 页面只可以在 Tag 标记结束后。才可以使用 variable 指令返回给 JSP 页面的对象。

下面的 variable 指令给出的对象的名字是 time，类型为 java.time.LocalDate，有效范围是 AT_END：

```
<%@ variable name-given = "time"
    variable-class = "java.time.LocalDate" scope = "AT_END" %>
```

❷ 对象的返回

Tag 文件为了给 JSP 页面返回一个对象，就必须将返回的对象的名字以及该对象的引用存储到 Tomcat 服务器提供的内置对象 jspContext 中。Tag 文件只有将对象的名字及其引用存储到 jspContext 中，JSP 页面才可以使用该对象。比如，Tag 文件的 variable 指令：

```
<%@ variable name-given="time"
    variable-class="java.time.LocalDate" scope="AT_END" %>
```

为 JSP 页面返回名字是 time 的 LocalDate 对象。那么 Tag 文件中必须让 jspContext 调用

```
setAttribute("对象名",对象的引用);
```

方法存储名字是 time 的对象以及该对象的引用,例如:

```
jspContext.setAttribute("time",LocalDate.now());
```

将名字是 time 的 LocalDate 对象存储到 jspContext 中。

下面的例 3_4 中,JSP 页面 example3_4.jsp 将 String 对象交给 Tag 文件 handleData.tag,handleData.tag 解析出 String 对象的字符序列中的全部数字,并计算出数字总和,将数字总和放在 Double 对象 price 中,然后返回给 JSP 页面 example3_4.jsp。example3_4.jsp 输出 price 对象中的数字总和。handleData.tag.tag 存放在 ch3\WEB-INF\tags\example4 目录中,example3_4.jsp 保存在 ch3 目录中。

例 3_4

example3_4.jsp(效果如图 3.4 所示)

```jsp
<%@ page contentType="text/html" %>
<%@ page pageEncoding="utf-8" %>
<%@ taglib tagdir="/WEB-INF/tags/example4" prefix="getPrice" %>
<HTML><body bgcolor=#FFCCFF>
<% String str="麻辣豆腐:20.6元,红烧肉:68.9元,烤鸭:199元";
%>
<getPrice:handleData mess="<%= str %>"/>   <%-- 使用 Tag 标记 --%>
<p style="font-family:宋体;font-size:36">
菜单:<br>"<%= str %>"<br>价格总和:
<%= price %>        <%-- 使用 Tag 标记返回的 Double 对象 price --%>
</p>
<% str="毛巾:2.6元,香皂:6.9元,牙刷:12.3元";
%>
<getPrice:handleData mess="<%= str %>"/>
<p style="font-family:黑体;font-size:36;color:blue">
购物小票:<br>"<%= str %>"<br>价格总和:
<%= price %>
</p>
</body></HTML>
```

handleData.tag

```jsp
<%@ attribute name="mess" required="true" type="java.lang.String" %>
<%@ tag import="java.util.regex.Pattern" %>
<%@ tag import="java.util.regex.Matcher" %>
<%@ variable name-given="price" variable-class="java.lang.Double"
   scope="AT_BEGIN" %>
<%!
public Double getPriceSum(String input){           //定义方法
    Pattern pattern;                                //模式对象
```

```
            Matcher matcher;                            //匹配对象
            String regex = "-?[0-9][0-9]*[.]?[0-9]*";   //匹配数字的正则表达式
            pattern = Pattern.compile(regex);           //初始化模式对象
            matcher = pattern.matcher(input);           //初始化匹配对象,用于检索 input
            double sum = 0;
            while(matcher.find()) {
                String str = matcher.group();
                sum += Double.parseDouble(str);
            }
            return new Double(sum);
        }
     %>
     <%  //将返回的 Double 对象放在 jspContext 中,用名字 price 返回给 JSP 页面
         jspContext.setAttribute("price",getPriceSum(mess));
     %>
```

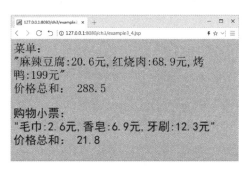

图 3.4　向 JSP 页面返回对象

注：在 JSP 页面中不可以再定义与 Tag 文件返回的对象具有相同名字的变量,否则 Tag 文件无法将 variable 指令给出的对象返回给 JSP 页面(并将出现编译错误)。如果 Tag 文件同时使用 variable 指令和 attribute 指令,那么 variable 指令中 name-given 和 attribute 指令中 name 给出的对象不能相同(否则将出现编译错误)。

▶ 3.3.5　taglib 指令

JSP 页面或 Tag 文件都可以使用 taglib 指令引入标记库(如前面各个例子所示)。taglib 指令格式如下：

`<%@ taglib tagdir = "自定义标记库的位置" prefix = "前缀">`

一个 Tag 文件也可以使用几个 taglib 指令标记引入若干个标记库,例如：

`<%@ taglib tagdir = "/WEB-INF/tags" prefix = "beijing" %>`
`<%@ taglib tagdir = "/WEB-INF/tags/tagsTwo" prefix = "dalian" %>`

3.4　上机实验

提供了详细的实验步骤要求,按步骤完成,提升学习效果,积累经验,不断提高 Web 设计能力。

视频讲解

3.4.1 实验1 解析单词

❶ 实验目的

掌握 JSP 页面中使用 Tag 文件，向 Tag 文件传递对象，Tag 文件负责处理对象中的数据。

❷ 实验要求

（1）JSP 页面 giveText.jsp 负责将 String 对象，比如 String str = "how are you"，传递给所调用的 Tag 文件 backWords.tag。

（2）backWords.tag 文件负责解析出 String 对象中的单词，并将这些单词返回给 JSP 页面 giveText.jsp。

（3）JSP 页面 giveText.jsp 负责显示 backWords.tag 返回给它的单词。

（4）在 Tomcat 服务器的 webapps 目录下（比如，D:\apache-tomcat-9.0.26\webapps）新建一个名字是 ch3_practice_one 的 Web 服务目录。把 giveText.jsp 文件保存到 ch3_practice_one 目录中。在 ch3_practice_one 目录下再建立目录结构：\WEB-INF\tags\ practice1，将 backWords.tag 保存在 practice1 目录中。

（5）用浏览器访问 JSP 页面 giveText.jsp。

❸ 参考代码

参考代码运行效果如图 3.5 所示。

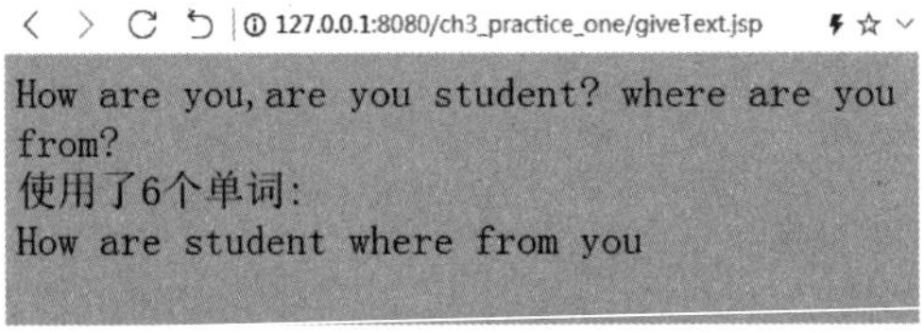

图 3.5 解析单词

giveText.jsp

```
<%@ page contentType = "text/html" %>
<%@ page pageEncoding = "utf - 8" %>
<%@ page import = "java.util.Iterator" %>
<%@ taglib tagdir = "/WEB - INF/tags/practice1" prefix = "getWords" %>
<HTML><body bgcolor = #CCCCFF>
<% String str = "How are you,are you student? where are you from? ";
%>
<getWords:backWords okString = "<% = str %>" /><% -- 使用 Tag 标记 -- %>
<p style = "font - family:宋体;font - size:26">
<% = str %><br>
<%
      Iterator<String> ite = words.iterator();    //使用 Tag 标记返回的对象 words
      out.print("使用了" + words.size() + "个单词:<br>");
      while(ite.hasNext()) {                       //遍历集合
         out.print(" " + ite.next());
      }
%>
</p></body></HTML>
```

backWords.tag

```
<%@ tag import = "java.util.HashSet" %>
<%@ tag import = "java.util.regex.Pattern" %>
<%@ tag import = "java.util.regex.Matcher" %>
<%@ attribute name = "okString" required = "true" type = "java.lang.String" %>
<%@ variable name-given = "words" variable-class = "java.util.HashSet"
      scope = "AT_BEGIN" %>
<%
   HashSet<String>    set = new HashSet<String>();   //集合不允许有相同的元素
   Pattern pattern;                                  //模式对象
   Matcher matcher;                                  //匹配对象
   String regex = "[a-zA-Z]+";                       //匹配英文单词
   pattern = Pattern.compile(regex);                 //初始化模式对象
   matcher = pattern.matcher(okString);              //初始化匹配对象,用于检索 okString
   while(matcher.find()) {
       String str = matcher.group();
       set.add(str);
   }
   //将返回的 set 对象放在 jspContext 中,用名字 words 返回给 JSP 页面
   jspContext.setAttribute("words",set);
%>
```

3.4.2　实验 2　显示日历

❶ 实验目的

和实验 1 的目的相同,进一步强化掌握 JSP 页面使用 Tag 文件,即向 Tag 文件传递对象,Tag 文件负责处理数据。

❷ 实验要求

(1) Tag 文件 calendar.tag 负责显示日历。

(2) 编写 JSP 页面 useCalendar.jsp,要求 useCalendar.jsp 使用 Tag 标记使用 calendar.tag,并将日期的年份和月份传递给 calendar.tag。

(3) 在 Tomcat 服务器的 webapps 目录下(比如,D:\apache-tomcat-9.0.26\webapps)新建一个名字是 ch3_practice_two 的 Web 服务目录。把 useCalendar.jsp 文件保存到 ch3_practice_two 目录中。在 ch3_practice_two 目录下再建立目录结构:\WEB-INF\tags\ practice2,将 calendar.tag 保存在 practice2 目录中。

(4) 用浏览器访问 JSP 页面 useCalendar.jsp。

❸ 参考代码

参考代码运行效果如图 3.6 所示。

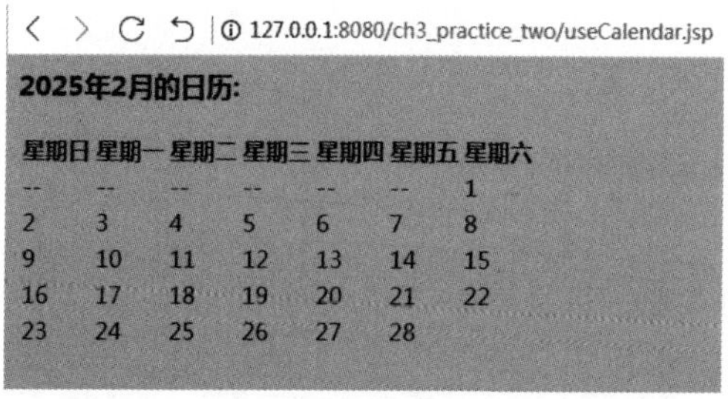

图 3.6　显示日历

useCalendar.jsp

```jsp
<%@ page contentType="text/html" %>
<%@ page pageEncoding="utf-8" %>
<%@ taglib tagdir="/WEB-INF/tags/practice2" prefix="getCalendar" %>
<HTML><body bgcolor=#CCCCFF>
<getCalendar:calendar year="2025" month="2" /><%-- 使用 Tag 标记 --%>
</body></HTML>
```

calendar.tag

```jsp
<%@ tag import="java.time.LocalDate" %>
<%@ tag import="java.time.DayOfWeek" %>
<%@ attribute name="year" required="true" type="java.lang.String" %>
<%@ attribute name="month" required="true" type="java.lang.String" %>
<%
    int y = Integer.parseInt(year);
    int m = Integer.parseInt(month);
    LocalDate date = LocalDate.of(y,m,1);
    int days = date.lengthOfMonth();              //得到该月有多少天
    int space = 0;                                 //存放空白字符的个数
    DayOfWeek dayOfWeek = date.getDayOfWeek();    //得到1号是星期几
    switch(dayOfWeek) {
            case SUNDAY:     space = 0;
                             break;
            case MONDAY:     space = 1;
                             break;
            case TUESDAY:    space = 2;
                             break;
            case WEDNESDAY:  space = 3;
                             break;
            case THURSDAY:   space = 4;
                             break;
            case FRIDAY:     space = 5;
                             break;
            case SATURDAY:   space = 6;
                             break;
    }
    String [] calendar = new String[space + days];  //用于存放日期和1号前面的空白
    for(int i = 0; i < space; i++)
        calendar[i] = " -- ";
    for(int i = space, n = 1; i < calendar.length; i++){
        calendar[i] = String.valueOf(n) ;
        n++;
    }
%>
<h3><%= year %>年<%= month %>月的日历:</h3>
<table border=0>
  <tr><th>星期日</th><th>星期一</th><th>星期二</th><th>星期三</th>
      <th>星期四</th><th>星期五</th><th>星期六</th>
  </tr>
<%
```

```
        int n = 0;
        while(n < calendar.length){
            out.print("<tr>");
            int increment = Math.min(7,calendar.length - n);
            for(int i = n; i < n + increment; i++) {
               out.print("<td>" + calendar[i] + "</td>");
            }
            out.print("</tr>");
            n = n + increment;
        }
%>
</table>
```

习题 3

1. 用户可以使用浏览器直接访问一个 Tag 文件吗？
2. Tag 文件应当存放在怎样的目录中？
3. Tag 文件中的 tag 指令可以设置哪些属性的值？
4. Tag 文件中的 attribute 指令有怎样的作用？
5. Tag 文件中的 variable 指令有怎样的作用？
6. 编写两个 Tag 文件 Rect.tag 和 Circle.tag。Rect.tag 负责计算并显示矩形的面积，Circle.tag 负责计算并显示圆的面积。编写一个 JSP 页面 lianxi6.jsp，该 JSP 页面使用 Tag 标记调用 Rect.tag 和 Circle.tag。调用 Rect.tag 时，向其传递矩形的两个边的长度；调用 Circle.tag 时，向其传递圆的半径。
7. 编写一个 Tag 文件：GetArea.tag 负责求出三角形的面积，并使用 variable 指令返回三角形的面积给调用该 Tag 文件的 JSP 页面。JSP 页面负责显示 Tag 文件返回的三角形的面积。JSP 在调用 Tag 文件时，使用 attribute 指令将三角形三边的长度传递给 Tag 文件。one.jsp 和 two.jsp 都使用 Tag 标记调用 GetArea.tag。one.jsp 返回的三角形的面积保留最多 3 位小数，two.jsp 返回的三角形的面积保留最多 6 位小数。

第 4 章　JSP内置对象

本章导读

　　主要内容
- request 对象
- response 对象
- session 对象
- out 对象
- application 对象

　　难点
- 理解 session 对象
- 使用 session 对象存储数据

　　关键实践
- 计算器
- 成绩与饼图
- 记忆测试

　　本章在 webapps 目录下新建一个 Web 服务目录 ch4，除非特别约定，本章例子中的 JSP 页面均保存在 ch4 目录中。

　　有些对象不用声明就可以在 JSP 页面的 Java 程序片和表达式部分使用，这就是 JSP 的内置对象。JSP 的常用内置对象有 request、response、session、application 和 out。

　　response 和 request 对象是 JSP 内置对象中较重要的两个，这两个对象提供了对服务器和浏览器通信方法的控制。直接讨论这两个对象前，要先对 HTTP 协议——Word Wide Web 底层协议进行简单介绍。

　　Word Wide Web 是怎样运行的呢？在浏览器上输入一个正确的网址后，若一切顺利，网页就出现了。例如，在浏览器输入栏中输入 http://www.renren.com，人人网的主页就出现在浏览器窗口中。这背后是什么在起作用？

　　使用浏览器从网站获取 HTML 页面时，实际在使用 Hypertext Transfer Protocol (HTTP)。HTTP 协议规定了信息在 Internet 上的传输方法，特别规定了浏览器与服务器的交互方法。

　　从网站获取页面时，浏览器在网站上打开了一个对网络服务器的连接，并发出请求。服务器收到请求后回应，所以 HTTP 协议被称作"请求和响应"协议。

　　浏览器请求有某种结构，HTTP 请求包括一个请求行、头域和可能的信息体。最普通的请求类型是对页面的一个简单请求，如下例：

```
get/hello.html HTTP/1.1
Host:www.sina.com.cn
```

这是对网站 www.sina.com.cn 上页面 hello.html 的 HTTP 请求的例子。首行是请求行，规定了请求的方法、请求的资源及使用的 HTTP 协议的版本。

上例中，请求的方法是 get 方法，此方法获取特定的资源。上例中 get 方法用来获取名为 hello.html 的网页。其他请求方法包括 post、head、delete、trace 及 put 方法等。

此例中的第二行是头（header）。Host 头给出了网站上 hello.html 文件的 Internet 地址。此例中，主机是 www.sina.com.cn。

一个典型请求通常包含许多头，称为请求的 HTTP 头。头提供了关于信息体的附加信息及请求的来源。其中有些头是标准的，有些和特定的浏览器有关。

一个请求还可能包含信息体，例如，信息体可包含 HTML 表单的内容。在 HTML 表单上单击 submit 键（提交键）时，该表单内容就由 post 方法或 get 方法在请求的信息体中发送。

服务器在收到请求时，返回 HTTP 响应。响应也有某种结构，每个响应都由状态行开始，可以包含几个头及可能的信息体，称为响应的 HTTP 头和响应信息体，这些头和信息体由服务器发送给用户的浏览器，信息体就是用户请求的网页的运行结果，对于 JSP 页面，就是网页的静态信息。状态行说明了正在使用的协议、状态代码及文本信息。例如，若服务器认为请求出错，则状态行返回错误及对错误的描述，比如 HTTP/1.1 404 Object Not Found。若服务器成功地响应了对网页的请求，返回包含 200 OK 的状态行。

4.1 request 对象

HTTP 通信协议是用户与服务器之间一种提交（请求）信息与响应信息（request/response）的通信协议。在 JSP 中，内置对象 request 封装了用户提交的信息，那么该对象调用相应的方法可以获取封装的信息，即使用该对象可以获取用户提交的信息。

内置对象 request 是实现了 ServletRequest 接口类的一个实例，可以在 Tomcat 服务器的 webapps\tomcat-docs\servletapi 中查找 ServletRequest 接口的方法。

用户通常使用 HTML 的 form 表单（也称 form 标记）请求访问服务器的某个 JSP 页面（或 servlet，见第 6 章），并提交必要信息给所请求的 JSP 页面（servlet）。表单的一般格式是：

```
< form action = "请求访问的页面或 Servlet" method = get | post >
    提交手段
</form>
```

其中，< form >是 form 表单的开始标签、</form >是结束标签，开始标签和结束标签之间是 form 表单的标记体的内容。action 是 form 表单的属性，其属性值给出表单请求访问的 JSP 页面或 servlet。form 表单中的 method 属性取值 get 或 post。get 方法和 post 方法的主要区别是：使用 get 方法提交的信息会在提交的过程中显示在浏览器的地址栏中，而 post 方法提交的信息不会显示在地址栏中。提交手段包括文本框、列表、文本区等，例如：

```
< form action = "tom.jsp" method = "post" >
    < input type = "text" name = "boy" value = "ok" />
    < input type = "submit" name = "submit" value = "提交" />
</form>
```

该 form 表单使用 post 方法向请求访问的 tom.jsp 页面提交信息，提交信息的手段是 text（文本框），文本框的名字由 name 属性的值指定（例如，名字是 boy）。在文本框输入信息

(其中默认信息是 ok),然后单击 submit 提交键向服务器的 JSP 页面 tom.jsp 提交信息。

form 表单请求访问的 JSP 页面可以让内置 request 对象调用 getParameter(String s)方法,并让 s 取值是 input 标记中 name 给出的文本框 text 的名字(比如 boy),获取该表单通过文本框 text 提交的文本信息(即用户在 GUI 控件文本框中输入的文本,默认值是 ok),比如:

```
request.getParameter("boy");
```

▶ 4.1.1 获取用户提交的信息

视频讲解

request 对象获取用户提交信息的最常用的方法是 getParameter(String s)。在下面例 4_1 中,example4_1.jsp 通过表单向 example4_1_computer.jsp 提交三角形三边的长度,example4_1_computer.jsp 负责计算并显示三角形的面积。

例 4_1

example4_1.jsp(效果如图 4.1(a)所示)

```
<%@ page contentType = "text/html" %>
<%@ page pageEncoding = "utf-8" %>
<HTML><body bgcolor = #ffccff>
   <form action = "example4_1_computer.jsp" method = post>
      <input type = "text" name = "sizeA" value = 9 size = 6/>
      <input type = "text" name = "sizeB" value = 8 size = 6/>
      <input type = "text" name = "sizeC" value = 8 size = 6/>
      <input type = "submit" name = "submit" value = "提交"/>
   </form>
</body></HTML>
```

example4_1_computer.jsp(效果如图 4.1(b)所示)

```
<%@ page contentType = "text/html" %>
<%@ page pageEncoding = "utf-8" %>
<HTML><body bgcolor = #ccffff>
<p style = "font-family:黑体;font-size:36;color:blue">
<% String sideA = request.getParameter("sizeA");
   String sideB = request.getParameter("sizeB");
   String sideC = request.getParameter("sizeC");
   try {  double a = Double.parseDouble(sideA);
          double b = Double.parseDouble(sideB);
          double c = Double.parseDouble(sideC);
          double p = (a + b + c)/2, area = 0;
          area = Math.sqrt(p * (p-a) * (p-b) * (p-c));
          String result = String.format("%.2f", area);
          out.println("<BR>三边:" + sideA + "," + sideB + "," + sideC);
          out.println("<BR>三角形面积(保留2位小数):" + result);
   }
   catch(NumberFormatException ee){
          out.println("<BR>请输入数字字符");
   }
%>
</p></body></HTML>
```

第4章 JSP内置对象

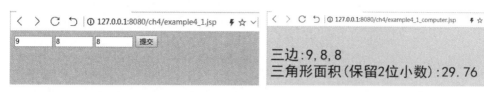

(a) 使用表单提交信息　　　　　　(b) 获取并处理表单提交的信息

图 4.1　计算三角形的面积

在下面的例 4_2 中，example4_2.jsp 通过表单向当前页面提交购物小票，当前页面负责计算购物小票的价格总和。如果 form 表单中的 action 请求的页面是当前页面，可以用双引号或单引号代替当前页面，即写成 action=""或 action=''，注意双引号或单引号中不能含有空格。也可省略 action 参数，即不显式写出 action 参数。

例 4_2

example4_2.jsp（效果如图 4.2 所示）

```
<%@ page contentType = "text/html" %>
<%@ page pageEncoding = "utf-8" %>
<%@ page import = "java.util.regex.Pattern" %>
<%@ page import = "java.util.regex.Matcher" %>
<style>
    #tom{
        font-family:宋体;font-size:22;color:blue
    }
</style>
<HTML><body id = "tom" bgcolor = #ffccff>
输入购物小票内容(显示的是默认内容):
<%
    String content = "牛奶:12.68元,面包:6.6元,"
                   + "苹果:28元,香皂:6.58元";
%>
<form action = "" method = "post" id = "tom">
    <textArea name = "shopping" rows = 5 cols = 32 id = "tom">
        <% = content %>
    </textArea>
    <input type = "submit" id = "tom" name = "submit" value = "提交"/>
</form>
<%  String shoppingReceipt = request.getParameter("shopping");
    if(shoppingReceipt == null) {
        shoppingReceipt = "0";
    }
    Pattern pattern;                              //模式对象
    Matcher matcher;                              //匹配对象
    String regex = "-?[0-9][0-9]*[.]?[0-9]*";     //匹配数字,整数或浮点数的正则表达式
    pattern = Pattern.compile(regex);             //初始化模式对象
    matcher =
    pattern.matcher(shoppingReceipt);             //matcher 检索 shoppingReceipt
    double sum = 0;
    while(matcher.find()) {
        String str = matcher.group();
        sum += Double.parseDouble(str);
```

```
        }
        out.print("购物小票消费总额:" + sum);
%>
</body></HTML>
```

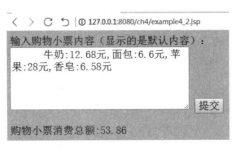

图 4.2　向当前页面提交信息

注：使用 request 对象获取当前页面提交的信息时要格外小心，在上面的例 4_2 中，当用户在浏览器中输入页面地址请求页面时，用户还没有机会提交数据，那么页面在执行

```
String shoppingReceipt = request.getParameter("shopping");
```

时得到的 shoppingReceipt 就是空对象。如果程序使用了空对象，Java 解释器就会提示出现 NullPointerException 异常。因此，在上述例 4_2 中为了避免在运行时出现 NullPointerException 异常，使用了如下代码：

```
String shoppingReceipt = request.getParameter("shopping");
if(shoppingReceipt == null) {
    shoppingReceipt = "0";
}
```

4.1.2　处理汉字信息

视频讲解

　　request 对象获取用户提交的信息中如果含有汉字字符或其他非 ASCII 字符，就必须进行特殊的处理方式，以防出现乱码现象。
　　JSP 页面文件的编码为 UTF-8 编码，只要让 request 对象在获取信息之前调用 setCharacterEncoding 方法设置编码为 UTF-8（默认是 ISO-8859-1）就可以避免乱码现象，代码如下：

```
request.setCharacterEncoding("utf-8");
```

例 4_3 中 example4_3.jsp 通过 form 表单向自己提交不同语言的问候语，request 对象在获取信息之前调用 setCharacterEncoding 方法设置编码为 UTF-8，然后再获取数据。

例 4_3

example4_3.jsp（效果如图 4.3 所示）

```
<%@ page contentType = "text/html" %>
<%@ page pageEncoding = "utf-8" %>
<style>
```

```
    #tom{
        font-family:宋体;font-size:26;color:blue
    }
</style>
<HTML><body id="tom" bgcolor=#ffccff>
可以输入各种语言的文字,单击提交键:
<%
    String content = "早上好,Good morning,อรุณสวัสดิ์คะ(泰语)," +
                     "おはよう,Доброе утро,좋은 아침";
%>
<form action="" method=post>
    <textArea name="language" id="tom" rows=3 cols=50>
        <%= content %>
    </textArea>
    <input type="submit" id="tom" name="submit" value="提交"/>
</form>
<%   request.setCharacterEncoding("utf-8");
    String variousLanguages = request.getParameter("language");
    out.print(variousLanguages);
%>
</p></body></HTML>
```

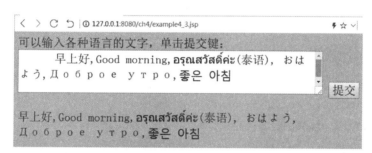

图 4.3 request 设置 UTF-8 避免乱码

4.1.3 常用方法举例

当用户访问一个页面时,会提交一个 HTTP 请求给 Tomcat 服务器,这个请求包括一个请求行、HTTP 头和信息体,例如:

视频讲解

```
post/example3_1.jsp/HTTP.1.1
host: localhost: 8080
accept-encoding: gzip, deflate
```

其中,首行叫请求行,规定了向访问的页面请求提交信息的方式,例如 post、get 等方式,以及请求的页面的名字和使用的通信协议。

第 2、3 行分别是两个头(Header):host 和 accept-encoding,称 host、accept-encoding 是头名字,而 localhost:8080 以及 gzip、deflate 分别是两个头的值。一个典型的请求通常包含很多的头,有些头是标准的,有些和特定的浏览器有关。

一个请求还包含信息体,即 HTML 标记组成的部分,可能包括各式各样用于提交信息的表单等,如:

```
<form action = "tom.jsp" method = "post" >
    <input type = "text" name = "boy" value = "ok" />
    <input type = "submit" name = "submit" value = "提交"/>
</form>
```

尽管服务器非常关心用户提交的 HTTP 请求中 form 表单的信息,如前面的例 4_1 和例 4_2 中使用 request 对象的 getParameter 方法获取 form 表单提交的有关信息,但实际上,request 对象调用相关方法可以获取请求的许多细节信息。request 对象常用方法如下:

(1) String getProtocol()获取用户向服务器提交信息所使用的通信协议,例如 http/1.1 等。

(2) String getServletPath()获取用户请求的 JSP 页面文件的名字(带目录符号\,例如\hello.jsp)。

(3) String getContextPath() 获取用户请求的当前 Web 服务目录(例如 ch4)。

(4) int getContentLength()获取用户提交的整个信息的长度。

(5) String getMethod()获取用户提交信息的方式,例如 post 或 get。

(6) String getHeader(String s)获取 HTTP 头文件中由参数 s 指定的头名字的值,一般来说参数 s 可取的头名有 accept、accept-language、content-type、accept-encoding、user-agent、host、content-length、connection、cookie 等,例如,s 取值 user-agent 将获取用户的浏览器的版本号等信息。

(7) Enumeration getHeaderNames()获取头名字的一个枚举。

(8) Enumeration getHeaders(String s)获取头文件中指定头名字的全部值的一个枚举。

(9) String getRemoteAddr()获取用户的 IP 地址。

(10) String getRemoteHost()获取用户机的名称(如果获取不到,就获取 IP 地址)。

(11) String getServerName()获取服务器的名称。

(12) String getServerPort()获取服务器的端口号。

(13) Enumeration getParameterNames()获取用户提交的信息体部分中各个 name 给出的参数的一个枚举。

下面的例 4_4 使用了 request 对象的一些常用方法。

例 4_4

example4_4.jsp(效果如图 4.4 所示)

```
<%@ page contentType = "text/html" %>
<%@ page pageEncoding = "utf-8" %>
<HTML><body bgcolor = #ffccff >
<p style = "font - family:宋体;font - size:36;color:blue">
<% request.setCharacterEncoding("utf - 8");
    String jsp = request.getServletPath();        //请求的 JSP 页面
    jsp = jsp.substring(1);                        //去掉 JSP 页面名称前面的目录符号/
    String webDir = request.getContextPath();     //获取当前 Web 服务目录的名称
    webDir = webDir.substring(1);                  //去掉 Web 服务目录的名称前面的目录符号/
    String clientIP = request.getRemoteAddr();    //用户的 IP 地址
    int serverPort = request.getServerPort();     // 服务器的端口号
%>
用户请求的页面:<% = jsp %>
<br>Web 服务目录的名字:<% = webDir %>
```

```
<br>用户的 IP 地址:<% = clientIP %>
<br>服务器的端口号:<% = serverPort %>
</p></body></HTML>
```

```
用户请求的页面：example4_4.jsp
Web服务目录的名字：ch4
用户的IP地址:127.0.0.1
服务器的端口号:8080
```

图 4.4　request 对象获取的信息

▶ 4.1.4　处理 HTML 标记

视频讲解

本节对经常用于提交信息的 HTML 标记进行简单介绍，有关细节建议读者查阅 HTML 资料。JSP 页面可以含有 HTML 标记，当用户通过浏览器请求一个 JSP 页面时，Tomcat 服务器将该 JSP 页面中的 HTML 标记直接发送到用户的浏览器，由用户的浏览器负责执行这些 HTML 标记。而 JSP 页面中的变量声明、程序片以及表达式由 Tomcat 服务器处理后，再将有关的结果用文本方式发送到用户的浏览器。

目前的 HTML 大约有 100 多个标记，这些标记可以描述数据的显示格式，如果读者对 HTML 语言比较陌生，建议补充这方面的知识。

❶ form 标记

习惯称 form 标记为 form 表单，由于用户经常需要使用 form 表单提交数据，所以有必要对 form 表单进行简明的介绍。

form 表单的一般格式是：

```
< form action = "请求访问的页面或 servlet" method = get | post >
    各种提交手段
    提交键
</form>
```

其中，< form >是 form 表单的开始标签，</form >是结束标签，开始标签和结束标签之间是 form 表单的标记体的内容。action 是 form 表单的属性，给出表单请求访问的 JSP 页面或 servlet。form 表单中的 method 属性取值 get 或 post。get 方法和 post 方法的主要区别是：使用 get 方法提交的信息会在提交的过程中显示在浏览器的地址栏中，而 post 方法提交的信息不会显示在地址栏中。提交手段包括文本框、列表、文本区等，例如：

```
< form action = "tom.jsp" method = "post" >
    < input type = "text" name = "boy" value = "ok" />
    < input type = "submit" name = "submit" value = "提交"/>
</form>
```

form 表单标记经常将下列标记作为 form 表单的子标记，以便提供提交数据的手段，这些标记都以 GUI 形式出现，方便用户输入或选择数据，例如，文本框、下拉列表、滚动列表等。

- < input .../>
- < select...></select >
- < option ...></option >

- < textArea ...></textArea >

❷ input 标记

在 form 表单将 input 标记作为子标记来指定 form 表单中数据的输入方式以及 form 表单的提交键。input 标记属于空标记,即没有标记体,所以 input 标记没有开始标签和结束标签(空标记的基本格式是<标记名称 属性列表 />)。<input>标记的基本格式为:

```
< input type = "GUI 对象" name = "GUI 对象的名字" value = "GUI 中的默认值"/>
```

input 中的 type 属性的值指定输入方式的 GUI 对象,name 属性的值指定这个 GUI 对象的名字。Tomcat 服务器的内置对象 request 通过 name 指定的名字来获取 GUI 对象中提交的数据。GUI 对象可以是 text(文本框)、checkbox(复选框)、submit(提交键)等。

(1) 文本框 text。

当 type 属性值指定输入方式的 GUI 是 text 时,除了用 name 为 text 指定名字外,还可以为 text 指定其他的一些值。比如:

```
< input type = "text" name = "m" value = "h" size = "8" algin = "left" maxlength = "9"/>
```

其中,value 的值是 text 中文本的初始值,size 是 text 的长度(单位是字符),algin 是 text 在浏览器窗体中的对齐方式,maxlength 指定 text 可输入的最多字符。request 对象通过 name 指定的名字来获取用户在 text 输入的字符串。如果用户没有在 text 输入任何信息,就单击 form 表单中的 submit 提交键,request 对象调用 getParameter 方法将获取由 value 指定的默认值(text 中显示的默认文本),如果 value 未指定任何值,getParameter 方法获取的字符串的长度为 0,即该字符串为""。

(2) 单选框 radio。

当 type 属性指定输入方式的 GUI 是 radio 时,除了用 name 为 radio 指定名字外,还可以为 radio 指定其他的一些值。比如:

```
< input type = "radio" name = "hi" value = "男" algin = "top" checked = "ok" />
< input type = "radio" name = "hi" value = "女" algin = "top"  />
```

其中,value 指定 radio 的值,algin 是 radio 在浏览器窗体中的对齐方式,如果几个单选键的 name 取值相同,那么同一时刻只能有一个被选中。request 对象调用 getParameter 方法获取被选中的 radio 中 value 属性指定的值。checked 如果取值是一个非空的字符串,那么该单选框的初始状态就是选中状态。

(3) 复选框 checkbox。

当 type 属性指定输入方式的 GUI 是 checkbox 时,除了用 name 为 checkbox 指定名字外,还可以为 checkbox 指定其他的一些值。比如:

```
< input type = "checkbox" name = "item" value = "A" algin = "top" checked = "ok" />
< input type = "checkbox" name = "item" value = "B" algin = "top"/>
< input type = "checkbox" name = "item" value = "C" algin = "top" checked = "ok" />
< input type = "checkbox" name = "item" value = "D" algin = "top"/>
```

其中,value 指定 checkbox 的值。复选框与单选框的区别就是可以多选,即如果几个 checkbox 的 name 取值相同,那么同一时刻可有多个 checkbox 被选中。这时,request 对象需调用 getParameterValues 方法(不是 getParameter 方法)获取被选中的多个 checkbox 中 value 属性指定的值。checked 如果取值是一个非空的字符串，那么该复选框的初始状态就是选中状态。

(4) 口令框 password。

当 type 属性指定输入方式的 GUI 是 password，用户输入的信息用"*"回显，即防止他人偷看口令。例如：

```
<input type = "password" name = "me" size = "12" maxlength = "30" />
```

用户在口令框中输入 tiger，单击提交键，tiger 将被提交给 form 表单请求的页面，请求的页面的内置对象 request 调用 getParameter 方法获取 password 提交的值 tiger(password 仅仅起着不让别人偷看的作用，不提供加密措施)。

(5) 隐藏 hidden。

当 type 属性指定输入方式是 hidden 时，input 没有可见的输入界面，form 表单会将 input 标记中 value 属性的值提交给所请求的页面。例如：

```
<input type = "hidden" name = "nogui" value = "hello" />
```

用户单击 form 表单中的 submit 提交键，那么 form 表单所请求的页面的 request 对象调用 getParameter 方法将获取由 value 指定的值 hello。

(6) 提交键 submit。

为了能把 form 表单的数据提交给服务器，一个 form 表单至少包含一个提交键(可以有多个提交键，见稍后的例 4_10)，例如：

```
<input type = "submit" name = "me" value = "确定" size = "12" />
```

单击提交键后，form 表单请求的页面才有机会获取 form 表单提交的各个数据。

(7) 重置键 reset。

重置键将表单中输入的数据清空，以便重新输入数据，例如：

```
<input type = "reset" value = "重置" />
```

例 4_5 中，JSP 页面 example4_5.jsp 用 form 表单向 example4_5_receive.jsp 提交数据，example4_5_receive.jsp 使用 request 对象获得 example4_5.jsp 提交的数据。用户在 example4_5.jsp 单击 form 表单 submit 提交键提交信息，所提交的信息包括通过 radio 选择的是否打开背景音乐的信息、通过 checkbox 选择回答的奥运项目信息、通过 hidden 隐藏的信息。调试例 4_5 时需要将名字是 back.mp3 的 mp3 文件存放到 Web 服务目录 ch4 的子目录 sound 中。

注：HTML embed 标记可以播放音乐，音乐文件可以是 .wav 和 .mp3 格式(见稍后的介绍)。

例 4_5

example4_5.jsp(效果如图 4.5(a)所示)

```
<%@ page contentType = "text/html" %>
<%@ page pageEncoding = "utf-8" %>
<style>
    #tom{
        font-family:宋体;font-size:26;color:blue
```

```
    }
</style>
<HTML><body id="tom" bgcolor=#ffccff>
<form action="example4_5_receive.jsp" method=post id=tom>
    <br>音乐:
    <input type="radio" name="R" value="on" />打开
    <input type="radio" name="R" value="off" checked="default">关闭
<br>哪些是奥运会项目:<br>
    <input type="checkbox" name="item" value="A"   algin="top"   />足球
    <input type="checkbox" name="item" value="B"   algin="top"   />围棋
    <input type="checkbox" name="item" value="C"   algin="top"   />乒乓球
    <input type="checkbox" name="item" value="D"   algin="top"   />篮球
    <input type="hidden" value="我是球迷,但不会踢球" name="secret"/>
    <br><input type="submit" id="tom" name="submit" value="提交"/>
    <input type="reset" id="tom" value="重置" />
</form>
</body></HTML>
```

example4_5_receive.jsp（效果如图 4.5(b)所示）

```
<%@ page contentType="text/html" %>
<%@ page pageEncoding="utf-8" %>
<%@ page import="java.util.Arrays" %>
<%! public boolean isSame(String []a,String [] b){
        Arrays.sort(a);
        Arrays.sort(a);
        return Arrays.equals(a,b);
    }
%>
<HTML><body bgcolor=white>
<p style="font-family:宋体;font-size:36;color:blue">
<%   String answer[] = {"A","C","D"};
    request.setCharacterEncoding("utf-8");
    String onOrOff = request.getParameter("R");              //获取 radio 提交的值
    String secretMess = request.getParameter("secret");       //获取 hidden 提交的值
    String itemName[] = request.getParameterValues("item");   //获取 checkbox 提交的值
    out.println("<br>是否打开音乐:" + onOrOff);
    out.println("<br>您的答案:");
    if(itemName == null) {
        out.print("没给答案");
    }
    else {
        for(int k = 0;k<itemName.length;k++) {
          out.print(" " + itemName[k]);
        }
        if(isSame(itemName,answer)){
            out.print("<br>回答正确.");
        }
    }
    out.println("<br>提交的隐藏信息:" + secretMess);
    if(onOrOff.equals("on")) {
```

```
%>              <br><embed src = "sound/back.mp3" />
<%      }
%>
</p></body></HTML>
```

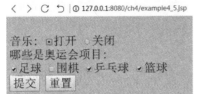

(a) 使用表单提交数据　　　　　　　(b) 用request获得数据

图 4.5　数据的提交与获取

❸ select、option 标记

下拉式列表和滚动列表通过 select 和 option 标记来定义，经常作为 form 的子标记，为表单提供选择数据的 GUI。select 标记将 option 作为子标记，形成下拉列表或滚动列表。

下拉列表的基本格式：

```
<select name = "myName">
  <option value = "item1">文本描述</option>
  <option value = "item2">文本描述</option>
    …
</select>
```

在 select 中增加 size 属性的值就变成滚动列表，size 的值是滚动列表的可见行的数目。滚动列表的基本格式：

```
<select name = "myName" size = "正整数">
  <option value = "item1">文本描述</option>
  <option value = "item2">文本描述</option>
    …
</select>
```

request 对象通过滚动列表的 name 指定的值来获取下拉列表或滚动列表中被选中的 option 的参数 value 指定的值。

例 4_6 中用户通过下拉列表为当前页面选择一首音乐，通过滚动列表为当前页面选择一幅图像。调试例 4_6 时需要将名字是 back1.mp3、back2.mp3 和 back3.mp3 的 mp3 文件存放到 Web 服务目录 ch4 的子目录 sound 中，需要将名字是 back1.jpg、back2.jpg 和 back3.jpg 的 jpg 文件存放到 Web 服务目录 ch4 的子目录 image 中。

例 4_6

example4_6.jsp(效果如图 4.6 所示)

```
<%@ page contentType = "text/html" %>
<%@ page pageEncoding = "utf-8" %>
<style>
    #tom{
```

```
        font-family:宋体;font-size:26;color:blue
     }
  </style>
  <%  String music = request.getParameter("music");
      String pic = request.getParameter("pic");
      String onOrOff = request.getParameter("R");
      if(music == null) music = "";
      if(pic == null)pic = "";
      if(onOrOff == null) onOrOff = "off";
  %>
  <HTML><body id = tom background = "image/<% = pic %>">
  <form action = ""method = post>
     <b>选择音乐:<br>
     <select id = tom name = "music">
        <Option selected value = "back1.mp3">绿岛小夜曲
        <Option value = "back2.mp3">我是一片云</option>
        <Option value = "back3.mp3">红河谷</option>
     </select>
     <input type = "radio" name = "R" value = "on" />打开
     <input type = "radio" name = "R" value = "off" />关闭
     <br><b>选择背景图像:<br>
     <select id = tom name = "pic" size = 2>
        <option value = "back1.jpg">荷花图</option>
        <option value = "back2.jpg">玫瑰图</option>
        <option value = "back3.jpg">校园图</option>
     </select><br>
     <input id = tom type = "submit"  name = "submit" value = "提交"/>
  </form>
  <%  if(onOrOff.equals("on")) {
  %>       <br><embed src = "sound/<% = music %>" height = 50 />
  <%  }
  %>
  </body></HTML>
```

图 4.6 下拉列表和滚动列表

❹ textArea 标记

<textArea>是一个能输入或显示多行文本的文本区,在 form 表单中使用<textArea>作为子标记可以提交多行文本给所请求的 JSP 页面。<textArea>的基本格式为:

```
<textArea name = "名字" rows = "文本可见行数"cols = "文本可见列数" >
</textArea>
```

❺ style 样式标记

style 标记可用于定义 HTML 其他标记中的字体样式,例如,style 标记给出样式:

```
<style>
    #textStyle{
        font-family:宋体;font-size:18;color:blue
    }
    #tom{
        font-family:黑体;font-size:16;color:black
    }
</style>
```

其中,♯字符之后的字符序列是样式名称,例如,♯textStyle 给出的样式名称是 textStyle(起一个自己喜欢且容易理解的名字),其他 HTML 标记可以让其 id 属性值是样式名称来使用这个样式。例如,段落标记 p 就可以如下使用 textStyle 样式:

```
<p id = "textStyle">你好</p>
```

textArea 标记如下使用 tom 样式:

```
<textArea name = "english" id = "tom" rows = 5 cols = 38>大家好</textArea>
```

submit 标记如下使用 textStyle 样式:

```
<input type = "submit"name = "submit" id = "textStyle" value = "提交"/>
```

❻ table 标记

表格以行列形式显示数据,不提供输入数据功能。经常将某些数据或 GUI 放置在表格的单元格中,让界面更加简练、美观。

表格由<table>标记定义,一般格式:

```
<table border = "边框的宽度">
    <tr width = "该行的宽度">
        <th width = "单元格的宽度" >单元格中的数据</th>
        ...
        <td width = "单元格的宽度" >单元格中的数据</td> ...
    </tr>
...
</table>
```

其中

```
<tr>....</tr>
```

定义表格的一个行,<th>或<td>标记定义这一行中的表格单元,二者的区别是<th>定义的单元着重显示,<td>称为普通单元,不着重显示。table 中增加选项 border 可给出该表格边框的宽度,当 border 取值是 0 时,相当于没有边框。

例 4_7 中用户在 example4_7.jsp 输入年份和月份提交给 example4_7_showCalendar.

jsp,example4_7_showCalendar.jsp 用 table 显示日历。

例 4_7

example4_7.jsp（效果如图 4.7(a)所示）

```jsp
<%@ page contentType="text/html" %>
<%@ page pageEncoding="utf-8" %>
<style>
    #textStyle{
        font-family:宋体;font-size:28;color:blue
    }
</style>
<HTML><body id=textStyle bgcolor=#ffccff>
<form action="example4_7_showCalendar.jsp" method=post>
输入日期的年份选择月份查看日历.<br>
年份：<input type="text" name="year" id=textStyle value=2022 size=12 />
月份 <select name="month" id=textStyle size=1>
    <option value="1">1 月</option>
    <option value="2">2 月</option>
    <option value="3">3 月</option>
    <option value="4">4 月</option>
    <option value="5">5 月</option>
    <option value="6">6 月</option>
    <option value="7">7 月</option>
    <option value="8">8 月</option>
    <option value="9">9 月</option>
    <option value="10">10 月</option>
    <option value="11">11 月</option>
    <option value="12">12 月</option>
</select><br>
<input type="submit" id=textStyle value="提交"/>
</form>
</body></HTML>
```

example4_7_showCalendar.jsp（效果如图 4.7(b)所示）

```jsp
<%@ page import="java.time.LocalDate" %>
<%@ page import="java.time.DayOfWeek" %>
<%
    request.setCharacterEncoding("utf-8");
    String year = request.getParameter("year");
    String month = request.getParameter("month");
    int y = Integer.parseInt(year);
    int m = Integer.parseInt(month);
    LocalDate date = LocalDate.of(y,m,1);
    int days = date.lengthOfMonth();        //得到该月有多少天
    int space = 0;                          //存放空白字符的个数
    DayOfWeek dayOfWeek = date.getDayOfWeek();  //得到 1 号是星期几
    switch(dayOfWeek) {
            case SUNDAY:    space = 0;
                            break;
            case MONDAY:    space = 1;
                            break;
```

```
            case TUESDAY:     space = 2;
                              break;
            case WEDNESDAY:   space = 3;
                              break;
            case THURSDAY:    space = 4;
                              break;
            case FRIDAY:      space = 5;
                              break;
            case SATURDAY:    space = 6;
                              break;
       }
       String [ ] calendar = new String[ space + days];
       for( int i = 0; i < space; i++ )
           calendar[ i ] = " -- ";
       for( int i = space, n = 1; i < calendar.length; i++){
           calendar[ i ] = String.valueOf(n) ;
           n++;
       }
%>
<HTML><body bgcolor = #ffccff>
<h3><% = year %>年<% = month %>月的日历:</h3>
<table border = 0>
   <tr><th>星期日</th><th>星期一</th><th>星期二</th><th>星期三</th>
       <th>星期四</th><th>星期五</th><th>星期六</th>
   </tr>
<%
   int n = 0;
   while(n < calendar.length){
       out.print("<tr>");
       int increment = Math.min(7, calendar.length - n);
       for( int i = n; i < n + increment; i++) {
          out.print("<td align = center>" + calendar[i] + "</td>");
       }
       out.print("</tr>");
       n = n + increment;
   }
%>
</table></body></HTML>
```

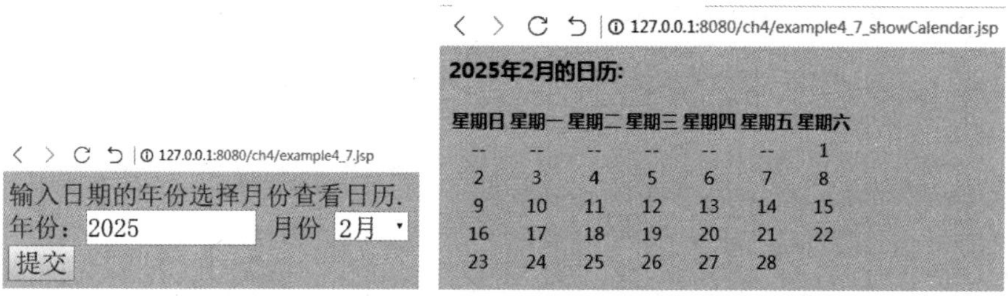

(a) 输入年份选择月份　　　　　(b) table显示日历

图 4.7　使用 table 显示数据

❼ <image>标记

使用 image 标记可以显示一幅图像，image 标记的基本格式为：

<image src="图像文件的 URL">描述文字</image>

如果图像文件和当前页面在同一 Web 服务目录中，图像文件的地址就是该图像文件的名字，如果图像文件在当前 Web 服务目录一个子目录中，比如 image 子目录中，那么"图像文件的 URL"就是"image/图像文件的名字"。

image 标记中可以使用 width 和 height 属性指定被显示的图像的宽和高，如果省略 width 和 height 属性，image 标记将按图像的原始宽度和高度来显示图像。

❽ embed 标记

使用 embed 标记可以播放音乐和视频，当浏览器执行该标记时，会把浏览器所在机器上的默认播放器嵌入到浏览器中，以便播放音乐或视频文件。embed 标记的基本格式为：

<embed src="音乐或视频文件的 URL">描述文字</embed>

或

<embed src="音乐或视频文件的 URL" />

如果音乐或视频文件和当前页面在同一 Web 服务目录中，embed 标记中 src 属性的值就是该文件的名字；如果视频文件在当前 Web 服务目录一个子目录中，比如 avi 子目录中，那么 embed 标记中 src 属性的值就是"avi/视频文件的名字"。

embed 标记中经常使用的属性及取值如下：

- autostart 属性取值 true 或 false，autostart 属性的值用来指定音乐或视频文件传送完毕后是否立刻播放，该属性的默认值是 false。
- loop 属性取值正整数指定音乐或视频文件重复播放的次数，取值-1 为无限循环播放。
- width 和 height 属性取值均为正整数，用 width 和 height 属性的值指定播放器的宽和高，如果省略 width 和 height 属性，将使用默认值。

例 4_8 中，example4_8.jsp 页面使用 image 标记显示一幅图像，用户使用下拉列表选择要播放视频提交给 example4_8_play_mp4.jsp，example4_8_play_mp4.jsp 页面使用 embed 标记播放用户选择的视频。其中图像文件 flower.jpg 和视频文件"高山.mp4"和"湖水.mp4"分别存放在当前 Web 服务目录 ch4 的子目录 image 和 video 中。

例 4_8

example4_8.jsp（效果如图 4.8(a)所示）

```
<%@ page contentType="text/html" %>
<%@ page pageEncoding="utf-8" %>
<HTML><body bgcolor=#ffccff>
<p style="font-family:宋体;font-size:18;color:blue">
<form action="example4_8_play_mp4.jsp" method=post name=form>
<b>选择视频:<br>
  <select name="mp4">
    <option value="高山.mp4">美丽的珠峰
    <option value="湖水.mp4">美丽的高原湖
```

```
    </select>
    <input type = "submit" name = "submit" value = "提交" />
</form><br>
<image src = "image/flower.jpg" width = 300 height = 190></image><br>
</p></body></HTML>
```

example4_8_play_mp4.jsp（效果如图 4.8(b)所示）

```
<%@ page contentType = "text/html" %>
<%@ page pageEncoding = "utf-8" %>
<%
    request.setCharacterEncoding("utf-8");
    String mp4 = request.getParameter("mp4");
%>
<HTML><body bgcolor = #ffccff>
<embed src = "video/<% = mp4 %>" width = 300 height = 180></embed>
</p></body></HTML>
```

(a) 显示图像　　　　　　　　　　　(b) 播放视频

图 4.8　显示图像与播放视频

❾ style 样式标记

style 标记可用于定义 HTML 其他标记中的字体样式，例如，style 标记给出样式：

```
<style>
    #textStyle{
        font-family:宋体;font-size:18;color:blue
    }
</style>
```

其中，♯字符之后的字符序列是样式名称，例如♯textStyle 给出的样式名称是 textStyle（起一个自己喜欢且容易理解的名字），其他 HTML 显示文本的标记可以让其 id 属性值是样式名称来使用这个样式。例如，段落标记 p 就可以如下使用 textStyle 样式：

```
<p id = "textStyle">你好</p>
```

textArea 标记如下使用 textStyle 样式：

```
<textArea name = "english" id = "textStyle" rows = 5 cols = 38>大家好</textArea>
```

submit 标记如下使用 textStyle 样式：

```
<input type = "submit" name = "submit" id = "textStyle" value = "提交"/>
```

❿ 属性值格式的说明

许多 HTML 标记都有属性,并可以指定属性的值,例如:

< input type = "text" name = "testAmount" value = 10 />

其中,type、name、value 都是 input 标记的属性,属性值可以用双引号括起,也可以用单引号括起或者不用任何符号,比如 type 属性的值可以用双引号括起"text",也可以用单引号括起'text',或者不用任何符号 text。

注:一个好的习惯是用单引号括起。例如,下面超链接标记中的 href 的属性值用单引号括起。

< a href = 'example4_1.jsp'>超链接</a>

▶ 4.1.5 处理超链接

视频讲解

HTML 的超链接标记

< a href = 链接的页面地址 >文字说明</a>

是一个常用标记。例如:

< a href = "example4_9_receive.jsp">购买</a>

用户单击超链接标记的文字说明,可以访问超链接给出的链接页面。使用超链接标记时还可以增加参数,以便向所链接的页面传递值,格式如下:

< a href = 链接的页面地址?参数1 = 字符串1&参数2 = 字符串2… >文字说明</a>

例如:

< a href = "example4_9_receive.jsp?id = A1001&price = 8765">购买</a>

超链接所链接的页面,使用 request 对象调用 getParameter("参数")方法获得超链接的参数传递过来的参数的值,即字符串。例如:

String idStr = request.getParameter("id");

需要注意的是,超链接标记向所链接的页面传递的参数的值,即字符串中不允许含有非 ASCII 字符(例如汉字等)。

例 4_9 中,example4_9.jsp 用超链接向 example4_9_receive.jsp 传递商品的编号和价格。

例 4_9

example4_9.jsp(效果如图 4.9(a)所示)

```
<%@ page contentType = "text/html" %>
<%@ page pageEncoding = "utf - 8" %>
< HTML >< body bgcolor = #ffccff >
<%
    double price = 98.78;
```

```
%>
<p style = "font-family:宋体;font-size:36;color:blue">
商品编号 A1001,价格 8765
<a href = "example4_9_receive.jsp?id = A1001&price = 8765">购买</a><br>
商品编号 A1002,价格<% = price %>
<a href = "example4_9_receive.jsp?id = A1002&price = <% = price %>">购买</a>
</p></body></HTML>
```

example4_9_receive.jsp(效果如图 4.9(b)所示)

```
<%@ page contentType = "text/html" %>
<%@ page pageEncoding = "utf-8" %>
<HTML><body bgcolor = #EEEEFF>
<p style = "font-family:宋体;font-size:36;color:blue">
<% String id = request.getParameter("id");
   String price = request.getParameter("price");
%>
<b>商品编号:<% = id %><br>
商品价格:<% = price %>
</p></body></HTML>
```

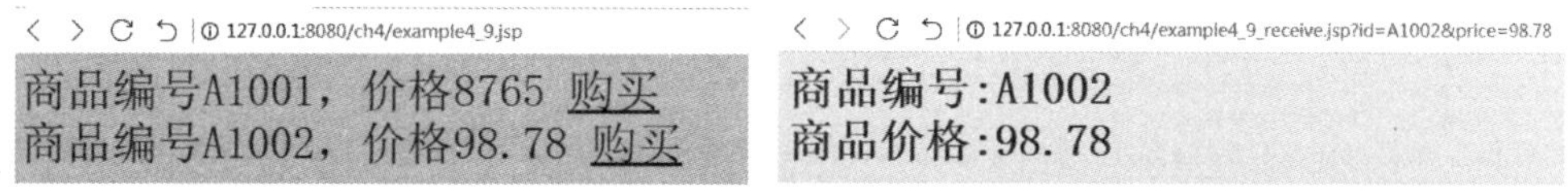

(a) 带参数的超链接　　　　　　　(b) 获取超链接传递的参数值

图 4.9　使用超链接

4.2　response 对象

当用户访问一个服务器的页面时,会提交一个 HTTP 请求,服务器收到请求时,返回 HTTP 响应。响应和请求类似,也有某种结构,每个响应都由状态行开始,可以包含几个头及可能的信息体(网页的结果输出部分)。

4.1 节学习了用 request 对象获取用户请求提交的信息,与 request 对象相对应的对象是 response 对象。可以用 response 对象对用户的请求作出动态响应,向用户端发送数据。比如,当一个用户请求访问一个 JSP 页面时,该页面用 page 指令设置页面的 contentType 属性的值是 text/html,那么 Tomcat 服务器将按照这种属性值响应用户对页面的请求,将页面的静态部分返回给用户,用户浏览器接收到该响应就会使用 HTML 解释器解释执行所收到的信息。

▶ 4.2.1　动态响应 contentType 属性

页面用 page 指令设置页面的 contentType 属性的值,那么 Tomcat 服务器将按照这种属性值作出响应,将页面的静态部分返回给用户,用户浏览器接收到该响应就会使用相应的手段处理所收到的信息。由于 page 指令只能为 contentType 指定一个值来决定响应的 MIME 类型,如果想动态地改变这个属性的值来响应用户,就需要使用

response 对象的 setContentType(String s)方法来改变 contentType 的属性值,该方法中的参数 s 可取值:text/html、text/plain、image/gif、image/x-xbitmap、image/jpeg、image/pjpeg、application/x-shockwave-flash、application/vnd.ms-powerpoint、application/vnd.ms-excel、application/msword 等。

当用 setContentType(String s)方法动态改变了 contentType 的属性值,即响应的 MIME 类型,Tomcat 服务器就会按照新的 MIME 类型将 JSP 页面的输出结果返回给用户。

例 4_10 中,用户在 example4_10.jsp 页面输入圆半径,然后单击"提交看面积"提交键,请求访问 example4_10_show.jsp 页面,该页面显示圆的面积。但是,如果用户输入圆半径,单击"提交看圆形"提交键,那么 example4_10_show.jsp 的 response 对象将默认的 MIME 类型 text/html 改变成 image/jpeg,以便用户的浏览器启用相应的图形解码器显示服务器发来的图形。例 4_10 中用户单击"提交看面积"提交键后的效果如图 4.10(b)所示,单击"提交看图形"提交键后的效果如图 4.10(c)所示。

(a) 输入半径　　　　(b) 看面积　　　　(c) 看图形

图 4.10　动态改变 MIME 类型

例 4_10
example4_10.jsp(效果如图 4.10(a)所示)

```
<%@ page contentType = "text/html" %>
<%@ page pageEncoding = "utf-8" %>
<style> #textStyle
    { font-family:宋体;font-size:36;color:blue
    }
    #tomStyle
    { font-family:黑体;font-size:26;color:black
    }
</style>
<HTML><body id = "textStyle" bgcolor = #ffccff>
<form action = "example4_10_show.jsp"  method = post>
输入圆半径:<br>
半径:<input type = "text" name = "radius"
            id = "textStyle" value = 100.8 size = 12 /><br>
<input type = "submit" name = "submit" id = "tomStyle" value = "提交看面积"/><br>
<input type = "submit" name = "submit" id = "tomStyle" value = "提交看圆形"/>
</form>
</body></HTML>
```

example4_10_show.jsp(效果如图 4.10(b)或(c)所示)

```
<%@ page contentType = "text/html" %>
<%@ page pageEncoding = "utf-8" %>
```

```jsp
<%@ page import="java.awt.*" %>
<%@ page import="java.io.OutputStream" %>
<%@ page import="java.awt.image.BufferedImage" %>
<%@ page import="java.awt.geom.*" %>
<%@ page import="javax.imageio.ImageIO" %>
<style>#textStyle
    {font-family:宋体;font-size:36;color:blue
    }
</style>
<%! void drawCircle(double r,HttpServletResponse response) {          //定义方法
        int width = 320, height = 300;
        BufferedImage image =
        new BufferedImage(width,height,BufferedImage.TYPE_INT_RGB);
        Graphics g = image.getGraphics();
        g.fillRect(0, 0, width, height);
        Graphics2D g_2d = (Graphics2D)g;
        Ellipse2D ellipse = new Ellipse2D.Double(160-r,150-r,2*r,2*r);
        g_2d.setColor(Color.blue);
        g_2d.draw(ellipse);
        try {
          OutputStream outClient = response.getOutputStream();
          boolean boo = ImageIO.write(image,"jpeg",outClient);
        }
        catch(Exception exp){}
    }
    double getArea(double r) {                              //定义求面积的方法
       return  Math.PI*r*r;
    }
%>
<%   request.setCharacterEncoding("utf-8");
    String submitValue = request.getParameter("submit");
    String radius = request.getParameter("radius");
    double r = Double.parseDouble(radius);
    if(submitValue.equals("提交看圆形")){
       response.setContentType("image/jpeg");          //response更改 MIME 类型
       drawCircle(r,response) ;                         //绘制圆
    }
%>
<HTML><body bgcolor=#EEEEFF>
<p id="textStyle">
<%
    double area = getArea(r);
    String result = String.format("%.2f",area);
%>
半径:<%= radius %><br>
<b>面积(保留2位小数)<br><%= result %>
</p></body></HTML>
```

▶ 4.2.2 response 对象的 HTTP 文件头

当用户访问一个页面时,会提交一个 HTTP 请求给 Tomcat 服务器,这个请求包括一个请求行、HTTP 头和信息体,例如:

视频讲解

```
post/example3_1.jsp/HTTP.1.1
host: localhost: 8080
accept-encoding: gzip, deflate
```

其中,首行叫请求行,规定了向请求访问的页面提交信息的方式,例如,post、get 等方式,以及请求的页面的名字和使用的通信协议。

第 2、3 行分别是两个头(Header):host 和 accept-encoding,称 host、accept-encoding 是头名字,而 localhost:8080 以及 gzip、deflate 分别是两个头的值。一个典型的请求通常包含很多的头,有些头是标准的,有些和特定的浏览器有关。

同样,响应也包括一些头。response 对象可以使用方法

```
addHeader(String head,String value);
```

或

```
setHeader(String head,String value)
```

动态添加新的响应头和头的值,将这些头发送给用户的浏览器。如果添加的头已经存在,则先前的头被覆盖。

在下面的例 4_11 中,response 对象添加一个响应头 refresh,其头值是 5。那么用户收到这个头之后,5 秒后将再次刷新该页面,导致该网页每 5 秒刷新一次。

例 4_11

example4_11.jsp

```
<%@ page contentType = "text/html" %>
<%@ page pageEncoding = "utf-8" %>
<%@ page import = "java.time.LocalTime" %>
<HTML><body bgcolor = #ffccff>
<p style = "font-family:宋体;font-size:36;color:blue">
现在的时间是:<br>
<%   out.println("" + LocalTime.now());
     response.setHeader("Refresh","5");
%>
</p></body></HTML>
```

▶ 4.2.3 response 对象的重定向

视频讲解

在某些情况下,当响应用户时,需要将用户重新引导至另一个页面。例如,如果用户输入的 form 表单信息不完整,就会再被引导到该 form 表单的输入页面。

可以使用 response 对象的 sendRedirect(URL url)方法实现用户的重定向,即让用户从一个页面跳转到 sendRedirect(URL url)中 url 指定的页面,即所谓的客户端跳转。需要注意的是当使用 sendRedirect(URL url)方法将用户从当前页面重定向另一个页面时,Tomcat 服务器还是要把当前 JSP 页面执行完毕后才实施重定向(跳转)操作,但 Tomcat 服务器不再给用户看当前页面的执行效果。如果在执行 sendRedirect(URL url)方法后,紧接着执行了 return 返回语句,那么 Tomcat 服务器会立刻结束当前 JSP 页面的执行。

在下面的例 4_12 中,用户在 example4_12.jsp 页面的 form 表单中输入姓名,提交给 example4_12_receive.jsp 页面。如果未输入姓名就提交 form 表单,则会重新定向到 example4_12.jsp 页面。

例 4_12

example4_12.jsp

```jsp
<%@ page contentType="text/html" %>
<%@ page pageEncoding="utf-8" %>
<style>#textStyle
    {font-family:宋体;font-size:36;color:blue
    }
</style>
<HTML><body bgcolor=#ffccff>
<p id="textStyle">
填写姓名(<%=(String)session.getAttribute("name")%>):<br>
<form action="example4_12_receive.jsp" method="post" name=form>
    <input type="text" id="textStyle"  name="name">
    <input type="submit" id="textStyle" value="确定"/>
</form>
</body></HTML>
```

example4_12_receive.jsp

```jsp
<%@ page contentType="text/html" %>
<%@ page pageEncoding="utf-8" %>
<HTML><body bgcolor=#DDEEFF>
<% request.setCharacterEncoding("utf-8");
   String name = request.getParameter("name");
   if(name == null||name.length() == 0) {
       response.sendRedirect("example4_12.jsp");
       String str = (String)session.getAttribute("name");      //这个仍然会被执行
       session.setAttribute("name","李四" + str);              //这个仍然会被执行
   }
%>
<b>欢迎<%= name %>访问网页。
</body></HTML>
```

4.3 session 对象

HTTP 协议是一种无状态协议。一个用户向服务器发出请求(request),然后服务器返回响应(response),在服务器端不保留用户的有关信息,因此当下一次发出请求时,服务器无法判断这一次请求和以前的请求是否属于同一用户。当一个用户访问一个 Web 服务目录时,可能会在这个服务目录的几个页面反复链接,反复刷新一个页面或不断地向一个页面提交信息等,服务器应当通过某种办法知道这是同一个用户。Tomcat 服务器可以使用内置对象 session(会话)记录用户的信息。内置对象 session 由 Tomcat 服务器负责创建,session 是实现了 HttpSession 接口类的一个实例,可以在 Toamcat 服务器的 webapps\tomcat-docs\servletapi 中查找 HttpSession 接口的方法。

▶ 4.3.1 session 对象的 id

视频讲解

当一个用户首次访问 Web 服务目录中的一个 JSP 页面时,Tomcat 服务器产生一个 session 对象,这个 session 对象调用相应的方法可以存储用户在访问该 Web 服务目录中各个页面期间提交的各种信息,比如姓名、号码等信息。这个 session 对象被分配了一个 String 类型的 id 号,Tomcat 服务器同时将这个 id 号发送到用户端,存放在用户(浏览器)的 Cookie 中。这样,session 对象和用户之间就建立起一一对应的关系,即每个用户都对应着一个 session 对象(称作用户的会话),不同用户(不同浏览器)的 session 对象互不相同,具有不同的 id 号码。当用户再访问该 Web 服务目录的其他页面时,或从该 Web 服务目录链接到其他 Web 服务器再回到该 Web 服务目录时,Tomcat 服务器不再分配给用户的新 session 对象,而是使用完全相同的一个,直到 session 对象达到了最大生存时间或用户关闭自己的浏览器或 Tomcat 服务器关闭,Tomcat 服务器将销毁用户的 session 对象,即和用户的会话对应关系消失。如果用户的 session 对象被销毁,当用户再请求访问该 Web 服务目录时,Tomcat 服务器将为该用户创建一个新的 session 对象。

简单地说,用户(浏览器)在访问一个 Web 服务目录期间,服务器为该用户分配一个 session 对象(称作和该用户的会话),服务器可以在各个页面使用这个 session 对象记录当前用户的有关信息。而且服务器保证不同用户的 session 对象互不相同。

注:同一个用户在不同的 Web 服务目录中的 session 对象是互不相同的。

例 4_13 中用户在服务器的某个 Web 服务目录中的 2 个页面 example4_13_a.jsp 和 example4_13_b.jsp 进行链接,2 个页面的 session 对象是完全相同的。

例 4_13

example4_13_a.jsp(效果如图 4.11(a)所示)

```
<%@ page contentType="text/html" %>
<%@ page pageEncoding="utf-8" %>
<style>#textStyle
    {font-family:宋体;font-size:36;color:blue
    }
</style>
<HTML><body bgcolor=#ffccff>
<p id="textStyle">
这是 example4_13_a.jsp 页面<br>单击提交键链接到 example4_13_b.jsp
<% String id = session.getId();
    out.println("<br>session 对象的 ID 是<br>" + id);
%>
<form action="example4_13_b.jsp" method=post>
    <input type="submit" id="textStyle" value="访问 example4_13_b.jsp" />
</form>
</body></HTML>
```

example4_13_b.jsp(效果如图 4.11(b)所示)

```
<%@ page contentType="text/html" %>
<%@ page pageEncoding="utf-8" %>
```

```
<style>#textStyle
   {font-family:黑体;font-size:36;color:red
   }
</style>
<HTML><body bgcolor=cyan>
<p id="textStyle">
这是example4_13_b.jsp 页面
<%   String id = session.getId();
   out.println("<br>session 对象的 ID 是<br>" + id);
%>
<br>链接到 example4_13_a.jsp 的页面.<br>
<a href = "example4_13_a.jsp"> example4_13_a.jsp </a>
</body></HTML>
```

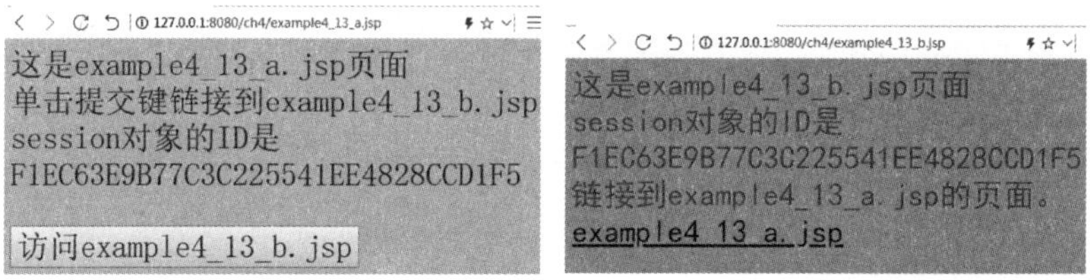

(a) example4_13_a中的ID   (b) example4_13_b中的ID

图 4.11 session 对象的 ID

▶ 4.3.2 session 对象与 URL 重写

视频讲解

session 对象能和用户建立起一一对应关系依赖于用户浏览器是否支持 Cookie。如果用户端浏览器不支持 Cookie,那么用户在不同网页之间的 session 对象可能是互不相同的,因为 Tomcat 服务器无法将 id 存放到用户端浏览器中,就不能建立 session 对象和用户的一一对应关系。用户将浏览器 Cookie 设置为禁止后(选择浏览器(IE)菜单→工具→Internet 选项→隐私→高级,将第三方 Cookie 设置成禁止),运行上述例 4_13 会得到不同的结果。也就是说,同一用户对应了多个 session 对象,这样 Tomcat 服务器就无法知道在这些页面上访问的实际上是同一个用户。

如果用户不支持 Cookie,JSP 页面可以通过 URL 重写来实现 session 对象的唯一性。所谓 URL 重写,就是当用户从一个页面重新链接到一个页面时,通过向这个新的 URL 添加参数,把 session 对象的 id 传带过去,这样就可以保障用户在该 Web 服务目录的各个页面中的 session 对象是完全相同的。可以让 response 对象调用 encodeURL()或 encodeRedirectURL()方法实现 URL 重写,比如,如果从 example4_13_a.jsp 页面链接到 example4_13_b.jsp 页面,首先在程序片中实现 URL 重写:

```
String str = response.encodeRedirectURL("example4_13_b.jsp");
```

然后将链接目标写成<%= str %> 即可。例如,将 example4_13_a.jsp 的代码:

```
<form action = "example4_13_b.jsp" method = post>
```

更改为:

```
<form action = <%= str %> method = post>
```

4.3.3 session 对象存储数据

视频讲解

session 对象驻留在服务器端,该对象调用某些方法保存用户在访问某个 Web 服务目录期间的有关数据。session 对象使用下列方法处理数据。

(1) public void setAttribute (String key,Object obj)。session 对象可以调用该方法将参数 Object 指定的对象 obj 添加到 session 对象中,并为添加的对象指定了一个索引关键字,如果添加的两个对象的关键字相同,则先前添加的对象被清除。

(2) public Object getAttribute(String key)。获取 session 对象索引关键字是 key 的对象。由于任何对象都可以添加到 session 对象中,因此用该方法取回对象时,应显式转化为原来的类型。

(3) public Enumeration getAttributeNames()。session 对象调用该方法产生一个枚举对象,该枚举对象使用 nextElemets()遍历 session 中的各个对象所对应的关键字。

(4) public void removeAttribute(String key)。session 对象调用该方法移掉关键字 key 对应的对象。

例 4_14 实现猜数字游戏。当用户访问 example4_14.jsp 时,随机分配给用户一个 1 到 100 之间的整数,然后将这个整数存在用户的 session 对象中。用户在 form 表单的 text 里输入自己的猜测。用户输入猜测后单击提交键,访问 example4_14_judge.jsp 页面,该页面负责判断用户给出的猜测是否和用户的 session 对象中存放的那个整数相同。如果相同,就将用户定向到 example4_14_success.jsp;如果不相同,就将用户定向到 example4_14_large.jsp 或 example4_14_small.jsp,然后,用户在这些页面再重新提交新的猜测给 example4_14_judge.jsp 页面。

例 4_14

example4_14.jsp(效果如图 4.12(a)所示)

```
<%@ page contentType = "text/html" %>
<%@ page pageEncoding = "utf-8" %>
<style>#textStyle
    {font-family:宋体;font-size:36;color:blue
    }
</style>
<style>#tomStyle
    {font-family:黑体;font-size:26;color:black
    }
</style>
<HTML><body bgcolor = #ccffff>
<p id = "textStyle">
随机分给了一个 1 到 100 之间的数,请猜!
<% int number = (int)(Math.random() * 100) + 1;
    session.setAttribute("count",new Integer(0));
    //保存需要猜测的数
    session.setAttribute("saveGuessNumber",new Integer(number));
%>
<br>输入猜测:
<form action = "example4_14_judge.jsp" method = "post">
    <input type = "text" id = "tomStyle" name = "guess">
    <input type = "submit" id = "tomStyle" value = "提交"/>
```

```
</form>
</p></body></HTML>
```

example4_14_judge.jsp

```jsp
<%   String str = request.getParameter("guess");
     if(str == null||str.length() == 0) {
         response.sendRedirect("example4_14.jsp");
     }
     else {
         int userGuessNumber = Integer.parseInt(str);     //用户的猜测
         session.setAttribute("userGuess",new Integer(userGuessNumber));
         Integer saveGuessNumber =
         (Integer)session.getAttribute("saveGuessNumber");
         if(userGuessNumber == saveGuessNumber.intValue()) {
             int n = ((Integer)session.getAttribute("count")).intValue();
             n = n + 1;
             session.setAttribute("count",new Integer(n));
             response.sendRedirect("example4_14_success.jsp");
         }
         else if(userGuessNumber > saveGuessNumber.intValue()){
             int n = ((Integer)session.getAttribute("count")).intValue();
             n = n + 1;
             session.setAttribute("count",new Integer(n));
             response.sendRedirect("example4_14_large.jsp");
         }
         else if(userGuessNumber < saveGuessNumber.intValue()) {
             int n = ((Integer)session.getAttribute("count")).intValue();
             n = n + 1;
             session.setAttribute("count",new Integer(n));
             response.sendRedirect("example4_14_small.jsp");
         }
     }
%>
```

example4_14_small.jsp(效果如图4.12(b)所示)

```jsp
<%@ page contentType = "text/html" %>
<%@ page pageEncoding = "utf-8" %>
<style>#textStyle
    { font-family:宋体;font-size:36;color:blue
    }
</style>
<style>#tomStyle
    { font-family:黑体;font-size:26;color:black
    }
</style>
<HTML><body bgcolor = white>
<%   Integer userGuess = (Integer)session.getAttribute("userGuess");
%>
<p id = "textStyle">
<% = userGuess %>数小了,请再猜:
```

```jsp
<form action = "example4_14_judge.jsp" method = "post" name = form>
    <input type = "text" id = "tomStyle" name = "guess" />
    <input type = "submit" id = "tomStyle" value = "送出" />
</form>
</p></body></HTML>
```

example4_14_large.jsp（效果如图 4.12(c)所示）

```jsp
<%@ page contentType = "text/html" %>
<%@ page pageEncoding = "utf-8" %>
<style>#textStyle
    {font-family:宋体;font-size:36;color:blue
    }
</style>
<style>#tomStyle
    {font-family:黑体;font-size:26;color:black
    }
</style>
<HTML><body bgcolor = white>
<%   Integer userGuess = (Integer)session.getAttribute("userGuess");
%>
<p id = "textStyle">
<% = userGuess %>数大了,请再猜:
<form action = "example4_14_judge.jsp" method = "post" name = form>
    <input type = "text" id = "tomStyle" name = "guess">
    <input type = "submit" id = "tomStyle" value = "送出" />
</form>
</p></body></HTML>
```

example4_14_success.jsp（效果如图 4.12(d)所示）

```jsp
<%@ page contentType = "text/html" %>
<%@ page pageEncoding = "utf-8" %>
<%   Integer count = (Integer)session.getAttribute("count");
     Integer num = (Integer)session.getAttribute("saveGuessNumber");
%>
<HTML><body bgcolor = pink>
<p style = "font-family:黑体;font-size:36;color:blue">
```

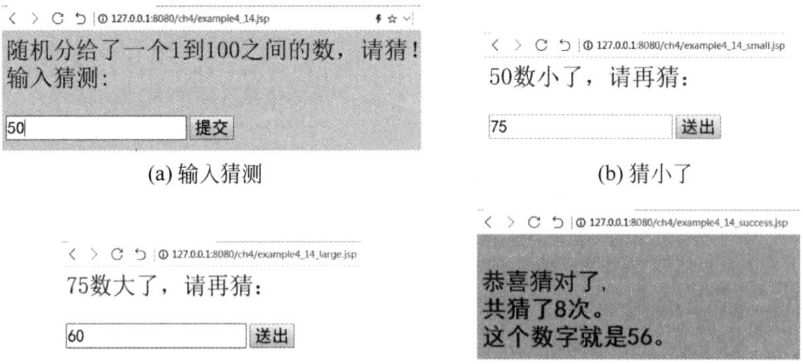

(a) 输入猜测

(b) 猜小了

(c) 猜大了

(d) 猜对了

图 4.12　猜数字游戏

```
<br>恭喜猜对了,
<br><b>共猜了<% = count %>次.
<br>这个数字就是<% = num %>.
</p></body></HTML>
```

4.3.4 session 对象的生存期限

视频讲解

一个用户在某个 Web 服务目录的 session 对象的生存期限依赖于 session 对象是否调用 invalidate()方法使得 session 无效或 session 对象达到了设置的最长的"发呆"状态时间以及用户是否关闭浏览器或服务器被关闭。所谓"发呆"状态时间是指用户对某个 Web 服务目录发出的两次请求之间的间隔时间(默认的发呆时间是 30 分钟)。比如,用户对某个 Web 服务目录下的 JSP 页面发出请求,并得到响应,如果用户不再对该 Web 服务目录发出请求(可能去请求其他的 Web 服务目录),那么用户对该 Web 服务目录进入"发呆"状态,直到用户再次请求该 Web 服务目录时,"发呆"状态结束。

可以修改 Tomcat 服务器下的 web.xml,重新设置各个 Web 服务目录下的 session 对象的最长"发呆"时间。打开 Tomcat 安装目录中 conf 文件下的配置文件 web.xml,找到

```
<session-config>
        <session-timeout>30</session-timeout>
</session-config>
```

将其中的 30 修改成所要求的值即可(单位为分钟)。

session 对象可以使用下列方法获取或设置和生存时间有关的信息:

(1) public long getCreationTime()获取 session 创建的时间,单位是毫秒(GMT 时间,1970 年 7 月 1 日午夜起至 session 创建时刻所走过的毫秒数)。

(2) public long getLastAccessedTime()获取 session 最后一次被操作的时间,单位是毫秒。

(3) public int getMaxInactiveInterval()获取 session 最长的"发呆"时间(单位是秒)。

(4) public void setMaxInactiveInterval(int interval)设置 session 最长的"发呆"时间(单位是秒)。

(5) public boolean isNew()判断 session 是否是一个新建的 session。

(6) invalidate()使 session 无效。

例 4_15 中,session 对象使用 setMaxInactiveInterval(int interval)方法设置最长的"发呆"状态时间为 6 秒。用户可以通过刷新页面检查是否达到了最长的"发呆"时间,如果两次刷新之间的间隔超过 6 秒,用户先前的 session 将被取消,用户将获得一个新的 session 对象。

例 4_15

example4_15.jsp(效果如图 4.13(a)(b)所示)

```
<%@ page import = "java.time.LocalTime" %>
<%@ page import = "java.time.temporal.ChronoUnit" %>
<%@ page contentType = "text/html" %>
<%@ page pageEncoding = "utf-8" %>
<HTML><body bgcolor = #ffccff>
<%   out.println("session 的 Id:" + session.getId());
```

```
        session.setMaxInactiveInterval(6);
        LocalTime time = LocalTime.now();
        out.print("<br>(时:分:秒:纳秒)" + time);
        out.println("<br>开始发呆");
    %>
</body></HTML>
```

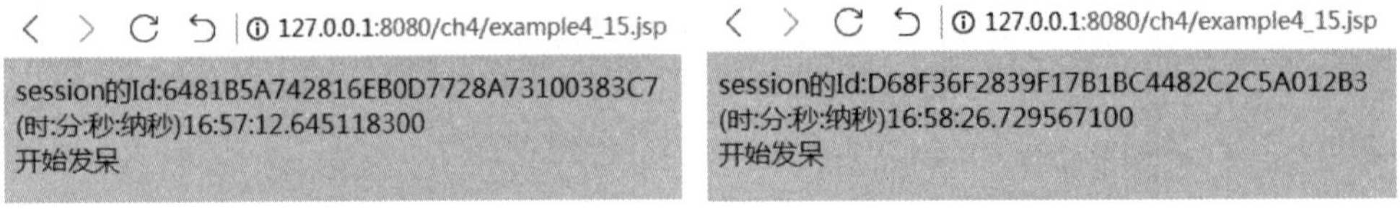

(a) 开始发呆　　　　　　　　　　　　　(b) 发呆超过6秒

图 4.13　session 对象的生存时间

4.4　application 对象

4.3 节学习了 session 对象,用户第一次访问 Web 服务目录时,Tomcat 服务器创建和该用户相对应的 session 对象,当用户在所访问的 Web 服务目录的各个页面之间浏览时,这个 session 对象都是同一个,而且不同用户的 session 对象是互不相同的。本节学习的 application 对象也是由 Tomcat 服务器负责创建,但与 session 对象不同的是,application 对象被访问该 Web 服务目录的所有的用户共享,但不同 Web 服务目录下的 application 互不相同。

▶ 4.4.1　application 对象的常用方法

视频讲解

(1) public void setAttribute(String key,Object obj)。application 对象可以调用该方法将参数 Object 指定的对象 obj 添加到 application 对象中,并为添加的对象指定了一个索引关键字,如果添加的两个对象的关键字相同,则先前添加对象被清除。

(2) public Object getAttitute(String key)。获取 application 对象含有的关键字是 key 的对象。由于任何对象都可以添加到 application 对象中,因此用该方法取回对象时,应显式转化为原来的类型。

(3) public Enumeration getAttributeNames()。application 对象调用该方法产生一个 Enumeration(枚举)对象,该 Enumeration 对象使用 nextElemets()遍历 application 中的各个对象所对应的关键字。

(4) public void removeAttribute(String key)。从当前 application 对象中删除关键字是 key 的对象。

(5) public String getServletInfo()。获取 Servlet 编译器的当前版本的信息。

由于 application 对象对所有的用户都是相同的,因此,在某些情况下,对该对象的操作需要实现同步(synchronized)处理。

需要注意的是,Tomcat 服务器启动后,application 对象中已经有 8 个关键字,因此,JSP 程序使用 setAttribute(String key,Object obj)在 application 对象中存储数据时,key 要避免使用这 8 个关键字。这 8 关键字的名字都很长,例如:javax.servlet.context.tempdir、org. apache.jasper.runtime.JspApplicationContextImpl 等。

▶ 4.4.2 application 留言板

在实际项目中一般是使用数据库来实现留言板,不会用 application 对象(除非是非常简单的告示留言板)。例 4_16 是为了练习使用 application 内置对象,让其担当留言板的角色,并设置留言板上最多可留言 99999 条,即让 application 对象用 1 至 99999 之间的一个整数作为关键字(key)存放一条留言。用户通过 example4_16.jsp 向 example4_16_pane.jsp 提交姓名、留言标题和留言内容,example4_16_pane.jsp 页面用 application 存放这些内容,查找这些内容所使用的关键字是 1 至 99999 之间的某个整数。当留言数目超过 99999 时,提示用户无法再留言。example4_16_show.jsp 负责显示全部留言。example4_16_delete.jsp 页面负责删除留言(删除留言时需要输入密码 123456,这个密码可以只让管理者知道)。

视频讲解

example4_16.jsp(效果如图 4.14(a)所示)

```jsp
<%@ page contentType = "text/html" %>
<%@ page pageEncoding = "utf-8" %>
<HTML>
<style>
    #textStyle{
        font-family:宋体;font-size:18;color:blue
    }
</style>
<body id = 'textStyle' bgcolor = #ffccff>
<form action = 'example4_16_pane.jsp' method = 'post'>
留言者:<input  type = 'text' name = 'peopleName' size = 40/>
<br>标题:<input type = 'text' name = 'title' size = 42/>
<br>留言:<br>
<textArea name = 'contents' id = 'textStyle' rows = '10' cols = 36 wrap = 'physical'>
</textArea>
<br><input type = 'submit' id = 'textStyle' value = '提交留言' name = 'submit'/>
</form>
<a href = 'example4_16_show.jsp'>查看留言</a>
<a href = 'example4_16_delete.jsp'>删除留言</a>
</body></HTML>
```

example4_16_pane.jsp

```jsp
<%@ page contentType = "text/html" %>
<%@ page pageEncoding = "utf-8" %>
<%@ page import = "java.time.LocalDateTime" %>
<%@ page import = "java.util.Stack" %>
<HTML><body>
<%!
    Stack<Integer> maxAmount = null;             //存放留言序号
    //向 application 添加对象的 synchronized 方法
    synchronized void addMess(ServletContext application,StringBuffer mess){
        int index = -1;                          //留言序号
        if(!maxAmount.empty()){
            index = maxAmount.pop();
            mess.insert(0,"No." + index + ".");
```

```jsp
            application.setAttribute("" + index,new String(mess));
        }
    }
%>
<%
    if(maxAmount == null){
        maxAmount = new Stack<Integer>();
        for(int i = 999999;i>=1;i--){            //最多可以有999999条留言
            maxAmount.push(i);
        }
    }
    boolean isSave = true;
    request.setCharacterEncoding("utf-8");
    String peopleName = request.getParameter("peopleName");
    String title = request.getParameter("title");
    String contents = request.getParameter("contents");
    if(peopleName.length() == 0||title.length() == 0||contents.length() == 0){
        isSave = false;
        out.print("<h2>" + "请输入留言者,标题和内容");
    }
    if(isSave) {
        LocalDateTime dateTime = LocalDateTime.now();
        StringBuffer message = new StringBuffer();
        message.append("留言者:" + peopleName + "#");
        message.append("<br>留言标题«" + title + "»#");
        message.append("<br>留言内容:<br>" + contents + "#");
        String timeFormat =
        String.format("%tY年%<tm月%<td日,%<tH:%<tM:%<tS",dateTime);
        message.append("<br>留言时间<br>" + timeFormat + "#");
        if(maxAmount.empty()){
            out.print("<h2>" + "留言板已满,无法再留言" + "</h2>");
        }
        else {
            addMess(application,message);          //信息存放到application(留言板角色)
        }
    }
%>
<br><a href = "example4_16.jsp">返回留言页面</a><br>
<a href = "example4_16_show.jsp">查看留言板</a>
</body></HTML>
```

example4_16_show.jsp(效果如图4.14(b)所示)

```jsp
<%@ page contentType = "text/html" %>
<%@ page pageEncoding = "utf-8" %>
<%@ page import = "java.util.Enumeration" %>
<HTML><body bgcolor = cyan>
<p style = "font-family:宋体;font-size:14;color:black">
<a href = "example4_16.jsp" >返回留言页面 </a><br><br>
<%
```

```jsp
        Enumeration<String> e = application.getAttributeNames();
        while(e.hasMoreElements()) {
            String key = e.nextElement();
            String regex = "[1-9][0-9]*";                //匹配用户的关键字
            if(key.matches(regex)){
                String message = (String)application.getAttribute(key);
                String [] mess = message.split("#");
                out.print(mess[0]);                       //留言者和序号
                out.print(mess[1]);                       //标题
                out.print(mess[2]);                       //留言内容
                out.print(mess[3]);                       //留言时间
                out.print("<br>-------------------------------------<br>");
            }
        }
%>
</p></body></HTML>
```

example4_16_delete.jsp(效果如图 4.14(c)所示)

```jsp
<%@ page contentType="text/html" %>
<%@ page pageEncoding="utf-8" %>
<%@ page import="java.util.Enumeration" %>
<HTML><body bgcolor=pink>
<p style="font-family:宋体;font-size:18;color:blue">
管理员删除留言.
<form action="" method=post>
输入密码:<input type="password" name="password"    size=12 /><br>
输入留言序号:<input type="text" name="index" size=6 />
<br><input type="submit" name="submit"   value="删除"/>
</form>
<%   String password = request.getParameter("password");
     String index = request.getParameter("index");
     if(password == null ) password = "";
     if(index == null ) index = "";
     if(password.equals("123456")){
       Enumeration<String> e = application.getAttributeNames();
       while(e.hasMoreElements()) {
            String key = e.nextElement();
            if(key.equals(index)){
                application.removeAttribute(key);
                out.print("<br>删除了第"+ index +"条留言<br>");
            }
        }
      }
%>
<a href="example4_16.jsp" >返回留言页面</a><br>
<a href="example4_16_show.jsp">查看留言板</a>
</p></body></HTML>
```

(a) 输入留言　　　　　　　　　(b) 查看留言

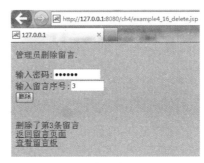

(c) 删除留言

图 4.14　留言板

4.5　out 对象

视频讲解

　　out 对象是一个输出流,用来向用户端输出数据。在前面的许多例子里曾多次使用 out 对象进行数据的输出。out 对象可调用如下的方法用于各种数据的输出,例如:

（1）out.print(boolean)或 out.println(boolean)用于输出一个布尔值。

（2）out.print(char)或 out.println(char)用于输出一个字符。

（3）out.print(double)或 out.println(double)用于输出一个双精度的浮点数。

（4）out.print(fload)或 out.println(float)用于输出一个单精度的浮点数。

（5）out.print(long)或 out.println(long)用于输出一个长整型数据。

（6）out.print(String)或 out.println(String)用于输出一个 String 对象的字符序列。

println 和 print 方法的区别是:println 会向缓存区写入一个换行,而 print 不写入换行。但是浏览器的显示区域目前不识别 println 写入的换行,如果希望浏览器显示换行,应当向浏览器写入"< br >"实现换行。

4.6　上机实验

视频讲解

　　提供了详细的实验步骤要求,按步骤完成,提升学习效果,积累经验,不断提高 Web 设计能力。

4.6.1 实验1 196算法之谜

❶ 实验目的

掌握怎样在JSP中使用内置对象request获取form表单的文本框text提交的数据。

❷ 实验要求

（1）编写inputNumber.jsp，该页面提供一个form表单，该form表单提供一个文本框text，用于用户输入一个正整数，用户在form表单中输入的数字，单击submit提交键将正整数提交给huiwenNumber.jsp页面。

（2）huiwenNumber.jsp获取inputNumber.jsp提交的正整数，然后huiwenNumber.jsp寻找回文数，也称196算法（一个数正读反读都一样，就把它称为"回文数"），对于正整数number，196算法如下：

① number加上把它反过来写之后得到的数reverseNumber得到resultNumber，如果resultNumber是回文数，进行③，否则②。

② 对新得到的数resultNumber重复上述①操作。

③ 结束。

例如对于59，59＋95＝154，154＋451＝605，605＋506＝1111。3步得到回文数1111。数字196是一个相当引人注目的例外，数学家们已经用计算机算到了3亿多位数，都没有产生回文数。从196出发，能否得到回文数？196究竟特殊在哪儿？至今仍是个谜。

（3）在Tomcat服务器的webapps目录下（比如，D:\apache-tomcat-9.0.26\webapps）新建名字是ch4_practice_one的Web服务目录。把JSP页面都保存到ch4_practice_one目录中。

（4）用浏览器访问JSP页面inputNumber.jsp。

❸ 参考代码

参考代码运行效果如图4.15所示。

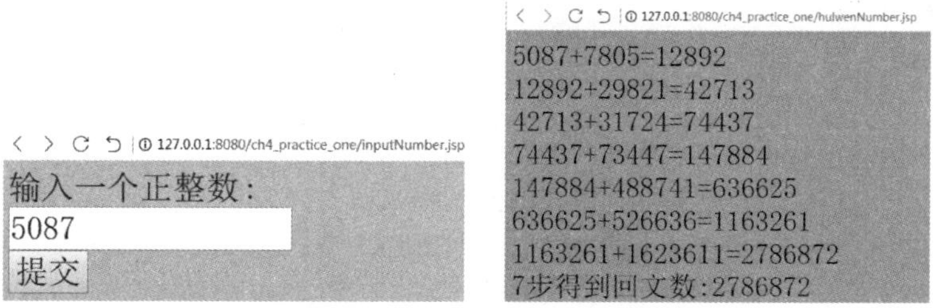

(a) 输入正整数　　　　(b) 产生回文数

图4.15　有趣的196算法

inputNumber.jsp（效果如图4.15(a)所示）

```
<%@ page contentType="text/html" %>
<%@ page pageEncoding="utf-8" %>
<style>
    #tomStyle{
        font-family:宋体;font-size:36;color:blue
    }
</style>
```

```html
<HTML><body id=tomStyle bgcolor=#ffccff>
<form action="huiwenNumber.jsp" method=post name=form>
输入一个正整数:<br>
  <input type=text name="number" id=tomStyle size=16 value=520 />
  <br><input type="submit" id=tomStyle value="提交" />
</form>
</body></HTML>
```

huiwenNumber.jsp（效果如图 4.15(b)所示）

```jsp
<%@ page contentType="text/html" %>
<%@ page pageEncoding="utf-8" %>
<%@ page import="java.math.BigInteger" %>
<HTML><body bgcolor=cyan>
<p style="font-family:宋体;font-size:26;color:black">
<%!
   public static String reverse(String s) {        //定义方法返回参数的反序
       StringBuffer buffer = new StringBuffer(s);
       StringBuffer reverseBuffer = buffer.reverse();
       return reverseBuffer.toString();
   }
%>
<%
   String regex = "[1-9][0-9]*";
   String startNumber = request.getParameter("number");
   if(startNumber == null || startNumber.length() == 0) {
       startNumber = "520";
   }
   if(!startNumber.matches(regex)) {
       response.sendRedirect("inputNumber.jsp");
       return;
   }
   long step = 1;
   BigInteger number = new BigInteger(startNumber);                    //开始的数
   BigInteger reverseNumber = new BigInteger(reverse(number.toString()));
   BigInteger resultNumber = number.add(reverseNumber);                //二者相加的结果
   out.print("<br>" + number + "+" + reverseNumber + "=" + resultNumber);  //计算过程
   BigInteger p = new BigInteger(reverse(resultNumber.toString()));
   while(!resultNumber.equals(p)) {                                    //判断是否是回文数
       number = new BigInteger(resultNumber.toString());
       reverseNumber = new BigInteger(reverse(number.toString()));
       resultNumber = number.add(reverseNumber);
       p = new BigInteger(reverse(resultNumber.toString()));
       out.print("<br>" + number + "+" + reverseNumber + "=" + resultNumber);
       step++;
   }
   out.print("<br>" + step + "步得到回文数:" + resultNumber);
%>
</p></body></HTML>
```

4.6.2 实验2 计算器

❶ **实验目的**

掌握怎样在 JSP 中使用 request 对象获取 form 表单提交的 text（文本框）以及 select（下拉列表）中的数据。掌握使用 response 对象实现重定向。

❷ **实验要求**

（1）编写 input.jsp，该页面提供一个 form 表单，该 form 表单提供两个 text 文本框，用于用户输入数字，提供一个 select 下拉列表，该下拉列表有加、减、乘、除四个选项，供用户选择运算符号。用户在 form 表单中输入的数字，选择运算符号，单击 submit 提交键将这些数据提交给 computer.jsp 页面。

（2）computer.jsp 页面获取 input.jsp 提交的数据，计算出相应的结果显示给用户。如果 computer.jsp 页面没有获取到数据，就将用户重新定向到 input 页面。

（3）在 Tomcat 服务器的 webapps 目录下（比如，D:\apache-tomcat-9.0.26\webapps）新建名字是 ch4_practice_two 的 Web 服务目录。把 JSP 页面都保存到 ch4_practice_two 目录中。

（4）用浏览器访问 JSP 页面 input.jsp。

❸ **参考代码**

参考代码运行效果如图 4.16 所示。

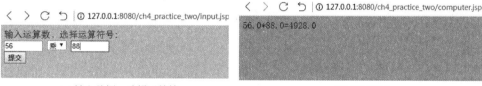

(a) 输入数据，选择运算符　　　　　　(b) 运算结果

图 4.16　计算器

input.jsp（效果如图 4.16(a) 所示）

```
<%@ page contentType = "text/html" %>
<%@ page pageEncoding = "utf-8" %>
<HTML><body bgcolor = #ffccff>
<form action = "computer.jsp" method = post name = form>
<p style = "font-family:宋体;font-size:18;color:blue">
输入运算数,选择运算符号:<br>
  <input type = text name = "numberOne" size = 6/>
     <select name = "operator">
        <option selected = "selected" value = " + ">加
        <option value = " - ">减
        <option value = " * ">乘
        <option value = "/">除
     </select>
  <input type = text name = "numberTwo" size = 6 />
  <br><input type = "submit" name = "submit" value = "提交" />
</form>
</p></body></HTML>
```

computer.jsp(效果如图 4.16(b)所示)

```jsp
<%@ page contentType="text/html" %>
<%@ page pageEncoding="utf-8" %>
<HTML><body bgcolor=cyan>
<p style="font-family:宋体;font-size:18;color:black">
<%
   String numberOne = request.getParameter("numberOne");
   String numberTwo = request.getParameter("numberTwo");
   String operator = request.getParameter("operator");
   if(numberOne == null||numberOne.length() == 0) {
       response.sendRedirect("input.jsp");
       return;
   }
   else if(numberTwo == null||numberTwo.length() == 0) {
       response.sendRedirect("input.jsp");
       return;
   }
   try{
       double a = Double.parseDouble(numberOne);
       double b = Double.parseDouble(numberTwo);
       double r = 0;
       if(operator.equals("+"))
           r = a + b;
       else if(operator.equals("-"))
           r = a - b;
       else if(operator.equals("*"))
           r = a * b;
       else if(operator.equals("/"))
           r = a/b;
       out.print(a + "" + operator + "" + b + "=" + r);
   }
   catch(Exception e){
       out.println("请输入数字字符");
   }
%>
</body></HTML>
```

▶ 4.6.3 实验3 单词的频率

❶ **实验目的**

掌握怎样在 JSP 中使用 request 对象获取 form 表单提交的 textArea 的数据。

❷ **实验要求**

(1) 编写 inputText.jsp,该页面提供一个 form 表单,该 form 表单提供一个 textArea 文本输入区,用于用户输入多行英文,单击 submit 提交键将英文提交给 findWord.jsp 页面。

(2) findWord.jsp 页面获取 input.jsp 提交的英文,分析出英文中出现的单词以及单词出现的频率。

(3) 在 Tomcat 服务器的 webapps 目录下(比如,D:\apache-tomcat-9.0.26\webapps)新建名字是 ch4_practice_three 的 Web 服务目录。把 JSP 页面都保存到 ch4_practice_three 目录中。

(4)用浏览器访问 JSP 页面 inputText.jsp。

❸ 参考代码

学生可按照实验要求,参考本代码编写自己的实验代码。参考代码运行效果如图 4.17 所示。

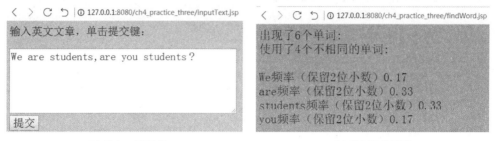

(a)输入一段英文　　　　　　　　　　　(b)单词以及频率

图 4.17　分析文章中的单词

inputText.jsp(效果如图 4.17(a)所示)

```jsp
<%@ page contentType = "text/html" %>
<%@ page pageEncoding = "utf-8" %>
<HTML>
<style>
    #textStyle{
        font-family:宋体;font-size:23;color:blue
    }
</style>
<body bgcolor = #ffccff>
<p id = "textStyle">
输入英文文章,单击提交键:
<%
    String content = "We are students,are you students?";
%>
<form action = "findWord.jsp" method = post name = form>
    <textArea name = "english" id = "textStyle" rows = 5 cols = 38><% = content %>
    </textArea>
<br><input type = "submit" name = "submit" id = "textStyle" value = "提交"/>
</form>
</p></body></HTML>
```

findWord.jsp(效果如图 4.17(b)所示)

```jsp
<%@ page contentType = "text/html" %>
<%@ page pageEncoding = "utf-8" %>
<%@ page import = "java.util.ArrayList" %>
<%@ page import = "java.util.Iterator" %>
<%@ page import = "java.util.regex.Pattern" %>
<%@ page import = "java.util.regex.Matcher" %>
<HTML><body bgcolor = cyan>
<p style = "font-family:宋体;font-size:25;color:black">
<%!
```

```
class Word  {                              //定义类,刻画单词
    String englishWord;                    //存放一个英文单词
    int count ;                            //存放单词出现的次数
    public boolean equals(Object o) {      //重写 Object 类的方法规定两个 Word 对象相等
        Word wd = (Word)o;
        return englishWord.equals(wd.englishWord);
    }
}
%>
<%  request.setCharacterEncoding("utf-8");
    String englishText = request.getParameter("english");
    ArrayList<Word> wordList = new ArrayList<Word>();  //存放 Word 对象的顺序表
    Pattern pattern;                                    //模式对象
    Matcher matcher;                                    //匹配对象
    String regex = "[a-zA-Z]+" ;                        //匹配英文单词
    pattern = Pattern.compile(regex);                   //模式对象
    matcher = pattern.matcher(englishText);             //匹配对象,用于检索 englishText
    int allWordAmount = 0;                              //全部单词的数量
    while(matcher.find()) {
        String str = matcher.group();
        allWordAmount++;
        Word wd = new Word();
        wd.englishWord = str;
        if(!wordList.contains(wd)) {
            wordList.add(wd);
            wd.count = 1;
        }
        else {
           int index = wordList.indexOf(wd);            //检索是否有和 wd 相等的 Word 对象
           wd = wordList.get(index);
           wd.count += 1;
        }
    }
    int diffrentWords = wordList.size();                //全部不相同的单词数量
    Iterator<Word> ite = wordList.iterator();
    out.print("出现了" + allWordAmount + "个单词:<br>");
    out.print("使用了" + diffrentWords + "个不相同的单词:<br>");
    while(ite.hasNext()) {                              //遍历
        Word wd = ite.next();
        double fq = (double)(wd.count)/allWordAmount;
        String frequency = String.format("%.2f",fq);
        out.print("<br>" + wd.englishWord + "频率(保留2位小数)" + frequency);
    }
%>
</p></body></HTML>
```

▶ **4.6.4 实验 4 成绩与饼图**

❶ 实验目的

掌握怎样在 JSP 页面中使用 response 对象改变 MIME 类型。

❷ 实验要求

(1) 用户在 inputStudents.jsp 页面输入考试人数,以及分数分别在 90～100 分,80～89 分,70～79 分,60～69 分和 60 分以下的人数,单击"看成绩分析"提交键,请求访问

scoreAnalysis.jsp，scoreAnalysis.jsp 页面按百分比分析成绩。单击"看成绩饼图"提交键，scoreAnalysis.jsp 页面用饼形状图显示成绩。如果总人数不等于参考人数之和，scoreAnalysis.jsp 将用户重定向到 inputStudents.jsp。

（2）scoreAnalysis.jsp 使用 response 对象，动态改变相应的 MIME 类型，即动态改变 contentType 属性的值。

（3）在 Tomcat 服务器的 webapps 目录下（比如，D:\apache-tomcat-9.0.26\webapps）新建名字是 ch4_practice_four 的 Web 服务目录。把 JSP 页面都保存到 ch4_practice_four 目录中。

（4）用浏览器访问 JSP 页面 inputStudents.jsp。

❸ 参考代码

参考代码运行效果如图 4.18 所示。

(a) 输入数据

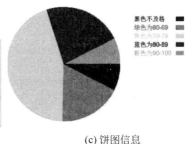
(b) 文字信息   (c) 饼图信息

图 4.18 成绩与饼图

inputStudents.jsp（效果如图 4.18(a)所示）

```
<%@ page contentType = "text/html" %>
<%@ page pageEncoding = "utf-8" %>
<HTML><body bgcolor = #ffccff>
<style>
    #textStyle{
        font-family:宋体;font-size:18;color:blue
    }
</style>
<p id = "textStyle">
<form action = "scoreAnalysis.jsp"   method = post>
输入考试有关数据：<br>
考试人数：<input type = "text" name = "totalStudents"value = 1 size = 12 /><br>
90～100 人数：<input type = "text" name = "student90"value = 1 size = 12 /><br>
80～89 人数：<input type = "text" name = "student80"value = 1 size = 12 /><br>
70～79 人数：<input type = "text" name = "student70"value = 1 size = 12 /><br>
60～69 人数：<input type = "text" name = "student60"value = 1 size = 12 /><br>
0～59 人数：<input type = "text" name = "student59"value = 1 size = 12 /><br>
<input type = "submit" name = "submit" id = "textStyle" value = "看成绩分析"/><br>
<input type = "submit" name = "submit" id = "textStyle" value = "看成绩饼图"/>
</form>
</p></body></HTML>
```

scoreAnalysis.jsp（效果如图 4.18(b)(c)所示）

```jsp
<%@ page contentType="text/html" %>
<%@ page pageEncoding="utf-8" %>
<%@ page import="java.awt.*" %>
<%@ page import="java.io.OutputStream" %>
<%@ page import="java.awt.image.BufferedImage" %>
<%@ page import="java.awt.geom.*" %>
<%@ page import="javax.imageio.ImageIO" %>
<% request.setCharacterEncoding("utf-8");
   String submitValue = request.getParameter("submit");
   String totalStudents = request.getParameter("totalStudents");
   String students90 = request.getParameter("student90");
   String students80 = request.getParameter("student80");
   String students70 = request.getParameter("student70");
   String students60 = request.getParameter("student60");
   String students59 = request.getParameter("student59");
   int students90_number = Integer.parseInt(students90);
   int students80_number = Integer.parseInt(students80);
   int students70_number = Integer.parseInt(students70);
   int students60_number = Integer.parseInt(students60);
   int students59_number = Integer.parseInt(students59);
   int totalStudents_number = Integer.parseInt(totalStudents);
   boolean isRight = totalStudents_number ==
     students90_number + students80_number +
     students70_number + students60_number + students59_number ;
   if(isRight == false) {
       response.sendRedirect(" inputStudents.jsp");
       return;
   }
   double percent90 = (double)students90_number/totalStudents_number;
   double percent80 = (double)students80_number/totalStudents_number;
   double percent70 = (double)students70_number/totalStudents_number;
   double percent60 = (double)students60_number/totalStudents_number;
   double percent59 = (double)students59_number/totalStudents_number;
   double angSt = 0;
   double angExt = 0;
   int x = 450, y = 50;
   if(submitValue.equals("看成绩饼图")){
     //response 对象更改 MIME 类型
       response.setContentType("image/jpeg");
       int width = 1000, height = 500;
       BufferedImage image =
       new BufferedImage(width, height, BufferedImage.TYPE_INT_RGB);
       Graphics g = image.getGraphics();
       g.fillRect(0, 0, width, height);
       Graphics2D g_2d = (Graphics2D)g;
       Arc2D arc = new  Arc2D.Double(0,0,400,400,0,360,Arc2D.PIE);
       arc.setAngleStart(angSt);                //设置饼弧的开始角度
       angExt = - percent59 * 360;
       arc.setAngleExtent(angExt);              //设置饼弧度数
       g_2d.setFont(new Font("",Font.PLAIN,22));
       g_2d.setColor(Color.black);
       g_2d.drawString("黑色不及格",x,y);
```

```
            g_2d.fillRect(x+150,35,30,15);
            g_2d.fill(arc);                       //绘制不及格的饼图
            angSt = - percent59 * 360;
            angExt = - percent60 * 360;
            arc.setAngleStart(angSt);             //设置饼弧的开始角度
            arc.setAngleExtent(angExt);           //设置饼弧度数
            g_2d.setColor(Color.green);
            g_2d.drawString("绿色为 60 - 69",x,y = y+30);
            g_2d.fillRect(x+150,65,30,15);
            g_2d.fill(arc);                       //绘制 60~69 分的饼图
            angSt = - (percent59 * 360 + percent60 * 360);
            angExt = - percent70 * 360;
            arc.setAngleStart(angSt);             //设置饼弧的开始角度
            arc.setAngleExtent(angExt);           //设置饼弧度数
            g_2d.setColor(Color.yellow);
            g_2d.drawString("黄色为 70 - 79",x,y = y+30);
            g_2d.fillRect(x+150,95,30,15);
            g_2d.fill(arc);                       //绘制 70~79 分的饼图
            angSt = - (percent59 * 360 + percent60 * 360 + percent70 * 360);
            angExt = - percent80 * 360;
            arc.setAngleStart(angSt);             //设置饼弧的开始角度
            arc.setAngleExtent(angExt);           //设置饼弧度数
            g_2d.setColor(Color.blue);
            g_2d.drawString("蓝色为 80 - 89",x,y = y+30);
            g_2d.fillRect(x+150,125,30,15);
            g_2d.fill(arc);                       //绘制 80~89 分的饼图
            angSt = - (percent59 * 360 + percent60 * 360 + percent70 * 360 + percent80 * 360);
            angExt = - percent90 * 360;
            arc.setAngleStart(angSt);             //设置饼弧的开始角度
            arc.setAngleExtent(angExt);           //设置饼弧度数
            g_2d.setColor(Color.pink);
            g_2d.drawString("粉色为 90 - 100",x,y = y+30);
            g_2d.fillRect(x+150,155,30,15);
            g_2d.fill(arc);                       //绘制 90~100 分的饼图
            try {                                 //指向用户端的输出流
                OutputStream outClient = response.getOutputStream();
                boolean boo = ImageIO.write(image,"jpeg",outClient);
            }
            catch(Exception exp){}
        }
%>
<HTML><body bgcolor = #EEEEFF>
<p style = "font-family:宋体;font-size:18;color:blue">
<%
    String percent90Result = String.format("%.3f%%",percent90 * 100);
    String percent80Result = String.format("%.3f%%",percent80 * 100);
    String percent70Result = String.format("%.3f%%",percent70 * 100);
    String percent60Result = String.format("%.3f%%",percent60 * 100);
    String percent59Result = String.format("%.3f%%",percent59 * 100);
%>
```

```
<b>成绩百分比分布(保留3位小数)<br>
90~100分:<%=percent90Result%><br>
80~89分:<%=percent80Result%><br>
70~79分:<%=percent70Result%><br>
60~69分:<%=percent60Result%><br>
不及格:<%=percent59Result%><br>
</p></body></HTML>
```

4.6.5 实验5 记忆测试

❶ 实验目的

掌握怎样在 JSP 中使用 session 对象存储用户的数据，以及使用 response 对象实现重定向和添加新的响应头。

❷ 实验要求

(1) 编写 choiceGrage.jsp，该页面中的 form 表单中使用 radio(单选按钮)提供选择记忆测试级别：初级、中级和高级。初级需要记忆一个长度为5个字符的字符序列(例如，☆□☆△○)，中级需要记忆一个长度为7个字符的字符序列，高级需要记忆一个长度为10个字符的字符序列。在 choiceGrage.jsp 页面选择级别后，单击 form 表单的提交键，提交给 giveTest.jsp 页面。

(2) 编写 giveTest.jsp 页面，该页面获取 choiceGrage.jsp 页面提交的级别后，根据级别随机显示一个字符序列，比如，对于中级，这个字符序列的长度为7个字符。然后提示用户在5秒内记住这个字符序列。5秒后，该页面将自动定向到 answerTest.jsp 页面。

(3) 编写 answerTest.jsp 页面，该页面的 form 表单提供用户给出答案的界面，即使用 radio 标记让用户选择字符序列中的各个字符，以此代表用户认为自己记住的字符序列。单击提交键，将选择提交给 judgeAnswer.jsp 页面。

(4) 编写 judgeAnswer.jsp 页面，该页面负责判断用户是否记住了 giveTest.jsp 页面给出的字符序列。

(5) 在 Tomcat 服务器的 webapps 目录下(比如，D:\apache-tomcat-9.0.26\webapps)新建名字是 ch4_practice_five 的 Web 服务目录。把 JSP 页面都保存到 ch4_practice_five 目录中。

(6) 用浏览器访问 JSP 页面 choiceGrade.jsp。

❸ 参考代码

参考代码运行效果如图 4.19 所示。

choiceGrade.jsp(效果如图 4.19(a)所示)

```
<%@ page contentType="text/html" %>
<%@ page pageEncoding="utf-8" %>
<HTML><body bgcolor=#ffccff>
<style>
   #textStyle{
      font-family:宋体;font-size:26;color:blue
   }
</style>
<form action="giveTest.jsp" id="textStyle" method=post name=form>
```

```
< input type = radio name = "grade" value = "5" />初级
< input type = radio name = "grade" value = "7" checked = "ok" />中级
< input type = radio name = "grade" value = "10" />高级
< br >< input type = "submit" name = "submit" id = "textStyle" value = "提交"/>
< input type = "reset" id = "textStyle" value = "重置" />
</form >
</body ></HTML >
```

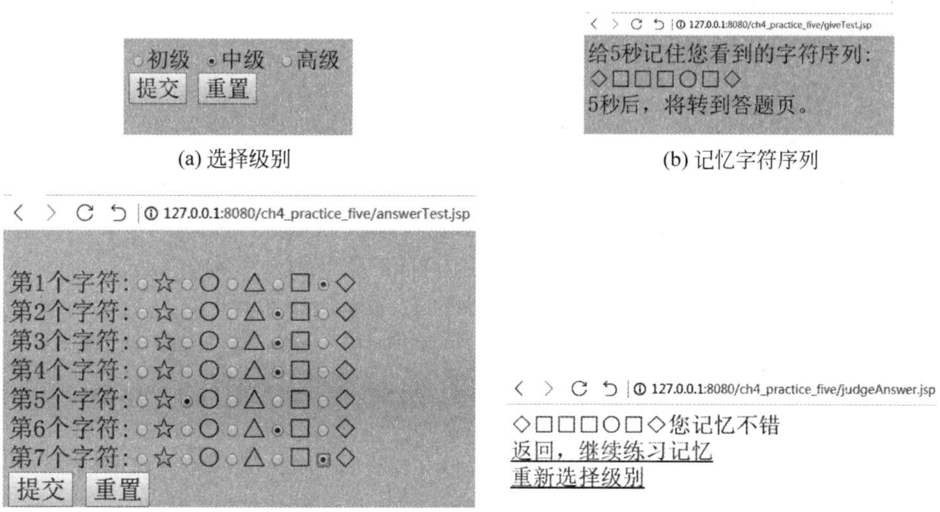

(a) 选择级别　　　　　　　　　　(b) 记忆字符序列

(c) 答题页面　　　　　　　　　　(d) 判断答案是否正确

图 4.19　记忆测试

giveTest.jsp（效果如图 4.19（b）所示）

```
<%@ page contentType = "text/html" %>
<%@ page pageEncoding = "utf - 8" %>
<%@ page import = "java.util.ArrayList" %>
<%@ page import = "java.util.Random" %>
< HTML >< body bgcolor = #ffccff >
< style >
    #textStyle{
        font - family:宋体;font - size:36;color:blue
    }
</style >
<%! static ArrayList < String > list = new ArrayList < String >();
    static {
        list.add("☆");
        list.add("○");
        list.add("△");
        list.add("□");
        list.add("◇");
    }
    String getNextTestString(int length) {
        StringBuffer buffer = new StringBuffer();
        Random random = new Random();
        for(int i = 0;i < length;i++) {
            int index = random.nextInt(list.size());
```

```
            String str = list.get(index);
            buffer.append(str);
         }
         return new String(buffer);
      }
%>
<% String grade = request.getParameter("grade");
   if(grade == null){
         grade = (String) session.getAttribute ("grade");
   }
   int number = Integer.parseInt(grade);
   session.setAttribute ("grade" ,grade);
   String testString = null;                         //存放测试题目,例如"☆△☆□◇○□"
   String yesORNo = null;                            //存放是否已经给了测试题目,例如"yes"
   yesORNo = (String)session.getAttribute ("yesORNo");
   if(yesORNo == null) {
         testString = getNextTestString(number);     //得到测试的题目
         session.setAttribute ("yesORNo" ,"yes");
         session.setAttribute ("testString" ,testString);
   }
   else if(yesORNo.equals("yes")){
         response.sendRedirect("answerTest.jsp");    //定向到答题页面
         return;
   }
   else if(yesORNo.equals("no")){
         testString = getNextTestString(number);     //得到测试的题目
         session.setAttribute ("yesORNo" ,"yes");
         session.setAttribute ("testString" ,testString);
   }
%>
<p id = "textStyle">给 5 秒记住您看到的字符序列:<br>
<% = testString %>
<br>5 秒后,将转到答题页.
<%   response.setHeader("refresh","5");
%>
</p></body></HTML>
```

answerTest.jsp(效果如图 4.19(c)所示)

```
<%@ page contentType = "text/html" %>
<%@ page pageEncoding = "utf - 8" %>
<HTML><body bgcolor = #ffccff>
<style>
   #textStyle{
       font - family:宋体;font - size:26;color:blue
   }
</style>
<form action = "judgeAnswer.jsp" id = textStyle method = post >
<%
   int n = Integer.parseInt((String)session.getAttribute ("grade"));
   session.setAttribute ("yesORNo" ,"no");
```

```
        for(int i = 1;i <= n;i++){
            out.print("<br>第" + i + "个字符:");
            out.print("<input type = radio name = R" + i + " value = '☆'/>☆" +
                "<input type = radio name = R" + i + " value = '○'/>○" +
                "<input type = radio name = R" + i + " value = '△'/>△" +
                "<input type = radio name = R" + i + " value = '□'/>□" +
                "<input type = radio name = R" + i + " value = '◇'/>◇");
        }
%>
<br><input type = "submit"  name = "submit" id = "textStyle" value = "提交"/>
<input type = "reset" id = "textStyle" value = "重置" />
</form>
</body></HTML>
```

judgeAnswer.jsp(效果如图 4.19(d)所示)

```
<%@ page contentType = "text/html" %>
<%@ page pageEncoding = "utf-8" %>
<HTML><body bgcolor = white >
<p style = "font-family:宋体;font-size:26;color:blue">
<%   session.setAttribute ("yesORNo","no");
     request.setCharacterEncoding("utf-8");
     int n = Integer.parseInt((String)session.getAttribute ("grade"));
     StringBuffer buffer = new StringBuffer();
     for(int i = 1;i <= n;i++){
        buffer.append(request.getParameter("R" + i));         //获取 radio 提交的值
        out.print("" + request.getParameter("R" + i));
     }
     String userAnswer = new String(buffer);
     String testString = (String)session.getAttribute ("testString");
     if(testString.equals(userAnswer)){
         out.print("您记忆不错");
     }
     else {
         out.print("您没记忆住额!答案是:<br>" + testString);
     }
%>
<br><a href = "giveTest.jsp">返回,继续练习记忆</a>
<br><a href = "choiceGrade.jsp">重新选择级别</a>
</p></body></HTML>
```

4.7　小结

- HTTP 通信协议是用户与服务器之间一种提交(请求)信息与响应信息(request/response) 的通信协议。在 JSP 中,内置对象 request 封装了用户提交的信息,request 对象获取用 户提交信息的最常用的方法是 getParameter(String s)。response 对象对用户的请求 作出动态响应,向用户端发送数据。
- HTTP 协议是一种无状态协议。一个用户向服务器发出请求(request),然后服务器返 回响应(response),但不记忆连接的有关信息。所以,Tomcat 服务器必须使用内置

session 对象(会话)记录有关连接的信息。同一个用户在某个 Web 服务目录中的 session 对象是相同的；同一个用户在不同的 Web 服务目录中的 session 对象是互不相同的；不同用户的 session 对象是互不相同的。

➢ 一个用户在某个 Web 服务目录的 session 对象的生存期限依赖于用户是否关闭浏览器，依赖于 session 对象是否调用 invalidate( )方法使得 session 无效或 session 对象达到了设置的最长的"发呆"状态时间。

➢ 内置对象 application 由服务器负责创建，每个 Web 服务目录下的 application 对象被访问该服务目录的所有的用户共享；不同 Web 服务目录下的 application 互不相同。

习题 4

1. 假设 JSP 使用的表单中有如下的 GUI(多选择框)：

```
< input type = "checkbox" name = "item" value = "bird" >鸟
< input type = "checkbox" name = "item" value = "apple" >苹果
< input type = "checkbox" name = "item" value = "cat" >猫
< input type = "checkbox" name = "item" value = "moon" >月亮
```

该表单所请求的 JSP 可以使用 request 对象获取该表单提交的数据，那么，下列哪些是 request 对象获取该表单提交的值的正确语句：

  A. String a＝request. getParameter("item")；

  B. String b＝request. getParameter("checkbox")；

  C. String c[ ]＝request. getParameterValues("item")；

  D. String d[ ]＝request. getParameterValues("checkbox")；

2. 如果表单提交的信息中有汉字，接收该信息的页面应做怎样的处理？

3. 编写两个 JSP 页面 inputString. jsp 和 computer. jsp，用户可以使用 inputString. jsp 提供的表单输入一个字符串，并提交给 computer. jsp 页面，该页面通过内置对象获取 inputString. jsp 页面提交的字符串，计算并显示该字符串的长度。

4. response 对象调用 sendRedirect(URL url)方法的作用是什么？

5. 对例 4_1 的代码进行改动，如果用户在 example4_1. jsp 页面提供的表单中输入了非数字字符，computer. jsp 就将用户重新定向到 example4_1. jsp。

6. 一个用户在不同 Web 服务目录中的 session 对象相同吗？

7. 一个用户在同一 Web 服务目录的不同子目录中的 session 对象相同吗？

8. 编写一个 JSP 页面 selectMusic. jsp，该页面使用 select(下拉列表)提供一些歌曲名，用户选择一个一个歌曲名，单击提交键提交给当前页面，然后当前页面播放用户选择的音乐(音频文件保存的 Web 服务目录的 \music 子目录中)。

第 5 章　JSP与JavaBean

本章导读

 主要内容

- ❖ 编写和使用 JavaBean
- ❖ 获取和修改 bean 的属性
- ❖ bean 的辅助类
- ❖ JSP 与 bean 结合的简单例子

 难点

- ❖ 获取和修改 bean 的属性

 关键实践

- ❖ 记忆测试

  在谈论组件之前让我们看一个常见的事——组装电视机。组装一台电视机时，人们可以选择多个组件，例如电阻、电容、显像管等，一个组装电视机的人不必关心显像管是怎么研制的，只要根据说明书了解其属性和功能就可以了。不同的电视机可以安装相同的显像管，显像管的功能完全相同。如果一台电视机的显像管发生了故障，并不会影响其他的电视机；如果两台电视机安装了一个共享的组件——天线，当天线发生了故障，两台电视机就会受到同样的影响。

  按照 Sun 公司的定义，JavaBean 是一个可重复使用的软件组件。实际上 JavaBean 是一种 Java 类，通过封装属性和方法成为具有某种功能或者处理某个业务的对象，简称 bean。由于 JavaBean 是基于 Java 语言的，因此 JavaBean 不依赖平台，具有以下特点：

- 可以实现代码的重复利用。
- 易编写、易维护、易使用。
- 可以在任何安装了 Java 运行环境的平台上的使用，而无须重新编译。

  JSP 页面可以将数据的处理过程指派给一个或几个 bean 来完成，即 JSP 页面调用这些 bean 完成数据的处理，并将有关处理结果存放到 bean 中，然后 JSP 页面负责显示 bean 中的数据。例如使用 Java 程序片或某些 JSP 指令标记显示 bean 中的数据（见 5.2 节），即 JSP 页面的主要工作是显示数据，不负责数据的逻辑业务处理，如图 5.1 所示。

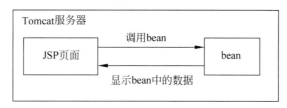

图 5.1　JSP+JavaBean

  本章在 webapps 目录下新建一个 Web 服务目录 ch5，除非特别约定，本章例子中涉及的 JSP 页面均保存在 ch5 目录中。创建 bean 的类的字节码文件须按要求存放，因此要在 ch5 目录下建立目录结构\ch5\WEB-INF\classes（WEB-INF 字母大写）。

5.1 编写和使用 JavaBean

▶ 5.1.1 编写 JavaBean

视频讲解

编写 JavaBean 就是编写一个 Java 的类,所以只要会写类就能编写一个 JavaBean。这个类创建的一个对象称为一个 JavaBean,简称 bean,分配给 bean 的变量(成员变量),也称 bean 的属性。为了能让使用 bean 的应用程序构建工具(比如 Tomcat 服务器)使用 JSP 动作标记知道 bean 的属性和方法,在类的命名上需要遵守以下规则:

(1) 如果类的成员变量(也称 bean 的属性)的名字是 xxx,那么为了获取或更改 bean 的属性的值,类中必须提供两个方法:

- getXxx(),用来获取属性 xxx。
- setXxx(),用来修改属性 xxx。

也就是方法的名字用 get 或 set 为前缀,后缀是将属性(成员变量)名字的首字母大写的字符序列。

(2) 类中定义的方法的访问权限都必须是 public 的。

(3) 类中必须有一个构造方法是 public、无参数的。

下面我们编写一个创建 bean 的 Java 类,并说明在 JSP 中怎样使用这个类创建一个 bean。要求创建 bean 的类带有包名,即 Java 源文件须使用 package 语句给出包名,例如:

```
package gping;
```

或

```
package tom.jiafei;
```

以下是用来创建 bean 的 Java 源文件。

Circle.java(负责创建 bean)

```
package tom.jiafei;
public class Circle {
    double radius;
    public Circle() {
        radius = 1;
    }
    public double getRadius() {
        return radius;
    }
    public void setRadius(double newRadius) {
        radius = newRadius;
    }
    public double circleArea() {
        return Math.PI * radius * radius;
    }
    public double circleLength() {
        return 2.0 * Math.PI * radius;
    }
}
```

第 5 章　JSP与JavaBean

将上述 Java 文件保存为 Circle.java。注意,保存 Java 源文件时,"保存类型"选择为"所有文件","将"编码"选择为"ANSI"。

▶ 5.1.2　保存 bean 的字节码

视频讲解

为了使 JSP 页面使用 bean,Tomcat 服务器必须使用相应的字节码文件创建一个对象,即创建一个 bean。为了让 Tomcat 服务器能找到字节码文件,字节码文件必须保存在特定的目录中。

ch5 \WEB-INF\classes 目录下,根据包名对应的路径,在 classes 目录下再建立相应的子目录。例如,包名 tom.jiafei 对应的路径是 tom\jiafei,那么在 classes 目录下建立子目录结构 tom\jiafei,如图 5.2 所示。

图 5.2　字节码文件的存放位置

将创建 bean 的字节码文件,例如 Circle.class,复制到\WEB-INF\classes\tom\jiafei 中。为了调试程序方便,可以直接按照 bean 的包名将 bean 的源文件(例如 Circle.java)保存在\WEB-INF\classes\tom\jiafei 目录中,然后用命令行进入 tom\jiafei 的父目录 classes(不要进入 tom 或 jiafei 目录)编译 Circle.java:

```
classes> javac tom\jiafei\Circle.java
```

▶ 5.1.3　创建与使用 bean

视频讲解

❶ 使用 bean

使用 JSP 动作标记 useBean 加载使用 bean,语法格式是:

```
< jsp:useBean id = "bean 的名字" class = "创建 bean 的类" scope = "bean 有效范围"/>
```

或

```
< jsp:useBean id = "bean 的名字" class = "创建 bean 的类" scope = "bean 有效范围">
</jsp:useBean >
```

例如:

```
< jsp:useBean id = "circle" class = "tom.jiafei.Circle" scope = "page" />
```

需要特别注意的是,其中的"创建 bean 的类"要带有包名,例如:

```
class = "tom.jiafei.Circle"
```

❷ bean 的加载原理

当 JSP 页面使用 JSP 动作标记 useBean 加载一个 bean 时,Tomcat 服务器首先根据 JSP

动作标记 useBean 中 id 给出的 bean 名字以及 scope 给出的使用范围（bean 生命周期），在 Tomcat 服务器管理的 pageContent 内置对象中查找是否含有这样的 bean（对象）。如果这样的 bean（对象）存在，Tomcat 服务器就复制这个 bean（对象）给 JSP 页面，就是常说的 Tomcat 服务器分配这样的 bean 给 JSP 页面。如果在 pageContent 中没有查找到 JSP 动作标记要求的 bean，就根据 class 指定的类创建一个 bean，并将所创建的 bean 添加到 pageContent 中。通过 Tomcat 服务器创建 bean 的过程可以看出，首次创建一个新的 bean 需要用相应类的字节码文件创建对象，当某些 JSP 页面再需要同样的 bean 时，Tomcat 服务器直接将 pageContent 中已经有的 bean 分配给 JSP 页面，从而提高 JSP 页面 bean 的使用效率。

注：如果修改了字节码文件，必须重新启动 Tomcat 服务器才能使用新的字节码文件。

❸ bean 的有效范围和生命周期

scope 的取值范围给出了 bean 的生命周期（存活时间），即 scope 取值决定了 Tomcat 服务器分配给用户的 bean 的有效范围和生命周期，因此需要理解 scope 取值的具体意义。下面就 JSP 动作标记 useBean 中 scope 取值的不同情况进行说明。

（1）page bean。scope 取值为 page 的 bean 称为 page bean，page bean 的有效范围是用户访问的当前页面，存活时间直到当前页面执行完毕。Tomcat 服务器分配给每个 JSP 页面的 page bean 是互不相同的。也就是说，尽管每个 JSP 页面的 page bean 的功能相同，但它们占有不同的内存空间。page bean 的有效范围是当前页面，当页面执行完毕，Tomcat 服务器取消分配的 page bean，即释放 page bean 所占有的内存空间。需要注意的是，不同用户（浏览器）的 page bean 也是互不相同的。也就是说，当两个用户同时访问同一个 JSP 页面时，一个用户对自己 page bean 的属性的改变，不会影响到另一个用户。

（2）session bean。scope 取值为 session 的 bean 称为 session bean，session bean 的有效范围是用户访问的 Web 服务目录下的各个页面，存活时间是用户的会话期（session）间，直到用户的会话消失（session 对象达到了最大生存时间或用户关闭自己的浏览器以及服务器关闭，见 4.3.1 节）。如果用户访问 Web 服务目录多个页面，那么每个页面 id 相同的 session bean 是同一个 bean（占有相同的内存空间）。因此，用户在某个页面更改了这个 session bean 的属性值，其他页面的这个 session bean 的属性值也将发生同样的变化。当用户的会话（session）消失，Tomcat 服务器取消所分配的 session bean，即释放 session bean 所占有的内存空间。需要注意的是，不同用户（浏览器）的 session bean 是互不相同的（占有不同的内存空间）。也就是说，当两个用户同时访问同一个 Web 服务目录，一个用户对自己 session bean 属性的改变，不会影响到另一个用户（一个用户在不同 Web 服务目录的 session bean 互不相同）。

（3）request bean。scope 取值为 request 的 bean 称为 request bean，request bean 的有效范围是用户请求的当前页面，存活时间是从用户的请求产生到请求结束。Tomcat 服务器分配给每个 JSP 页面的 request bean 是互不相同的。Tomcat 服务器对请求作出响应之后，取消分配给这个 JSP 页面的 request bean。简单地说，request bean 只在当前页面有效，直到响应结束。request bean 存活时间略长于 page bean 的存活时间，原因是 Tomcat 服务器认为页面执行完毕后，响应才算结束。需要注意的是，不同用户的 request bean 的也是互不相同的。也就是说，当两个用户同时请求同一个 JSP 页面时，一个用户对自己 request bean 属性的改变，不会影响到另一个用户。

（4）application bean。scope 取值为 application 的 bean 称为 application bean，application bean 的有效范围是当前 Web 服务目录下的各个页面，存活时间直到 Tomcat 服务器关闭。Tomcat 服务器为访问 Web 服务目录的所有用户分配一个共享的 bean，即不同用户的 application bean 也都是相同的一个。也就是说，任何一个用户对自己 application bean 属性的改变，都会影响到其他用户（不同 Web 服务目录的 application bean 互不相同）。

图 5.3 是 bean 的有效期的示意图，图 5.4 是有效范围示意图。

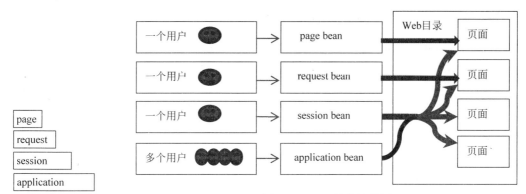

图 5.3　bean 的有效期　　　　图 5.4　bean 的有效范围

注：当使用 session bean 时，要保证用户端支持 Cookie。

例 5_1 中负责创建 page bean 的类是上述的 Circle 类，page bean 的名字是 circle。

例 5_1
example5_1.jsp（效果如图 5.5 所示）

```
<%@ page contentType = "text/html" %>
<%@ page pageEncoding = "utf-8" %>
<HTML><body bgcolor = #ffccff>
<style>
    #textStyle{
        font-family:宋体;font-size:36;color:blue
    }
</style>
<HTML><body bgcolor = #ffccff>
<p id = "textStyle">
<jsp:useBean id = "circle" class = "tom.jiafei.Circle" scope = "page" />
<%-- 通过 useBean 标记，获得名字是 circle 的 page bean --%>
圆的初始半径是:<% = circle.getRadius()%>
<%  double newRadius = 100;
    circle.setRadius(newRadius);              //修改半径
%>
<br>修改半径为<% = newRadius %>
<br><b>圆的半径是:<% = circle.getRadius()%>
<br><b>圆的周长是:<% = circle.circleLength()%>
<br><b>圆的面积是:<% = circle.circleArea()%>
</p></body></HTML>
```

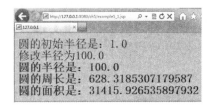

图 5.5　page bean

例 5_2 使用 id 为 girl 的 session bean,创建 session bean 的类仍然是上述的 Circle.class。在例 5_2 的 example5_2_a.jsp 页面中,session bean 的半径 radius 的值是 1(如图 5.6(a)所示),然后链接到 example5_2_b.jsp 页面,显示 session bean 的半径 radius 的值,然后将 session bean 的半径 radius 的值更改为 1.618(如图 5.6(b)所示)。用户再刷新 example5_2_a.jsp 或 example5_2_b.jsp 时看到的 session bean 的 radius 的值就都是 1.618 了(如图 5.6(c)所示)。

(a) 初次使用session bean　　(b) 修改bean的属性值　　(c) 同一个session bean

图 5.6　session bean

例 5_2

example5_2_a.jsp(效果如图 5.6(a)(c)所示)

```
<%@ page contentType = "text/html" %>
<%@ page pageEncoding = "utf - 8" %>
<HTML><body bgcolor = cyan>
<p style = "font - family:宋体;font - size:36;color:blue">
<% -- 通过 JSP 标记,用户获得一个 id 是 girl 的 session bean: -- %>
<jsp:useBean id = "girl" class = "tom.jiafei.Circle" scope = "session" />
<br>这里是 example5_2_a.jsp 页面.
<br>圆的半径是<% = girl.getRadius() %>
<br>单击超链接,到其他页面看圆的半径.
<a href = "example5_2_b.jsp"><br> example5_2_b.jsp </a>
</p></body></HTML>
```

example5_2_b.jsp(效果如图 5.6(b)所示)

```
<%@ page contentType = "text/html" %>
<%@ page pageEncoding = "utf - 8" %>
<HTML><body bgcolor = #ffccff>
<p style = "font - family:黑体;font - size:36;color:blue">
<% -- 用户的 id 是 girl 的 session bean: -- %>
<jsp:useBean id = "girl" class = "tom.jiafei.Circle" scope = "session"/>
<i><br>这里是 example5_2_b.jsp 页面
<br>当前圆的半径是:<% = girl.getRadius() %>
<% girl.setRadius(1.618);
```

```
<br>修改后的圆的半径是<%=girl.getRadius()%></i>
<br>单击超链接,到其他页面看圆的半径.
<a href="example5_2_a.jsp"><br>example5_2_a.jsp</a>
</p></body></HTML>
```

例5_3 中使用了 id 为 boy 的 application bean。当第一个用户访问这个页面时,显示 application bean 的 radius 的值,然后把 application bean 的 radius 的值更改为 2.718(如图 5.7(a) 所示)。当其他用户访问这个页面时,看到的 application bean 的 radius 的值都是 2.718(如图 5.7(b)所示)。

 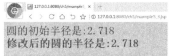

(a) 第一个用户访问的效果　　(b) 后续用户访问的效果

图 5.7　application bean

例 5_3

example5_3.jsp(效果如图 5.7(a)(b)所示)

```
<%@ page contentType="text/html" %>
<%@ page pageEncoding="utf-8" %>
<HTML><body bgcolor=#ffccff>
<p style="font-family:宋体;font-size:36;color:blue">
<jsp:useBean id="boy" class="tom.jiafei.Circle" scope="application" />
圆的初始半径是:<%=boy.getRadius()%>
<% boy.setRadius(2.718);
%>
<br><b>修改后的圆的半径是:<%=boy.getRadius()%>
</p></body></HTML>
```

5.2　获取和修改 bean 的属性

使用 useBean 动作标记获得一个 bean 后,在 Java 程序片或表达式中 bean 就可以调用方法产生行为,如前面的例 5_1~例 5_3 所示,这种情况下,不要求创建 bean 的类遵守 setXxx 和 getXxx 等规则(见 5.1.1 节)。获取或修改 bean 的属性还可以使用 JSP 动作标记 getProperty、setProperty,这种情况下,要求创建 bean 的类遵守 setXxx 和 getXxx 等规则,当 JSP 页面使用 getProperty、setProperty 标记获取或修改属性 xxx 时,必须保证 bean 有相应的 getXxx 和 setXxx 方法,即对方法名字的命名有特殊的要求。下面讲述怎样使用 JSP 的动作标记 getProperty、setProperty 去获取和修改 bean 的属性。

▶ 5.2.1　getProperty 动作标记

使用 getProperty 动作标记可以获得 bean 的属性值,并将这个值用串的形式发送给用户的浏览器。使用 getProperty 动作标记之前,必须使用 useBean 动作标记获得相应的 bean。

视频讲解

getProperty 动作标记的语法格式是：

```
<jsp:getProperty name = "bean 的 id" property = "bean 的属性" />
```

或

```
<jsp:getProperty name = "bean 的 id" property = "bean 的属性">
</jsp:getProperty>
```

其中，name 取值是 bean 的 id，用来指定要获取哪个 bean 的属性的值，property 取值是该 bean 的一个属性的名字。

注：让 request 调用 setCharacterEncoding 方法设置编码为 UTF-8，以避免显示 bean 的属性值出现乱码现象。

▶ 5.2.2 setProperty 动作标记

视频讲解

使用 setProperty 动作标记可以设置 bean 的属性值。使用这个标记之前，必须使用 useBean 标记得到一个相应的 bean。

setProperty 动作标记可以通过两种方式设置 bean 属性值。

（1）将 bean 属性值设置为一个表达式的值或字符序列。

```
<jsp:setProperty name = "bean 的 id" property = "bean 的属性"
value = "<% = expression %>"/>
<jsp:setProperty name = "bean 的 id" property = "bean 的属性"
value = "字符序列"/>
```

value 给出的值的类型要和 bean 的属性的类型一致。

（2）通过 HTTP 表单的参数值来设置 bean 的相应属性值。

① 用 form 表单的所有参数值设置 bean 相对应属性值的使用格式如下：

```
<jsp:setProperty name = "bean 的 id 的名字" property = " * " />
```

在 setProperty 标记的上述用法中不具体指定 bean 属性值将对应 form 表单中哪个参数指定的值，系统会自动根据名字进行匹配对应，但要求 bean 属性的名字必须在 form 表单中有名称相同的参数名字相对应，Tomcat 服务器会自动将参数的字符串值转换为 bean 相对应的属性值

② 用 form 表单的某个参数的值设置 bean 的某个属性值的使用格式如下：

```
<jsp:setProperty name = "bean 的名字" property = "属性名" param = "参数名" />
```

setProperty 标记的上述用法具体指定了 bean 属性值将对应表单中哪个参数名（param）指定的值，这种设置 bean 的属性值的方法，不要求 property 给出的 bean 属性的名字和 param 给出的参数名一致，即不要求 bean 属性的名字必须和表单中某个参数名字相同。

当把字符序列设置为 beans 的属性值时，这个字符序列会自动被转化为 bean 的属性类型。Java 语言将字符序列转化为其他数值类型的方法如下。

- 转化到 int：Integer.parseInt(Sting s)

- 转化到 long：Long.parseLong(Sting s)
- 转化到 float：Float.parseFloat(Sting s)
- 转化到 double：Double.parseDouble(Sting s)

这些方法都可能发生 NumberFormatException 异常，例如，当试图将字符序列 ab23 转化为 int 型数据时就发生了 NumberFormatException。

注：用 form 表单设置 bean 的属性值时，只有提交了表单，对应的 setProperty 标记才会被执行。

例 5_4 使用 Goods 类创建 request bean。example5_4_a.jsp 通过 form 表单指定 example5_4_a.jsp 和 example5_4_b.jsp 中的 request bean 的 name 和 price 属性值。example5_4_a.jsp 和 example5_4_b.jsp 使用 getProperty 动作标记以及 bean 调用方法两种方式显示 request bean 的 name 和 price 属性值。

例 5_4

> **JavaBean**

将 Goods.java 保存在 \ch5\WEB-INF\classes\tom\jiafei 中，用命令行进入 tom\jiafei 的父录 classes，按如下格式编译 Goods.java：

```
classes> javac tom\jiafei\Goods.java
```

Goods.java（负责创建 bean）

```java
package tom.jiafei;
public class Goods {
    String name = "无名";
    double price = 0;
    public String getName() {
        return name;
    }
    public void setName(String newName){
        name = newName;
    }
    public double getPrice() {
        return price;
    }
    public void setPrice(double newPrice) {
        price = newPrice;
    }
}
```

> **JSP 页面**

example5_4_a.jsp（效果如图 5.8(a) 所示）

```
<%@ page contentType = "text/html" %>
<%@ page pageEncoding = "utf-8" %>
<% request.setCharacterEncoding("utf-8");
%>
<jsp:useBean id = "phone" class = "tom.jiafei.Goods" scope = "request"/>
<HTML><body bgcolor = #ffccff>
```

```
<style>
    #textStyle{
        font-family:宋体;font-size:20;color:blue
    }
</style>
<p id="textStyle">
<form action="" Method="post">
手机名称:<input type=text id=textStyle name="name">
<br>手机价格:<input type=text id=textStyle name="price"/>
<br><input type=submit id=textStyle value="提交给本页面"/>
</form>
<form action="example5_4_b.jsp" Method="post">
手机名称:<input type= text name="name" id=textStyle>
<br>手机价格:<input type=text name="price" id=textStyle>
<br><input type=submit id=textStyle value="提交给example5_4_b.jsp页面">
</form>
<jsp:setProperty name="phone" property="name" param="name"/>
<jsp:setProperty name="phone" property="price" param="price"/>
<br><b>名称:<jsp:getProperty name="phone" property="name"/>
<br><b>名称:<%= phone.getName() %><br>
<br><b>价格:<jsp:getProperty name="phone" property="price"/>
<br><b>价格:<%= phone.getPrice() %>
</p></body></HTML>
```

example5_4_b.jsp(效果如图 5.8(b)所示)

```
<%@ page contentType="text/html" %>
<%@ page pageEncoding="utf-8" %>
<% request.setCharacterEncoding("utf-8");
%>
<jsp:useBean id="phone" class="tom.jiafei.Goods" scope="page"/>
<HTML><body bgcolor= cyan>
<p style="font-family:黑体;font-size:20;color:red">
<jsp:setProperty name="phone" property="name" param="name"/>
<jsp:setProperty name="phone" property="price" param="price"/>
```

(a) example5_4_a.jsp

(b) example5_4_b.jsp

图 5.8　设置与获取 bean 的属性值

```
<br><b>名称：<jsp:getProperty name = "phone" property = "name"/>
<br><b>名称：<% = phone.getName()%><br>
<br><b>价格：<jsp:getProperty name = "phone" property = "price"/>
<br><b>价格：<% = phone.getPrice()%>
</p></body></HTML>
```

注：setProperty 和 getProperty 动作标记适合在处理数据和显示数据不是很复杂的情况下使用，如果业务逻辑或显示数据比较复杂，可能在 Java 程序片中用 bean 调用方法更加方便（创建 bean 的类也不用遵循方法命名的 setXXX 和 getXXX 规则）。

5.3 bean 的辅助类

视频讲解

在写一个创建 bean 的类时，除了需要用 import 语句引入 JDK 提供的类，可能还需要自己编写一些其他的类，只要将这样类的包名和 bean 类的包名一致即可（也可以和创建 bean 的类写在一个 Java 源文件中）。

在下面的例 5_5 中，使用一个 bean 列出 Tomcat 服务器驻留的计算机上某目录中特定扩展名的文件。创建 bean 的 ListFile 类，需要一个实现 FilenameFilter 接口的辅助类 FileExtendName，该类可以帮助 bean 列出指定扩展名的文件（把 ListFile.java 编译生成的字节码 ListFile.class 和 FileExtendName.class 复制到\ch4\WEB-INF\classes\tom\jiafei 中）。

例 5_5 使用 ListFile 类创建 request bean。例 5_5 中用户通过表单设置 request bean 的 extendsName 属性值，request bean 列出目录中由 extendsName 属性值指定的扩展名的文件。

例 5_5

➢ **JavaBean**

用命令行进入 tom\jiafei 的父目录 classes，编译 ListFile.java（约定见 5.1.2 节）：

```
classes > javac tom\jiafei\ListFile.java
```

ListFile.java（负责创建 bean）

```java
package tom.jiafei;
import java.io.*;
class FileExtendName implements FilenameFilter {
    String str = null;
    FileExtendName (String s) {
        str = "." + s;
    }
    public boolean accept(File dir,String name) {
        return name.endsWith(str);
    }
}
public class ListFile {
    String extendsName = null;
    String [] allFileName = null;
    String dir = null;
```

```java
    public void setDir(String dir) {
        this.dir = dir;
    }
    public String getDir() {
        return dir;
    }
    public void setExtendsName(String s) {
        extendsName = s;
    }
    public String getExtendsName() {
        return extendsName;
    }
    public String [] getAllFileName() {
        if(dir!= null) {
           File mulu = new File(dir);
           FileExtendName help = new FileExtendName(extendsName);
           allFileName = mulu.list(help);
        }
        return allFileName;
    }
}
```

> **JSP 页面**

example5_5.jsp(效果如图 5.9 所示)

```jsp
<%@ page contentType="text/html" %>
<%@ page pageEncoding="utf-8" %>
<% request.setCharacterEncoding("utf-8");
%>
<style>
    #textStyle{
        font-family:宋体;font-size:36;color:blue
    }
</style>
<jsp:useBean id="file" class="tom.jiafei.ListFile" scope="request"/>
<HTML><body id=textStyle bgcolor=#ffccff>
<form action="" Method="post">
输入目录名(例如 D:/2000)<input type=text name="dir" id=textStyle size=15/><br>
输入文件的扩展名(例如 java)
<input type=text name="extendsName" id=textStyle size=6>
<input type=submit id=textStyle value="提交"/>
</form>
<jsp:setProperty name="file" property="dir" param="dir"/>
<jsp:setProperty name="file" property="extendsName" param="extendsName"/>
<br><b>目录 <jsp:getProperty name="file" property="dir"/>中
扩展名是 <jsp:getProperty name="file" property="extendsName"/> 的文件有:
<% String [] fileName = file.getAllFileName();
      if(fileName!= null) {
        for(int i=0;i<fileName.length;i++) {
           out.print("<br>"+fileName[i]);
        }
      }
%>
</body></HTML>
```

图 5.9 特定扩展名的文件

5.4 JSP 与 bean 结合的简单例子

JSP 页面中调用 bean 可以将数据的处理从页面中分离出来,实现代码复用,以便更有效地维护一个 Web 应用。本节将结合一些实际问题,进一步熟悉掌握 bean 的使用方法。在本节中,创建 bean 类的包名都是 red.star,使用的 Web 服务目录仍然是 ch5,因此,需要在 ch5 目录下建立目录结构:ch5\WEB-INF\classes\red\star,将创建 bean 的字节码文件都保存在该目录中。为了调试程序方便,可以直接按照创建 bean 类的包名将相应的 Java 源文件保存在 Web 服务目录的相应目录中,例如将 Java 源文件保存在 Web 服务目录 ch5 的 WEB-INF\classes\red\star 目录中,然后使用 MS-DOS 命令行进入\red\star 的父目录 classes,按如下格式编译 Java 源文件:

classes > javac red\star\源文件名

▶ 5.4.1 三角形 bean

例 5_6 使用 request bean(Triangle 类负责创建)完成三角形的有关数据的处理。例子中的 JSP 页面提供一个 form 表单,用户可以通过 form 表单将三角形三边的长度提交给该页面。用户提交 form 表单后,JSP 页面将计算三角形面积的任务交给一个 request bean 去完成。

视频讲解

例 5_6

> **JavaBean**

用命令行进入 red\star 的父目录 classes,编译 Triangle.java(约定见 5.1.2 节):

classes > javac red\star\Triangle.java

Triangle.java(负责创建 request bean)

```
package red.star;
public class Triangle {
   double sideA = -1,sideB = -1,sideC = -1;
   String area;
   boolean isTriangle;
   public void setSideA(double a) {
      sideA = a;
   }
   public double getSideA() {
```

```
        return sideA;
    }
    public void setSideB(double b) {
        sideB = b;
    }
    public double getSideB() {
        return sideB;
    }
    public void setSideC(double c) {
        sideC = c;
    }
    public double getSideC() {
        return sideC;
    }
    public String getArea() {
        double p = (sideA + sideB + sideC)/2.0;
        if(isTriangle){
            double result = Math.sqrt(p * (p - sideA) * (p - sideB) * (p - sideC));
            area = String.format("%.2f",result);        //保留2位小数
        }
        return area;
    }
    public boolean getIsTriangle(){
        if(sideA < sideB + sideC&&sideB < sideA + sideC&&sideC < sideA + sideB)
            isTriangle = true;
        else
            isTriangle = false;
        return isTriangle;
    }
}
```

➢ JSP 页面

example5_6.jsp(效果如图 5.10 所示)

```
<%@ page contentType = "text/html" %>
<%@ page pageEncoding = "utf-8" %>
<style>
    #textStyle{
        font-family:宋体;font-size:36;color:blue
    }
</style>
<% request.setCharacterEncoding("utf-8");
%>
<jsp:useBean id = "triangle" class = "red.star.Triangle" scope = "request"/>
<HTML><body id = textStyle bgcolor = #ffccff>
<form action = "" method = "post">
输入三角形三边:
边 A:<input type = text name = "sideA" id = textStyle value = 0 size = 5/>
边 B:<input type = text name = "sideB" id = textStyle value = 0 size = 5/>
边 C:<input type = text name = "sideC" id = textStyle value = 0 size = 5/>
<br><input type = submit id = textStyle value = "提交"/>
```

```
</form>
<jsp:setProperty name = "triangle" property = "*"/>
三角形的三边是：
<jsp:getProperty name = "triangle" property = "sideA"/>,
<jsp:getProperty name = "triangle" property = "sideB"/>,
<jsp:getProperty name = "triangle" property = "sideC"/>.
<br><b>这三个边能构成一个三角形吗?<jsp:getProperty name = "triangle" property =
"isTriangle"/>
<br>面积是:<jsp:getProperty name = "triangle" property = "area"/></b>
</body></HTML>
```

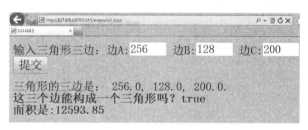

图 5.10　用 bean 计算三角形面积

5.4.2　四则运算 bean

例 5_7 使用 session bean(ComputerBean 类负责创建)完成四则运算。例子中的 JSP 页面提供一个 form 表单,用户可以通过 form 表单输入两个数,选择四则运算符号提交给该页面。用户提交 form 表单后,JSP 页面将计算任务交给 session bean 去完成。

视频讲解

例 5_7

➢ **JavaBean**

用命令行进入 red\star 的父目录 classes,编译 ComputerBean.java(约定见 5.1.2 节):

```
javac red\star\ComputerBean.java
```

ComputerBean.java(负责创建 session bean)

```
package red.star;
public class ComputerBean {
    double numberOne, numberTwo, result;
    String operator = " + ";
    public void setNumberOne(double n) {
        numberOne = n;
    }
    public double getNumberOne() {
        return numberOne;
    }
    public void setNumberTwo(double n) {
        numberTwo = n;
    }
    public double getNumberTwo() {
        return numberTwo;
    }
```

```
        public void setOperator(String s) {
            operator = s.trim();;
        }
        public String getOperator() {
            return operator;
        }
        public double getResult() {
            if(operator.equals("+"))
                result = numberOne + numberTwo;
            else if(operator.equals("-"))
                result = numberOne - numberTwo;
            else if(operator.equals("*"))
                result = numberOne * numberTwo;
            else if(operator.equals("/"))
                result = numberOne/numberTwo;
            return result;
        }
}
```

➢ **JSP 页面**

example5_7.jsp（效果如图 5.11 所示）

```
<%@ page contentType="text/html" %>
<%@ page pageEncoding="utf-8" %>
<style>
    #textStyle{
        font-family:宋体;font-size:36;color:blue
    }
</style>
<% request.setCharacterEncoding("utf-8");
%>
<jsp:useBean id="computer" class="red.star.ComputerBean" scope="session"/>
<HTML><body id=textStyle bgcolor=#ffccff >
<jsp:setProperty name="computer" property="*"/>
<form action="" method=post >
    <input type=text name="numberOne" id=textStyle size=6/>
    <select name="operator" id=textStyle >
        <option value="+" id=textStyle >+
        <option value="-" id=textStyle >-
        <option value="*" id=textStyle >*
        <option value="/" id=textStyle >/
    </select>
    <input type=text name="numberTwo" id=textStyle size=6/>
     <br><input type="submit" value="提交" id=textStyle "/>
</form>
<b>
<jsp:getProperty name="computer" property="numberOne"/>
<jsp:getProperty name="computer" property="operator"/>
<jsp:getProperty name="computer" property="numberTwo"/> =
<jsp:getProperty name="computer" property="result"/></b>
</body></HTML>
```

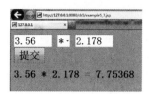

图 5.11　用 bean 完成四则运算

▶ 5.4.3　浏览图像 bean

视频讲解

例 5_8 中的 JSP 页面通过单击"下一张"或"上一张"超链接浏览图像，JSP 页面将获取图像名字的任务交给 session bean 去完成，JSP 页面根据 bean 获得的图像名字显示图像。例子中使用的图像文件保存在当前 Web 服务目录的子目录 image 中(图像文件的名字中不能含有空格)。

例 5_8

➢ **JavaBean**

用命令行进入 red\star 的父目录 classes，编译 Play.java(约定见 5.1.2 节)：

```
classes> javac red\star\Play.java
```

Play.java(负责创建 session bean)

```java
package red.star;
import java.io.*;
public class Play {
    String pictureName[];            //存放全部图片文件名字的数组
    String showImage;                //存放当前要显示的图片
    String webDir = "";              //Web 服务目录的名字,例如 ch5
    String tomcatDir;                //Tomcat 的安装目录,例如 apache-tomcat-9.0.26
    int index = 0;                   //存放图片文件的序号
    public Play() {
        File f = new File("");       //该文件认为在 Tomcat 服务器启动的目录中,即 bin 目录中
        String path = f.getAbsolutePath();
        int index = path.indexOf("bin");    //bin 是 Tomcat 的安装目录下的子目录
        tomcatDir = path.substring(0,index);  //得到 Tomcat 的安装目录的名字
    }
    public void setWebDir(String s) {
        webDir = s;
        File dirImage = new File(tomcatDir + "/webapps/" + webDir + "/image");
        pictureName = dirImage.list();
    }
    public String getShowImage() {
        showImage = pictureName[index];
        return showImage;
    }
    public void setIndex(int i) {
        index = i;
        if(index >= pictureName.length)
            index = 0;
        if(index < 0)
```

```
            index = pictureName.length - 1;
    }
    public int getIndex() {
        return index ;
    }
}
```

➢ **JSP 页面**

example5_8.jsp(效果如图 5.12 所示)

```
<%@ page contentType = "text/html" %>
<%@ page pageEncoding = "utf - 8" %>
<style>
    #textStyle{
        font - family:宋体;font - size:36;color:blue
    }
</style>
<% request.setCharacterEncoding("utf - 8");
%>
<jsp:useBean id = "play" class = "red.star.Play" scope = "session" />
<%
    String webDir = request.getContextPath();    //获取当前 Web 服务目录的名称
    webDir = webDir.substring(1);                //去掉名称前面的目录符号:/
%>
<jsp:setProperty name = "play" property = "webDir" value = "<% = webDir %>"/>
<jsp:setProperty name = "play" property = "index"  param = "index" />
<HTML><body bgcolor = cyan><p id = textStyle>
<image src =
image/<jsp:getProperty name = "play" property =
"showImage"/> width = 300 height = 200 ></image><br>
<a href = "?index = <% = play.getIndex() + 1 %>">下一张</a>
<a href = "?index = <% = play.getIndex() - 1 %>">上一张</a>
</p></body></HTML>
```

图 5.12 浏览图像

视频讲解

▶ **5.4.4 日历 bean**

例 5_9 使用 session bean(Calendar 类负责创建)显示某月的日历。用户单击"下一月""上一月"超链接可以翻阅日历。也可以输入年份,选择月份,单击 form 表单中的提交键查看日历。

例 5_9

> **JavaBean**

用命令行进入 red\star 的父目录 classes,编译 Calendar.java(约定见 5.1.2 节):

```
classes > javac red\star\Calendar.java
```

Calendar.java(负责创建 session bean)

```java
package red.star;
import java.time.LocalDate;
import java.time.DayOfWeek;
public class Calendar {
    int year ,month ;
    String saveCalender;                                    //存放日历
    public Calendar(){
        year = LocalDate.now().getYear();
        month = LocalDate.now().getMonthValue();
    }
    public void setYear(int y){
        year = y;
    }
    public int getYear(){
        return year;
    }
    public void setMonth(int m){
        month = m;
        if(month > 12){
            year++;
            month = 1;
        }
        if(month < 1){
            month = 12;
            year -- ;
        }
    }
    public int getMonth(){
        return month;
    }
    public String getSaveCalender(){
        LocalDate date = LocalDate.of(year,month,1);
        int days = date.lengthOfMonth();                    //得到该月有多少天
        int space = 0;                                      //存放空白字符的个数
        DayOfWeek dayOfWeek = date.getDayOfWeek();          //得到1号是星期几
        switch(dayOfWeek) {
            case SUNDAY:      space = 0;
                              break;
            case MONDAY:      space = 1;
                              break;
            case TUESDAY:     space = 2;
                              break;
            case WEDNESDAY:   space = 3;
                              break;
```

```
                    case THURSDAY:    space = 4;
                                      break;
                    case FRIDAY:      space = 5;
                                      break;
                    case SATURDAY:    space = 6;
                                      break;
            }
            String [] c = new String[space + days];
            for(int i = 0;i < space;i++)
                c[i] = " -- ";
            for(int i = space,n = 1;i < c.length;i++){
                c[i] = String.valueOf(n) ;
                n++;
            }
            String head =
            "<tr><th>星期日</th><th>星期一</th><th>星期二</th><th>星期三</th>" +
            "<th>星期四</th><th>星期五</th><th>星期六</th></tr>";
            StringBuffer buffer = new StringBuffer();
            buffer.append("<table border = 0>");
            buffer.append(head);
            int n = 0;
            while(n < c.length){
                buffer.append("<tr>");
                int increment = Math.min(7,c.length - n);
                for(int i = n;i < n + increment;i++) {
                    buffer.append("<td align = center>" + c[i] + "</td>");
                }
                buffer.append("</tr>");
                n = n + increment;
            }
            buffer.append("</table>");
            saveCalender = new String(buffer);
            return saveCalender;
        }
}
```

➢ JSP 页面

example5_9.jsp(效果如图 5.13 所示)

```
<%@ page contentType = "text/html" %>
<%@ page pageEncoding = "utf-8" %>
<style>
    #textStyle{
        font-family:宋体;font-size:18;color:blue
    }
</style>
<% request.setCharacterEncoding("utf-8");
%>
<HTML><body id = textStyle bgcolor = #ffccff>
<jsp:useBean id = "calendar" class = "red.star.Calendar" scope = "session" />
<jsp:setProperty name = "calendar" property = "year"    param = "year"/>
```

```
<jsp:setProperty name = "calendar" property = "month" param = "month" />
<jsp:getProperty name = "calendar" property = "year" />年
<jsp:getProperty name = "calendar" property = "month" />月的日历：<br>
<jsp:getProperty name = "calendar" property = "saveCalender" />
<br><a href = "?month = <% = calendar.getMonth() + 1 %>">下一月</a>
<a href = "?month = <% = calendar.getMonth() - 1 %>">上一月</a>
<form action = "" method = get>
输入年份<input type = text name = "year" id = textStyle size = 6 />
选择月份 <select name = "month" id = textStyle size = 1>
  <option value = "1">1 月</option>
  <option value = "2">2 月</option>
  <option value = "3">3 月</option>
  <option value = "4">4 月</option>
  <option value = "5">5 月</option>
  <option value = "6">6 月</option>
  <option value = "7">7 月</option>
  <option value = "8">8 月</option>
  <option value = "9">9 月</option>
  <option value = "10">10 月</option>
  <option value = "11">11 月</option>
  <option value = "12">12 月</option>
</select><br>
<input type = "submit" value = "提交" id = textStyle />
</form>
</body></HTML>
```

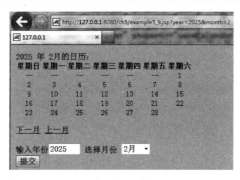

图 5.13　查看日历

5.4.5　计数器 bean

例 5_10 使用 application bean（ComputerCount 类负责创建）记录 Web 服务目录（通常所说的网站）被访问的次数。只要用户第 1 次访问（用户的 session 被创建）Web 服务目录，那么当前 Web 服务目录的访问计数就增加 1。如果用户的 session 没有消失，用户再访问当前 Web 服务目录，访问的计数不再增 1。

视频讲解

例 5_10

➢ **JavaBean**

用命令行进入 red\star 的父目录 classes，编译 ComputerCount.java（约定见 5.1.2 节）：

```
classes > javac red\star\ComputerCount.java
```

ComputerCount.java(负责创建 application bean)

```java
package red.star;
import java.io.*;
public class ComputerCount {
    int number = 0;
    public synchronized void addCount() {
        number++;
    }
    public int getNumber(){
        return number;
    }
}
import java.time.DayOfWeek;
```

> **JSP 页面**

example5_10_a.jsp(效果如图 5.14(a)所示)

```jsp
<%@ page contentType="text/html" %>
<%@ page pageEncoding="utf-8" %>
<jsp:useBean id="count" class="red.star.ComputerCount"
                scope="application"/>
<% if(session.isNew()) {
      count.addCount();
   }
%>
<HTML><body bgcolor=cyan>
<h1>这是网站的 example5_10_a.jsp 页面.
<br>网站访问量:<jsp:getProperty name="count" property="number"/>
<br>
<a href="example5_10_b.jsp">欢迎去 example5_10_b.jsp 参观</a>
</body></HTML>
```

example5_10_b.jsp(效果如图 5.14(b)所示)

```jsp
<%@ page contentType="text/html" %>
<%@ page pageEncoding="utf-8" %>
<jsp:useBean id="count" class="red.star.ComputerCount"
                scope="application"/>
<HTML><body bgcolor=#ffccff>
<% if(session.isNew()) {
      count.addCount();
   }
%>
<h1>这是网站的 example5_10_b.jsp 页面
<br>网站访问量:
<jsp:getProperty name="count" property="number"/>
<br>
<a href="example5_10_a.jsp">欢迎去 example5_10_a.jsp 参观</a>
</body></HTML>
```

第 5 章　JSP与JavaBean

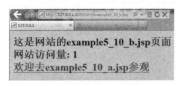
(a) 一个用户看到的计数

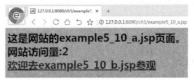
(b) 另一个用户看到的计数

图 5.14　两个用户访问 Web 服务目录

5.5　上机实验

提供了详细的实验步骤要求,按步骤完成,提升学习效果,积累经验,不断提高 Web 设计能力。

视频讲解

▶ 5.5.1　实验 1　小数表示为分数

❶ 实验目的

掌握怎样使用 request bean。

❷ 实验要求

(1) 编写 inputNumber.jsp,该页面提供一个 form 表单,该 form 表单提供一个 text 文本框,用于用户输入一个纯小数(例如 0.618)。用户在 form 表单中输入纯小数后,单击 submit 提交键将纯小数提交给 getFraction.jsp 页面。

(2) getFraction.jsp 使用 request bean,并使用 setProperty 动作标记让 request bean 将纯小数转换为分数,把分子和分母存放在 request bean 的名字是 numerator 和 denominator 的属性(变量)中。然后 getFraction.jsp 使用 getProperty 动作标记获取 request bean 的 numerator 和 denominator 的属性值。

(3) 在 Tomcat 服务器的 webapps 目录下(例如 D:\apache-tomcat-9.0.26\webapps)新建一个名字是 ch5_practice_one 的 Web 服务目录。把 JSP 页面都保存到 ch5_practice_one 目录中。在 ch5_practice_one 下建立子目录 WEB-INF(字母大写),然后在 WEB-INF 下再建立子目录 classes。将创建 request bean 的类的 Java 源文件保存在 classes 的相应子目录中(见5.1.2 节)。

(4) 用浏览器访问 JSP 页面 inputNumber.jsp。

❸ 参考代码

参考代码运行效果如图 5.15 所示。

(a) 输入纯小数

(b) 得到分数

图 5.15　把纯小数表示成分数

JavaBean 用命令行进入 sea\water 的父目录 classes，编译 Fraction.java（约定见 5.1.2 节）：

classes > javac sea\water\Fraction.java

Fraction.java（负责创建 request bean）

```java
package sea.water;
public class Fraction {
    public double number ;                              //存放小数
    public long numerator ;                             //存放分子
    public long denominator;                            //存放分母
    public double getNumber(){
        String numberString = String.valueOf(number);
        String xiaoshuPart =
        numberString.substring(numberString.indexOf(".") + 1);  //得到纯小数部分
        return Double.parseDouble("0." + xiaoshuPart);
    }
    public long getNumerator(){
        return numerator;
    }
    public long getDenominator(){
        return denominator;
    }
    public void setNumber(double number){
        this.number = number;
        String numberString = String.valueOf(number);
        String xiaoshuPart =
        numberString.substring(numberString.indexOf(".") + 1);  //得到小数部分
        int m = xiaoshuPart.length();                   //m 的值就是小数的小数位数
        numerator = Long.parseLong(xiaoshuPart);        //分子
        denominator = (long)Math.pow(10,m);             //分母
        long greatCommonDivisor = f(numerator,denominator) ;  //最大公约数
        numerator = numerator/greatCommonDivisor;
        denominator = denominator/greatCommonDivisor;
    }
    private long f(long a,long b) {                     //求 a 和 b 的最大公约数
      if(a == 0) return 1;
      if(a < b) {
         long c = a;
         a = b;
         b = c;
      }
      long r = a % b;
      while(r!= 0) {
         a = b;
         b = r;
         r = a % b;
      }
      return b;
    }
}
```

第 5 章　JSP与JavaBean

> **JSP 页面**

inputNumber.jsp（效果如图 5.15(a)所示）

```
<%@ page contentType="text/html" %>
<%@ page pageEncoding="utf-8" %>
<style>
    #tomStyle{
        font-family:宋体;font-size:36;color:blue
    }
</style>
<HTML><body bgcolor=#ffccff>
<form action="getFraction.jsp" id=tomStyle method=post>
输入一个纯小数:<br>
<input type=text name="number" id=tomStyle size=16 value=0.618 />
<br><input type="submit" id=tomStyle  value="提交" /><br>
看小数的分数表示.
</form>
</body></HTML>
```

getFraction.jsp（效果如图 5.15(b)所示）

```
<%@ page contentType="text/html" %>
<%@ page pageEncoding="utf-8" %>
<HTML><body bgcolor=cyan>
<p style="font-family:宋体;font-size:36;color:red">
<jsp:useBean id="fraction" class="sea.water.Fraction" scope="request" />
<jsp:setProperty name="fraction" property="number" param="number" />
<jsp:getProperty name="fraction" property="number" />
表示成分数是:<br>
<jsp:getProperty name="fraction" property="numerator" />
<jsp:getProperty name="fraction" property="denominator" />
</p></body></HTML>
```

▶ 5.5.2　实验 2　记忆测试

❶ 实验目的

掌握使用 session bean 存储用户的数据,和 4.6.5 节的记忆测试实验进行比对,体会使用 bean 的方便和好处。

❷ 实验要求

(1) 编写 choiceGrage.jsp,该页面中的 form 表单中使用 radio 标记选择记忆测试级别: 初级、中级和高级。初级需要记忆一个长度为 5 个字符的字符序列(例如★■★▲●),中级需要记忆一个长度为 7 个字符的字符序列,高级需要记忆一个长度为 10 个字符的字符序列。在 choiceGrage.jsp 页面选择级别后,单击 form 表单的提交键提交给 giveTest.jsp 页面。

(2) 编写 giveTest.jsp 页面,该页面获取 choiceGrage.jsp 页面提交的级别后,使用 session bean 显示 testString 的属性值(例如属性值是长度为 7 个字符的字符序列),然后提示用户在 5 秒内记住这个字符序列。5 秒后,该页面将自动定向到 answerTest.jsp 页面。

(3) 编写 answerTest.jsp 页面,该页面的 form 表单提供用户给出答案的界面,即使用 radio 标记让用户选择字符序列中的各个字符,以此代表用户认为自己记住的字符序列。单击

133

提交键，将选择提交给 judgeAnswer.jsp 页面。

（4）编写 judgeAnswer.jsp 页面，该页面负责判断有户是否记住了 giveTest.jsp 页面给出的字符序列。

（5）在 Tomcat 服务器的 webapps 目录下新建名字是 ch5_practice_two 的目录，即新建 Web 服务目录 ch5_practice_two。把 JSP 页面都保存到 ch5_practice_two 目录中。在 ch5_practice_two 目录下建立子目录 WEB-INF(字母大写)，然后在 WEB-INF 目录下再建立子目录 classes，即在 ch5_practice_two 目录下建立\WEB-INF\classes 目录结构。将创建 session bean 类的 Java 源文件按照包名保存在 classes 的相应子目录中(见 5.1.2 节)。

（6）用浏览器访问 JSP 页面 choiceGrage.jsp。

（7）将本实验与 4.6.5 节的实验进行对比，感受使用 session bean 的好处。

❸ 参考代码

请比较 4.6.5 节的记忆测试实验，体会使用 bean 的方便和好处。参考代码运行效果如图 5.16 所示。

(a) 选择级别的页面　　　　　　　　(b) 记忆字符序列的页面

(c) 进行答题的页面　　　　　　　　(d) 判断答题是否正确的页面

图 5.16　记忆测试

> **JavaBean**

用命令行进入 sea\water 的父目录 classes，编译 Memory.java(约定见 5.1.2 节)：

```
classes > javac sea\water\Memory.java
```

Memory.java(负责创建 session bean)

```java
package sea.water;
import java.util.ArrayList;
import java.util.Random;
public class Memory {
    static ArrayList<String> list = new ArrayList<String>();
    static {
        list.add("★");
        list.add("●");
        list.add("▲");
        list.add("■");
        list.add("◆");
    }
```

```java
        int grade = 5 ;              //存放级别,例如初级 grade 存放的值是 5,中级是 7,高级是 10
        String testString;           //存放需要记忆的字符序列,例如,★■★▲●
        boolean isGivenTestString = false;      //存放是否已经给了测试题目
        public void setGrade(int n){
            grade = n;
        }
        public int getGrade(){
            return grade;
        }
        public void giveTestString(){
            StringBuffer buffer = new StringBuffer();
            Random random = new Random();
            for(int i = 0;i < grade;i++) {
                int index = random.nextInt(list.size());
                String str = list.get(index);         //从 list 中得到一个字符,例如★
                buffer.append(str);
            }
            testString = new String(buffer);
        }
        public void setIsGivenTestString(boolean b){
            isGivenTestString = b;
        }
        public boolean getIsGivenTestString(){
            return isGivenTestString;
        }
        public String getTestString(){
            return testString;
        }
}
```

➢ **JSP 页面**

choiceGrade.jsp(效果如图 5.16(a)所示)

```jsp
<%@ page contentType = "text/html" %>
<%@ page pageEncoding = "utf-8" %>
<HTML><body bgcolor = #ffccff>
<style>
    #textStyle{
        font-family:宋体;font-size:26;color:blue
    }
</style>
<form action = "giveTest.jsp" id = "textStyle" method = post>
<input type = radio name = "grade" value = "5" />初级
<input type = radio name = "grade" value = "7" checked = "ok" />中级
<input type = radio name = "grade" value = "10" />高级
<br><input type = "submit" id = "textStyle" value = "提交"/>
<input type = "reset" id = "textStyle" value = "重置" />
</form>
</body></HTML>
```

giveTest.jsp（效果如图 5.16（b）所示）

```jsp
<%@ page contentType="text/html" %>
<%@ page pageEncoding="utf-8" %>
<%@ page import="java.util.ArrayList" %>
<%@ page import="java.util.Random" %>
<jsp:useBean id="memory" class="sea.water.Memory" scope="session" />
<HTML><body bgcolor=#ffccff>
<style>
    #tomStyle{
        font-family:宋体;font-size:36;color:blue
    }
</style>
<%   String grade = request.getParameter("grade");
     String testString = "";                          //存放测试题目,例如★■★▲●
     if(grade == null){
         memory.setGrade(memory.getGrade());
     }
     else {
         memory.setGrade(Integer.parseInt(grade));
     }
     if(memory.getIsGivenTestString() == false) {
         memory.giveTestString();
         testString = memory.getTestString();         //得到测试的题目
         memory.setIsGivenTestString(true);
     }
     else if(memory.getIsGivenTestString() == true){
         response.sendRedirect("answerTest.jsp");     //定向到答题页面
     }
%>
<p id=tomStyle>给 5 秒记住您看到的字符序列:<br>
<%= testString %>
<br>5 秒后,将转到答题页.
<%   response.setHeader("refresh","5");
%>
</p></body></HTML>
```

answerTest.jsp（效果如图 5.16（c）所示）

```jsp
<%@ page contentType="text/html" %>
<%@ page pageEncoding="utf-8" %>
<jsp:useBean id="memory" class="sea.water.Memory" scope="session" />
<HTML><body bgcolor=#ffccff>
<style>
    #tomStyle{
        font-family:宋体;font-size:26;color:blue
    }
</style>
<form action="judgeAnswer.jsp" id=tomStyle method=post>
您记住的字符序列是怎样的,请选择:
<%
   int n = memory.getGrade();
```

第 5 章　JSP与JavaBean

```
        memory.setIsGivenTestString(false);
    for(int i = 1;i <= n;i++){
        out.print("<br>第" + i + "个字符:");
        out.print
(       "<input type = radio id = tomStyle name = R" + i + " value = '★'/>★" +
        "<input type = radio id = tomStyle name = R" + i + " value = '●'/>●" +
        "<input type = radio id = tomStyle name = R" + i + " value = '▲'/>▲" +
        "<input type = radio id = tomStyle name = R" + i + " value = '■'/>■" +
        "<input type = radio id = tomStyle name = R" + i + " value = '◆'/>◆");
    }
%>
<br><input type = "submit"    id = tomStyle value = "提交"/>
<input type = "reset" id = tomStyle value = "重置" />
</form>
</body></HTML>
```

judgeAnswer.jsp（效果如图 5.16(d)所示）

```
<%@ page contentType = "text/html" %>
<%@ page pageEncoding = "utf-8" %>
<jsp:useBean id = "memory" class = "sea.water.Memory" scope = "session" />
<HTML><body bgcolor = white >
<p style = "font-family:宋体;font-size:26;color:blue">
<%  memory.setIsGivenTestString(false);
    request.setCharacterEncoding("utf-8");
    int n = memory.getGrade();
    StringBuffer buffer = new StringBuffer();
    for(int i = 1;i <= n;i++){
        buffer.append(request.getParameter("R" + i));       //获取 radio 提交的值
        out.print("" + request.getParameter("R" + i));
    }
    String userAnswer = new String(buffer);
    String testString = memory.getTestString();             //得到测试的题目
    if(testString.equals(userAnswer)){
        out.print("您记忆不错");
    }
    else {
        out.print("您没记忆住!答案是:<br>" + testString);
    }
%>
<br><a href = "giveTest.jsp">返回,继续练习记忆</a>
<br><a href = "choiceGrade.jsp">重新选择级别</a>
</p></body></HTML>
```

▶ 5.5.3　实验3　成语接龙

❶ 实验目的

掌握使用 application bean 存储所有用户共享的数据。

❷ 实验要求

（1）编写 inputIdioms.jsp，该页面使用 application bean 显示目前成语接龙的信息。提供 form 表单,用户根据目前成语接龙的信息输入一个成语,单击提交键提交给当前页面,如果输入成语符合成语接龙规则(例如输入的成语的首字和成语接龙中的最后一个成语的末字相同),application bean 就将用户输入的成语添加到成语接龙中。

(2) 在 Tomcat 服务器的 webapps 目录下新建名字是 ch5_practice_three 的 Web 服务目录。把 JSP 页面都保存到 ch5_practice_three 目录中。在 ch5_practice_three 下建立子目录 WEB-INF(字母大写),然后在 WEB-INF 目录下再建立子目录 classes,即在 ch5_practice_three 目录下建立\WEB-INF\classes 目录结构。将创建 application bean 的类的 Java 源文件按照包名保存在 classes 的相应子目录中(见 5.1.2 节)。

(3) 用浏览器访问 JSP 页面 inputIdioms.jsp。

❸ 参考代码

参考代码运行效果如图 5.17 所示。

图 5.17 成语接龙

➢ **JavaBean**

用命令行进入 sea\water 的父目录 classes,编译 ContinueIdioms.java(约定见 5.1.2 节):

```
classes > javac sea\water\ ContinueIdioms.java
```

ContinueIdioms.java(负责创建 application bean)

```java
package sea.water;
import java.util.LinkedList;
import java.util.Iterator;
import java.util.NoSuchElementException;
public class ContinueIdioms {
    LinkedList < String > listIdioms ;              //存放成语的链表
    public String nowIdioms;                        //当前参与接龙的成语
    public ContinueIdioms(){
        listIdioms = new LinkedList < String >();
    }
    public synchronized void setNowIdioms(String s){
        nowIdioms = s;
        try{
            String previous = listIdioms.getLast();  //得到上次添加的成语
            //上一个成语的最后一个字符
            char endChar = previous.charAt(previous.length() - 1);
            char startChar = nowIdioms.charAt(0);    //当前成语的第一个字符
            if(startChar  ==  endChar)
                    listIdioms.add(nowIdioms);
        }
        catch(NoSuchElementException exp){
            listIdioms.add(nowIdioms);
            System. out. println(exp);
```

```
            }
        }
        public String getAllIdioms(){
            StringBuffer buffer = new StringBuffer();
            Iterator<String> iterator = listIdioms.iterator();
            if(iterator.hasNext() == false)
                buffer.append("→");
            while(iterator.hasNext()){
                buffer.append(iterator.next() + "→");
            }
            return new String(buffer);
        }
}
```

➢ **JSP 页面**

inputIdioms.jsp（效果如图 5.17(a)(b)所示）

```
<%@ page contentType = "text/html" %>
<%@ page pageEncoding = "utf-8" %>
<jsp:useBean id = "idioms" class = "sea.water.ContinueIdioms"
                        scope = "application" />
<style>
#tomStyle{
    font-family:宋体;font-size:26;color:blue
    }
</style>
<% request.setCharacterEncoding("utf-8");
%>
<jsp:setProperty name = "idioms" property = "nowIdioms" param = "nowIdioms" />
<HTML><body bgcolor = #ffccff>
<p id = tomStyle>
目前的接龙情景:<br>
<textArea id = tomStyle rows = 5 cols = 38>
<% = idioms.getAllIdioms() %>
</textArea><br>
<form   action = "" id = tomStyle method = post>
继续接龙,输入成语:<text   name = "nowIdioms" value = 10 />
<br><input type = "text" name = "nowIdioms"   id = tomStyle />
<input type = "submit"   id = tomStyle value = "提交"/>
</form>
</p></body></HTML>
```

5.6　小结

- JavaBean 是一个可重复使用的软件组件,是遵循一定标准、用 Java 语言编写的一个类,该类的一个实例称作一个 JavaBean。
- 一个 JSP 页面可以将数据的处理过程指派给一个或几个 bean 来完成,我们在 JSP 页面中调用这些 bean 即可。在 JSP 页面中调用 bean 可以将数据的处理代码从页面中分离出来,实现代码复用,更有效地维护一个 Web 应用。
- bean 的生命周期分为 page、request、session 和 application。

习题 5

1. 假设 Web 服务目录 mymoon 中的 JSP 页面要使用一个 bean，该 bean 的包名为 blue.sky。请说明，应当怎样保存 bean 的字节码文件。

2. 假设 Web 服务目录是 mymoon，star 是 mymoon 的一个子目录，JSP 页面 a.jsp 保存在 star 中，并准备使用一个 bean，该 bean 的包名为 tom.jiafei。下列哪个叙述是正确的？
   A. 创建 bean 的字节码文件保存在 \mymoon\WEB-INF\classes\tom\jiafei 中
   B. 创建 bean 的字节码文件保存在 \mymoon\star\WEB-INF\classes\tom\jiafei 中
   C. 创建 bean 的字节码文件保存在 \mymoon\WEB-INF\star\classes\tom\jiafei 中
   D. 创建 bean 的字节码文件保存在 \mymoon\WEB-INF\classes\start\tom\jiafei 中

3. tom.jiafei.Circle 是创建 bean 的类，下列哪个标记是正确创建 session bean 的标记？
   A. `<jsp:useBean id="circle" class="tom.jiafei.Circle" scope="page" />`
   B. `<jsp:useBean id="circle" class="tom.jiafei.Circle" scope="request" />`
   C. `<jsp:useBean id="circle" class="tom.jiafei.Circle" scope="session" />`
   D. `<jsp:useBean id="circle" type="tom.jiafei.Circle" scope="session" />`

4. 假设创建 bean 的类有一个 int 型的属性 number，下列哪个方法是设置该属性值的正确方法？
   A. `public void setNumber(int n){`
      `    number = n;`
      `}`
   B. `void setNumber(int n){`
      `    number = n;`
      `}`
   C. `public void SetNumber(int n) {`
      `    number = n;`
      `}`
   D. `public void Setnumber(int n){`
      `    number = n;`
      `}`

5. 假设 JSP 页面使用标记

   `<jsp:useBean id="moon" class="tom.jiafei.AAA" scope="page"/>`

   创建了一个名字为 moon 的 bean，该 bean 有一个 String 类型、名字为 number 的属性。如果创建 moon 的 Java 类 AAA 没有提供 public String getNumber() 方法，JSP 页面是否允许使用如下 getProperty 标记获取 moon 的 number 属性值？

   `<jsp:getProperty name="moon" property="number"/>`

6. 编写一个 JSP 页面，该页面提供一个表单，用户可以通过表单输入梯形的上底、下底和高的值，并提交给本 JSP 页面，该 JSP 页面将计算梯形面积的任务交给一个 page bean 去完成。JSP 页面使用 getProperty 动作标记显示 page bean 中的数据，例如梯形的面积。

7. 编写两个 JSP 页面 a.jsp 和 b.jsp，a.jsp 页面提供一个表单，用户可以通过表单输入矩形的两个边长提交给 b.jsp 页面，b.jsp 调用一个 request bean 去完成计算矩形面积的任务。b.jsp 页面使用 getProperty 动作标记显示矩形的面积。

第 6 章　Java Servlet基础

本章导读

主要内容
- ❖ servlet 的部署、创建与运行
- ❖ servlet 的工作原理
- ❖ 通过 JSP 页面访问 servlet
- ❖ 共享变量
- ❖ doPost 和 doGet 方法
- ❖ 重定向与转发
- ❖ 使用 session

难点
- ❖ servlet 的工作原理
- ❖ 重定向与转发

关键实践
- ❖ 绘制多边形
- ❖ 双色球福利彩票

在第 1 章学习了 JSP 页面的运行原理：当用户请求一个 JSP 页面时，Tomcat 服务器自动生成和编译 Java 文件，并用编译得到的字节码文件在 Tomcat 服务器端，创建一个对象来响应用户的请求。JSP 的根基是 Java Servlet 技术，该技术的核心就是在 Tomcat 服务器端创建响应用户请求的对象，被创建的对象习惯上称为一个 servlet。在 JSP 技术出现之前，Web 应用开发人员就是自己编写创建 servlet 的类，并负责编译生成字节码文件，复制这个字节码文件到 Tomcat 服务器的特定目录中，以便 Tomcat 服务器使用这个字节码文件创建一个 servlet 来响应用户的请求。

有些 Web 应用可能只需要 JSP+JavaBean 就能设计得很好，但是有些 Web 应用就可能需要 JSP+JavaBean+servlet 来完成，即需要服务器再创建一些 servlet，配合 JSP 页面来完成整个 Web 应用程序的工作。关于这一点将在第 7 章的 MVC 模式中讲述。

本章在 webapps 目录下新建一个 Web 服务目录 ch6，除非特别约定，本章例子中涉及的 JSP 页面均保存在 ch6 目录中。创建 servlet 的类的字节码文件需要按要求存放，因此在 ch6 目录下建立目录结构\ch6\WEB-INF\classes（WEB-INF 字母大写）。

本章使用的 javax.servlet 和 javax.servlet.http 包中的类不在 JDK 提供的核心类库中，为了方便编译 Java 源文件，请事先将 Tomcat 安装目录 lib 子目录中的 servlet-api.jar 文件复制（不要剪贴）到\ch6\WEB-INF\classes 中。

6.1 servlet 的部署、创建与运行

Java Servlet 的核心思想是在 Tomcat 服务器端创建响应用户请求的 servlet 对象,简称 servlet。因此,学习 Java Servlet 的首要任务是掌握怎样编写创建 servlet 的类,怎样在 Tomcat 服务器上保存这个类所得到的字节码文件,怎样编写 web.xml 部署文件,怎样请求 Tomcat 服务器创建一个 servlet。有关 servlet 的工作原理以及使用细节将在后续内容中讲述。

▶ 6.1.1 源文件及字节码文件

视频讲解

❶ Servlet 类

写一个创建 servlet 的类就是编写一个特殊类的子类,这个特殊的类就是 javax.servlet.http 包中的 HttpServlet 类。HttpServlet 实现了 Servlet 接口,实现了响应用户的方法(这些方法将在后续内容中讲述)。HttpServlet 的子类被习惯地称作一个 Servlet 类,这样的类创建的对象习惯地被称作一个 servlet。

下面的例 6_1 中的 Example6_1.java 的 Example6_1 类是一个简单的 Servlet 类(为了便于 Web 应用程序的管理,Servlet 类应该具有包名),该类创建的 servlet 可以响应用户的请求,即用户请求这个 servlet 时,会在浏览器看到 hello servlet 这样的响应信息(见稍后的图 6.3)。

例 6_1

Example6_1.java(负责创建 servlet,servlet 运行效果见稍后的图 6.3)

```java
package moon.sun;
import java.io.*;
import javax.servlet.*;
import javax.servlet.http.*;
public class Example6_1 extends HttpServlet{
    public void init(ServletConfig config) throws ServletException{
        super.init(config);
    }
    public void service(HttpServletRequest request,
                        HttpServletResponse response) throws IOException{
      //设置响应的 MIME 类型
        response.setContentType("text/html;charset=utf-8");
        PrintWriter out = response.getWriter();         //获得向用户发送数据的输出流
        out.println("<html><body bgcolor = #ffccff>");
        out.println("<h1>hello servlet,你好 servlet</h1>");
        out.println("</body></html>");
    }
}
```

❷ 字节码文件的保存

为了能让 Tomcat 服务器使用 Example6_1 类创建一个 servlet,需要将例 6_1 中的 Java 源文件产生的 Example6_1.class 字节码文件按照类的包名对应的目录路径保存到 Web 服务目录中特定子目录中。包名 moon.sun 对应的目录路径是 moon\sun,因此,需要把 Example6_1.class 保存到\ch6\WEB-INF\classes\moon\sun 中,如图 6.1 所示。

图 6.1 字节码文件的存放位置

为了方便调试代码,可以事先将创建 servlet 的 Servlet 类的 Java 源文件,按照其包名(例如包名为 moon.sun),保存到\ch6\WEB-INF\classes\moon\sun 中。注意,保存 Java 源文件时,"保存类型"选择为"所有文件",将"编码"选择为"ANSI"。

然后用命令行进入 moon\sun 的父目录 classes(不要进入 moon 或 sun 目录)编译 Example6_1.java:

```
classes> javac -cp servlet-api.jar moon\sun\Example6_1.java
```

注意,编译时不要忘记使用-cp 参数,该参数指定使用非核心类库中的类,例如 servlet-api.jar 文档中给出的类,-cp 参数只要给出 jar 文档的位置即可(如果 jar 文档在当前目录,就直接输入 jar 文档的名字即可)。本章使用的 javax.servlet 和 javax.servlet.http 包中的类不在 JDK 提供的核心类库中,均在 servlet-api.jar 文档中,因此可以将 Tomcat 安装目录 lib 子目录中的 servlet-api.jar 文件复制(不要剪贴)到\ch6\WEB-INF\classes\中。否则,-cp 参数必须显式给出 servlet-api.jar 的位置,例如:

```
classes> javac -cp D:\tomcat\lib\servlet-api.jar moon\sun\Example6_1.java
```

当然,也可以将 Servlet 类的源文件,例如 Example6_1.java,按照其包名 moon.sun 保存到自己喜欢的目录中,例如 D:\geng\moon\sun。将 Tomcat 安装目录 lib 子目录中的 servlet-api.jar 文件复制(不要剪贴)到 D:\geng,然后用命令行进入 moon\sun 的父目录 geng(不要进入 moon 或 sun 目录)编译 Example6_1.java:

```
geng> javac -cp servlet-api.jar moon\sun\Examle6_1.java
```

然后将得到的字节码文件复制到\ch6\WEB-INF\classes\moon\sun 即可。

▶ 6.1.2 编写部署文件 web.xml

视频讲解

Servlet 类的字节码文件保存到指定的目录后,必须为 Tomcat 服务器编写一个部署文件,只有这样,Tomcat 服务器才会用 Servlet 类创建 servlet 对象。

该部署文件是一个 XML 文件,名字必须是 web.xml。web.xml 由 Tomcat 服务器负责管理,在这里,不需要深刻理解 XML 文件,只需要知道 XML 文件是由标记(也称元素)组成的文本文件,使用该 XML 文件的系统应用程序(例如 Tomcat 服务器)配有内置的解析器,可以解析 XML 文件的标记中的数据。可以在 Tomcat 服务器的 webapps 目录中的 root 目录找到一个 web.xml 文件,参照它编写自己的 web.xml 文件。

编写的 web.xml 文件必须保存到 Web 服务目录的 WEB-INF 子目录中,例如 ch6\WEB-INF 中(不要保存在 ch6 目录中)。根据例 6_1 给出的 Servlet 类,web.xml 文件的内容如下(需要用纯文本编辑器编辑 web.xml):

web.xml

```xml
<?xml version = "1.0" encoding = "utf-8" ?>
<web-app>
    <servlet>
        <servlet-name>hello</servlet-name>
        <servlet-class>moon.sun.Example6_1</servlet-class>
    </servlet>
    <servlet-mapping>
        <servlet-name>hello</servlet-name>
        <url-pattern>/lookHello</url-pattern>
    </servlet-mapping>
</web-app>
```

一个 XML 文件应当以 XML 声明作为文件的第一行,在其前面不能有空白、其他的处理指令或注释。XML 声明以"<? xml"标识开始、以"? >"标识结束。注意"<?"和"xml"之间,以及"?"和">"之间不要有空格。如果在 XML 声明中没有显式地指定 encoding 属性的值,那么该属性的默认值为 UTF-8 编码。如果 encoding 属性的值为 UTF-8,XML 文件按照 UTF-8 编码保存,如图 6.2 所示(如果是 GB 2312 或 ISO 8859-1 编码,按照 ANSI 编码保存)。XML 使用 UTF-8 编码,那么标记以及标记的内容除了可以使用 ASCII 字符外,还可以使用汉字、日文中的片假名、平假名等字符,Tomcat 服务器中的 XML 解析器会识别这些标记并正确解析标记中的内容。

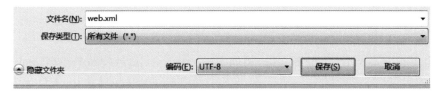

图 6.2 encoding 值是 UTF-8 时 XML 文件的保存

现在让我们看看 web.xml 文件中标记的具体内容及其作用。

❶ 根标记

xml 文件必须有一个根标记,web.xml 文件的根标记是 web-app,根标记 web-app 的开始标签是<web-app>,结束标签是</web-app>,开始标签和结束标签之间的内容称作根标记的内容。

❷ servlet 标记及子标记

web-app 根标记里可以有若干个 servlet 标记(称作根标记的子标记,该标记的开始标签是<servlet>,结束标签是</servlet>,开始标签和结束标签之间是 servlet 标记的内容。servlet 标记需要有两个子标记:servlet-name 和 servlet-class,其中 servlet-name 标记的内容是 Tomcat 服务器创建的 servlet 的名字,servlet-class 标记的内容告知 Tomcat 服务器用哪个 Servlet 类来创建 servlet。在我们给出的 web.xml 文件中,让 Tomcat 服务器使用例 6_1 给出的 Example6_1 类创建的名字是 hello 的 servlet。web.xml 文件可以有若干个 servlet 标记(一个 servlet 标记部署一个 servlet 对象),但要求它们的<servlet-name>子标记的内容互不相同,即服务器可以创建多个 servlet,它们的名字必须互不相同。

❸ servlet-mapping 标记及子标记

web.xml 文件中出现一个 servlet 标记就会对应地出现一个或多个 servlet-mapping 标记,该标记和 servlet 标记都是根标记的直接子标记。不同的是,servlet-mapping 标记需要有两个子标记：servlet-name 和 url-pattern，其中，servlet-name 标记的内容是 Tomcat 服务器创建的 servlet 的名字（该名字必须和对应的 servlet 标记的子标记 servlet-name 的内容相同）。url-pattern 标记用来指定用户用怎样的 URL 格式来请求 servlet，例如，url-pattern 标记的内容是/lookHello，那么用户必须在浏览器的地址栏中输入：

```
http://127.0.0.1:8080/ch6/lookHello
```

来请求名字是 hello 的 servlet。

Web 服务目录的 WEB-INF 子目录下的 web.xml 文件负责管理当前 Web 服务目录下的全部 servlet，当该 Web 服务目录需要提供更多的 servlet 时，只要在 web.xml 文件中增加 servlet 和 servlet-mapping 子标记即可。

对于 webapps 下的 Web 服务目录，如果修改并重新保存 web.xml 文件，Tomcat 服务器就会立刻重新读取 web.xml 文件，因此，修改 web.xml 文件不必重新启动 Tomcat 服务器。但是，如果修改导致 web.xml 文件出现错误，Tomcat 服务器就会关闭当前 Web 服务目录下的所有 servlet 的使用权限。所以必须保证 web.xml 文件正确无误，才能成功启动 Tomcat 服务器。但是，对于不是 webapps 下的 Web 服务目录，如果新建或修改了相应的 web.xml 文件，要重新启动 Tomcat 服务器。

本章中涉及的 JSP 页面存放在 Web 服务目录 ch6 中，负责创建 servlet 的 Servlet 类的字节码文件存放在 ch6\WEB-INF\classes\moon\sun 中（本章 Servlet 类的包名均为 moon.sun）。每当 Web 服务目录增加新的 servlet 时，都需要为 web.xml 文件添加 servlet 和 servlet-mapping 标记。

▶ 6.1.3 servlet 的创建与运行

servlet 由 Tomcat 服务器负责创建，Web 设计者只需为 Tomcat 服务器预备好 Servlet 类，编写好相应的配置文件 web.xml，用户就可以根据 web.xml 部署文件来请求 Tomcat 服务器创建并运行一个 servlet。如果 Tomcat 服务器没有名字为 hello 的 servlet，就会根据 web.xml 文件中 servlet 标记的子标记 servlet-class 指定的 Servlet 类创建一个名字为 hello 的 servlet。因此，如果名字是 hello 的 servlet 被创建之后，修改 Java 源文件、编译得到新的 Servlet 类，并希望 Tomcat 服务器用新的 Servlet 类创建 servlet，那么就要重新启动 Tomcat 服务器。

当用户请求 Tomcat 服务器运行一个 servlet 时，必须根据 web.xml 文件中标记 url-pattern 指定的格式输入请求。

现在，用户就可以请求例 6_1 给出的 Servlet 类创建的名字是 hello 的 servlet 了，根据 6.1.2 节中的 web.xml 文件，用户在浏览器输入：

```
http://127.0.0.1:8080/ch6/lookHello
```

请求 Tomcat 服务器运行名字是 hello 的 servlet，效果如图 6.3 所示。

图 6.3 用户请求 servlet 的效果

6.1.4 向 servlet 传递参数的值

视频讲解

在请求一个 servlet 时,可以在请求的 url-pattern 中额外加入参数及其值,格式是:

url - pattern?参数 1 = 值 & 参数 2 = 值 … 参数 n = 值

那么被请求的 servlet 就可以使用 request 对象获取参数的值,例如:

request.getParameter(参数 n)

例 6_2 中,用户请求 servlet 绘制一个椭圆,请求时将椭圆的宽(横轴长)和高(纵轴长)通过参数 width 和 height 传递给 servlet。

例 6_2

Example6_2.java(负责创建 servlet,servlet 运行效果如图 6.4 所示)

```java
package moon.sun;
import java.awt.image.BufferedImage;
import java.awt.*;
import java.awt.geom.*;
import javax.imageio.ImageIO;
import java.io.*;
import javax.servlet.*;
import javax.servlet.http.*;
public class Example6_2 extends HttpServlet{
    public void init(ServletConfig config) throws ServletException{
        super.init(config);
    }
    public void service(HttpServletRequest request,
                        HttpServletResponse response) throws IOException{
        request.setCharacterEncoding("utf - 8");
        double width = Double.parseDouble(request.getParameter("width"));
        double height = Double.parseDouble(request.getParameter("height"));
        response.setContentType("image/jpeg");
        Ellipse2D ellipse = new
        Ellipse2D.Double(400 - width/2, 300 - height/2, width, height);
        BufferedImage image = getImage(ellipse);
        try {
            OutputStream outClient = response.getOutputStream();
            boolean boo = ImageIO.write(image,"jpeg",outClient);   //发送到客户端
        }
        catch(Exception exp){}
    }
    BufferedImage getImage(Shape shape){                            //返回图形的图像
        int width = 800, height = 600;
        BufferedImage image =
        new BufferedImage(width,height,BufferedImage.TYPE_INT_RGB); //图像
        Graphics g = image.getGraphics();
        g.fillRect(0, 0, width, height);                            //图像的底色
        Graphics2D g_2d = (Graphics2D)g;
        g_2d.setColor(Color.blue);
```

```
            g_2d.fill(shape);                    //在图像上绘制图形
            return image;
    }
}
```

图 6.4 通过参数向 servlet 传递值

用命令行进入 moon\sun 的父目录 classes,编译 Example6_2.java(有关约定见 6.1.1 节):

```
classes> javac -cp servlet-api.jar moon\sun\Example6_2.java
```

向 ch6\WEB\INF\下的部署文件 web.xml 添加如下的 servlet 和 servlet-mapping 标记(见 6.1.2 节),部署的 servlet 的名字是 lookPic,访问 servlet 的 url-pattern 是/lookPic。

web.xml

```
<?xml version="1.0" encoding="utf-8"?>
<web-app>
    <!-- 以下是 web.xml 文件新添加的内容 -->
    <servlet>
        <servlet-name>lookPic</servlet-name>
        <servlet-class>moon.sun.Example6_2</servlet-class>
    </servlet>
    <servlet-mapping>
        <servlet-name>lookPic</servlet-name>
        <url-pattern>/lookPic</url-pattern>
    </servlet-mapping>
</web-app>
```

根据 web.xml 文件给出的 servlet 的 url-pattern,用户可以在请求 servlet 的 url-pattern 中额外加入 width、height 参数及其值,以便 servlet 获取这些值绘制椭圆,即在浏览器输入:

```
http://127.0.0.1:8080/ch6/lookPic?width=500&height=200
```

效果如图 6.4 所示。

6.2 servlet 的工作原理

servlet 由 Tomcat 服务器负责管理,Tomcat 服务器通过读取 web.xml,然后创建并运行 servlet。本节将详细讲解 servlet 的运行原理。

▶ 6.2.1 servlet 对象的生命周期

servlet 是 javax.servlet 包中 HttpServlet 类的子类的一个实例,由 Tomcat 服务器负责创

建并完成初始化工作。当多个用户请求同一个 servlet 时,服务器为每个用户分别启动一个线程而不是共用一个进程,这些线程由 Tomcat 服务器来管理,与传统的 CGI 为每个用户启动一个进程相比较,效率要高得多。

一个 servlet 的生命周期主要由下列三个过程组成:

(1) 初始化 servlet。servlet 第一次被请求加载时,服务器初始化这个 servlet,即创建一个 servlet,这 servlet 调用 init 方法完成必要的初始化工作。

(2) 新诞生的 servlet 再调用 service 方法响应用户的请求。

(3) 当服务器关闭时,调用 destroy 方法销毁 servlet。

init 方法只被调用一次,即在 servlet 第一次被请求加载时调用该方法。当后续的用户请求 servlet 服务时,Tomcat 服务器将启动一个新的线程,在该线程中,servlet 调用 service 方法响应用户的请求。也就是说,每个用户的每次请求都导致 service 方法被调用执行,其执行过程分别运行在不同的线程中。

▶ 6.2.2　init 方法

该方法是 HttpServlet 类中的方法,可以在子类中重写这个方法。init 方法的声明格式是:

```
public void init(ServletConfig config) throws ServletException
```

servlet 第一次被请求加载时,服务器创建一个 servlet,这个对象调用 init 方法完成必要的初始化工作。该方法在执行时,服务器会把一个 SevletConfig 类型的对象传递给 init 方法,这个对象就被保存在 servlet 中,直到 servlet 被销毁。这个 ServletConfig 对象负责向 servlet 传递服务设置信息,如果传递失败就会发生 ServeletException,servlet 就不能正常工作。

▶ 6.2.3　service 方法

该方法是 HttpServlet 类中的方法,可以在子类中直接继承该方法或重写这个方法。service 方法的声明格式是:

```
public void service(HttpServletRequest request HttpServletResponse response)throw ServletException,
IOException
```

当 servlet 成功创建和初始化之后,调用 service 方法来处理用户的请求并返回响应。Tomcat 服务器将两个参数传递给该方法。一个是 HttpServletRequest 类型的对象,该对象封装了用户的请求信息,另外一个参数对象是 HttpServletResponse 类型的对象,该对象用来响应用户的请求。和 init 方法不同的是,init 方法只被调用一次,而 service 方法可能被多次的调用。也就是说,当后续的用户请求该 servlet 时,Tomcat 服务器将启动一个新的线程,在该线程中 servlet 调用 service 方法响应用户的请求,即每个用户的请求都导致 service 方法被调用执行,调用过程运行在不同的线程中,互不干扰。因此,不同线程的 service 方法中的局部变量互不干扰,一个线程改变了自己的 service 方法中局部变量的值不会影响其他线程的 service 方法中的局部变量。

▶ 6.2.4　destroy 方法

该方法是 HttpServlet 类中的方法,子类可直接继承这个方法,一般不需要重写。destroy 方法的声明格式是:

```
public destroy()
```

当 Tomcat 服务器终止服务时,例如关闭 Tomcat 服务器等,destroy()方法会被执行,销毁 servlet。

6.3 通过 JSP 页面访问 servlet

按照部署文件 web.xml 给出的 servlet 的 url-pattern,用户除了可以在浏览器输入 url-pattern 请求运行一个 servlet 外,也可以通过 JSP 页面来请求一个 servlet。

需要特别注意的是,如果 web.xml 文件中给出的 servlet 的 url-pattern 是/lookHello,那么 JSP 页面请求 servlet 时,必须要写成 lookHello,不可以写成/lookHello,否则将变成请求 root 服务目录下的某个 servlet。

❶ 通过表单向 servlet 提交数据

Web 服务目录下的 JSP 页面都可以通过 form 表单请求该 Web 服务目录下的某个 servlet。如果 web.xml 文件中给出的 servlet 的 url-pattern 是/computeBill,那么 form 表单中 action 给出的值就是 computeBill,如下所示:

```
<form action = "computeBill">
</form>
```

当请求一个 servlet 时,也可以在请求的 url-pattern 中额外加入参数及其值,格式是:

```
url-pattern?参数1=值&参数2=值…参数n=值
```

例如:

```
<form action = "computeBill?sideA = 10.66&sideB = 23.9&sideC = 897">
</form>
```

通过 JSP 页面访问 servlet 的好处是,JSP 页面可以负责页面的信息显示,信息的有关处理交给 servlet 去完成。

例 6_3 中,JSP 页面 example6_3.jsp 通过 form 表单请求名字是 computeBill 的 servlet,向所请求的 computeBill 提交一个账单明细,例如"剁椒鱼头:62.9 元,烤鸭:199 元,红焖大虾:289.9 元"(效果如图 6.5(a)所示)。另外,在请求 computeBill 的 url-pattern 中额外加入一个参数 discount,其值代表优惠额度,例如 discount 值是 6 时,表示打 6 折,computeBill 将账单的消费额乘以 0.6(效果如图 6.5(b)所示)。

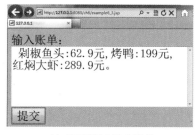
(a) JSP 页面中输入账单

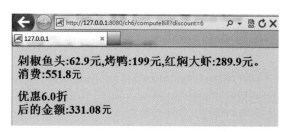
(b) computeBill 负责计算账单的消费

图 6.5 JSP 页面请求 servlet

例 6_3
➢ JSP 页面
example6_3.jsp(效果如图 6.5(a)所示)

```
<%@ page contentType = "text/html" %>
<%@ page pageEncoding = "utf-8" %>
<style>
    #textStyle{
        font-family:宋体;font-size:36;color:blue
    }
</style>
<HTML><body bgcolor = #ffccff>
<form action = "computeBill?discount = 6" id = textStyle method = post>
  输入账单:<br>
<textArea name = 'billMess' id = textStyle rows = 5 cols = 30>
  剁椒鱼头:62.9 元,烤鸭:199 元,红焖大虾:289.9 元.
</textArea>
<br><input type = submit id = textStyle value = "提交">
</form>
</body></HTML>
```

➢ Servlet 类
用命令行进入 moon\sun 的父目录 classes,编译 Example6_3_Servlet.java(约定见 6.1.1 节):

```
classes> javac -cp servlet-api.jar moon\sun\Example6_3_Servlet.java
```

Example6_3_Servlet.java(负责创建 servlet,servlet 运行效果如图 6.5(b)所示)

```
package moon.sun;
import java.io.*;
import javax.servlet.*;
import javax.servlet.http.*;
import java.util.regex.Pattern;
import java.util.regex.Matcher;
public class Example6_3_Servlet extends HttpServlet{
    public void init(ServletConfig config) throws ServletException{
        super.init(config);
    }
    public void service(HttpServletRequest request,
                    HttpServletResponse response) throws IOException{
        request.setCharacterEncoding("utf-8");
        response.setContentType("text/html;charset = utf-8");
        PrintWriter out = response.getWriter();
        out.println("<html><body bgcolor = yellow>");
        String discountMess = request.getParameter("discount");
        double discount = Double.parseDouble(discountMess);
        String billMess = request.getParameter("billMess");
        if(billMess == null) {
            out.print("没有账单");
            return;
        }
```

```
            double bill = getPriceSum(billMess);              //账单金额
            double billDiscount = bill * (discount/10);       //优惠后的金额
            out.print("<h2>" + billMess + "<br>消费:" + bill + "元");
            out.print("优惠" + discount + "折<br>后的金额:" + billDiscount + "元");
            out.print("</h2></body></html>");
        }
        public double getPriceSum(String input){             //定义方法
            Pattern pattern;                                  //模式对象
            Matcher matcher;                                  //匹配对象
            String regex = "-?[0-9][0-9]*[.]?[0-9]*";        //匹配数字的正则表达式
            pattern = Pattern.compile(regex);                 //初始化模式对象
            matcher = pattern.matcher(input);                 //初始化匹配对象,用于检索 input
            double sum = 0;
            while(matcher.find()) {
                String str = matcher.group();
                sum += Double.parseDouble(str);
            }
            return sum;
        }
    }
```

➤ **web.xml 文件**

向 ch6\WEB\INF\下的部署文件 web.xml 添加如下的 servlet 和 servlet-mapping 标记（知识点见 6.1.2 节），部署的 servlet 的名字是 computeBill，访问 servlet 的 url-pattern 是 /computeBill。

web.xml

```xml
<?xml version="1.0" encoding="utf-8"?>
<web-app>
    <!-- 以下是 web.xml 文件新添加的内容 -->
    <servlet>
        <servlet-name>computeBill</servlet-name>
        <servlet-class>moon.sun.Example6_3_Servlet</servlet-class>
    </servlet>
    <servlet-mapping>
        <servlet-name>computeBill</servlet-name>
        <url-pattern>/computeBill</url-pattern>
    </servlet-mapping>
</web-app>
```

❷ 通过超链接访问 servlet

JSP 页面可以使用超链接去请求某个 servlet。如果 web.xml 文件中给出的请求 servlet 的 url-pattern 是/circle，那么超链接标记中 href 的值是 circle（不要写成/circle）：

```
<a href="circle"></a>
```

例 6_4 使用超链接请求 url-pattern 是/circle 的 servlet，在请求的 url-pattern 中额外加入参数 radius、mess 及其值。如果参数 mess 的值是 area，servlet 将求半径是 radius 的圆面积，如果参数 mess 的值是 geometry，servlet 将绘制半径是 radius 的圆的图形。

例 6_4

➢ **JSP 页面**

example6_4.jsp(效果如图 6.6(a)所示)

```jsp
<%@ page contentType = "text/html" %>
<%@ page pageEncoding = "utf-8" %>
<style>
    #textStyle{
        font-family:宋体;font-size:36;color:blue
    }
</style>
<HTML><body bgcolor = #ffccff>
<% double r = 100.8;
%>
<p id = textStyle>
<a href = "circle?mess = area&radius = <% = r%>">看半径<% = r%>圆的面积.</a><br>
<a href = "circle?mess = geometry&radius = <% = r%>">看半径<% = r%>圆的图形.</a>
</p></body></HTML>
```

➢ **Servlet 类**

用命令行进入 moon\sun 的父目录 classes,编译 Example6_4_Servlet.java(约定见 6.1.1 节):

```
classes> javac -cp servlet-api.jar moon\sun\Example6_4_Servlet.java
```

Example6_4_Servlet.java(负责创建 servlet,servlet 运行效果如图 6.6(b)(c)所示)

```java
package moon.sun;
import java.io.*;
import javax.servlet.*;
import javax.servlet.http.*;
import java.awt.image.BufferedImage;
import java.awt.*;
import java.awt.geom.*;
import javax.imageio.ImageIO;
public class Example6_4_Servlet extends HttpServlet{
    HttpServletRequest request;
    HttpServletResponse response;
    public void init(ServletConfig config) throws ServletException{
        super.init(config);
    }
    public void service(HttpServletRequest request,
                        HttpServletResponse response)throws IOException{
        this.request = request;
        this.response = response;
        request.setCharacterEncoding("utf-8");
        String mess = request.getParameter("mess");
        String radius = request.getParameter("radius");
        if(mess.equals("area")) {
            getArea(Double.parseDouble(radius));
        }
        else if(mess.equals("geometry")){
```

第6章 Java Servlet基础

```java
            getGeometry(Double.parseDouble(radius));
        }
    }
    void getArea(double r) throws IOException{
        response.setContentType("text/html;charset=utf-8");
        PrintWriter out = response.getWriter();
        double area = Math.PI * r * r;
        String result = String.format("%.2f",area);
        out.print("<h2>半径"+r+"圆的面积(保留2位小数):<br>");
        out.print(result);
        out.print("</h2></body></html>");
    }
    void getGeometry(double r) throws IOException{
        response.setContentType("image/jpeg");
        Ellipse2D ellipse = new Ellipse2D.Double(30,30,2*r,2*r);
        BufferedImage image = getImage(ellipse);
        OutputStream outClient = response.getOutputStream();
        boolean boo = ImageIO.write(image,"jpeg",outClient);
    }
    BufferedImage getImage(Shape shape){         //得到图形的图像
        int width = 800, height = 600;
        BufferedImage image =
          new BufferedImage(width,height,BufferedImage.TYPE_INT_RGB);
        Graphics g = image.getGraphics();
        g.fillRect(0, 0, width, height);
        Graphics2D g_2d = (Graphics2D)g;
        g_2d.setColor(Color.blue);
        g_2d.fill(shape);
        return image;
    }
}
```

 (a) 使用超链接请求servlet   (b) servlet求面积   (c) servlet绘制圆

图 6.6 通过超链接访问 servlet

➢ web.xml 文件

向 ch6\WEB\INF\ 下的部署文件 web.xml 添加如下的 servlet 和 servlet-mapping 标记（见 6.1.2 节），部署的 servlet 的名字是 circle，访问 servlet 的 url-pattern 是 / circle。

web.xml

```xml
<?xml version = "1.0" encoding = "utf-8"?>
<web-app>
  <!-- 以下是web.xml文件新添加的内容 -->
  <servlet>
      <servlet-name>circle</servlet-name>
```

```
        <servlet-class>moon.sun.Example6_4_Servlet</servlet-class>
    </servlet>
    <servlet-mapping>
        <servlet-name>circle</servlet-name>
        <url-pattern>/circle</url-pattern>
    </servlet-mapping>
</web-app>
```

注：如果JSP页面，例如example6_4.jsp存放在了Web服务目录的子目录（例如ch6\subDir）中，那么web.xml给出的访问servlet的url-pattern中，必须有一个<url-pattern>含有子目录的名字，例如：

```
<servlet-mapping>
        <servlet-name>circle</servlet-name>
        <url-pattern>/subDir/circle</url-pattern>
</servlet-mapping>
```

6.4 共享变量

视频讲解

Servlet类是HttpServlet的一个子类，在编写子类时就可以声明某些成员变量，那么，请求servlet的用户将共享该servlet的成员变量。

数学上有一个计算π的公式：

$$\frac{\pi}{4}=1-\frac{1}{3}+\frac{1}{5}-\frac{1}{7}+\frac{1}{9}-\frac{1}{11}\cdots$$

例6_5利用servlet的成员变量被所有用户共享这一特性实现多用户计算π的值，即任何用户请求访问servlet时都参与了一次π的计算。通过单击example6_5.jsp页面的超链接访问名字为computePI的servlet，该servlet负责计算π的近似值。

例6_5

➢ **JSP页面**

example6_5.jsp（效果如图6.7(a)所示）

```
<%@ page contentType="text/html" %>
<%@ page pageEncoding="utf-8" %>
<HTML><body bgcolor=#ffccff>
<p style="font-family:宋体;font-size:36;color:blue">
<a href="computePI">参与计算 PI 的值</a>
</p></body></HTML>
```

➢ **Servlet类**

用命令行进入moon\sun的父目录classes，编译Example6_5_Servlet.java（约定见6.1.1节）：

```
classes> javac -cp servlet-api.jar moon\sun\Example6_5_Servlet.java
```

Example6_5_Servlet.java（负责创建servlet，servlet运行效果如图6.7(b)所示）

```
package moon.sun;
```

第6章 Java Servlet基础

```
import java.io.*;
import javax.servlet.*;
import javax.servlet.http.*;
public class Example6_5_Servlet extends HttpServlet{
    double sum = 0, i = 1, j = 1;           //被所有用户共享
    int number = 0;                          //被所有用户共享
    public void init(ServletConfig config) throws ServletException{
        super.init(config);
    }
    public synchronized void service(HttpServletRequest request,
                    HttpServletResponse response) throws IOException{
        response.setContentType("text/html;charset = utf - 8");
        PrintWriter out = response.getWriter();
        out.println("< html >< body bgcolor = cyan >");
        number++;
        sum = sum + i/j;
        j = j + 2;
        i = - i;
        out.println("< h1 > servlet:" + getServletName() +
                    "已经被请求了" + number + "次");
        out.println("< br >现在 PI 的值是:");
        out.println(4 * sum);
        out.println("</h1></body></html>");
    }
}
```

(a) 请求servlet计算PI

(b) servlet计算PI的近似值

图 6.7 共享 servlet 的成员变量

➢ web.xml 文件

向 ch6\WEB\INF\下的部署文件 web.xml 添加如下的 servlet 和 servlet-mapping 标记（知识点见 6.1.2 节），部署的 servlet 的名字是 computePI，访问 servlet 的 url-pattern 是 /computePI。

web.xml

```
<?xml version = "1.0" encoding = "utf - 8"?>
< web - app >
  <!-- 以下是 web.xml 文件新添加的内容 -->
    < servlet >
        < servlet - name > computePI </servlet - name >
        < servlet - class > moon.sun.Example6_5_Servlet </servlet - class >
    </servlet >
    < servlet - mapping >
```

```
          <servlet-name>computePI</servlet-name>
          <url-pattern>/computePI</url-pattern>
     </servlet-mapping>
</web-app>
```

6.5　doGet 和 doPost 方法

视频讲解

　　HttpServlet 类除了 init、service、destroy 方法外,该类还有两个很重要的方法:doGet 和 doPost,用来处理用户的请求并作出响应。当 Tomcat 服务器创建 servlet 后,该 servlet 会调用 init 方法初始化自己,以后每当 Tomcat 服务器再接受一个对该 servlet 请求时,就会产生一个新线程,并在这个线程中让该 servlet 调用 service 方法。实际上 HttpServlet 类所给出的 service 方法的功能是检查 HTTP 请求类型(get、post 等),并在 service 方法中根据用户的请求方式,在 service 方法中对应地再调用 doGet 或 doPost 方法。因此,在编写的 Servlet 类(HttpServlet 类的一个子类)时,也可以不重写 service 方法来响应用户,直接继承 scrvicc 方法即可。

　　如果不重写 service 方法,就需要在 Servlet 类中重写 doPost 或 doGet 方法来响应用户的请求。如果不论用户请求类型是 post 还是 get,Tomcat 服务器的处理过程完全相同,那么可以只在 doPost 方法中编写处理过程,而在 doGet 方法中再调用 doPost 方法即可,或只在 doGet 方法中编写处理过程,而在 doPost 方法中再调用 doGet 方法。如果根据请求的类型进行不同的处理,就要在两个方法中编写不同的处理过程(这一点比 service 方法更为灵活)。

　　例 6_6 的 example6_5.jsp 页面有两个 form 表单,每个 form 表单都请求名字为 sumORproduct 的 servlet,并提交一串数字给 sumORproduct。一个 form 表单的请求方式是 post,另一个 form 表单的请求方式是 get。当 form 表单的请求方式是 post 时,sumORproduct 计算各数字的代数和,当 form 表单的请求方式是 get 时,sumORproduct 计算各数字的乘积。

例 6_6

➢ **JSP 页面**

example6_6.jsp(效果如图 6.8(a)所示)

```
<%@ page contentType="text/html" %>
<%@ page pageEncoding="utf-8" %>
<style>
   #textStyle{
      font-family:宋体;font-size:36;color:blue
   }
</style>
<HTML><body bgcolor=#ffccff>
<form action="sumORproduct" id=textStyle method=post>
   输入数字,用逗号分隔提交给 servlet(post 方式):
   <br><input type=text id=textStyle name="number" />
   <input type=submit id=textStyle value="提交" />
</form>
<form action="sumORproduct" id=textStyle method=get>
   输入数字,用逗号分隔提交给 servlet(get 方式):
```

第6章 Java Servlet基础

```
  <br><input type=text id = textStyle name = "number" />
  <input type=submit id = textStyle value = "提交" />
</form>
</body></HTML>
```

➢ **Servlet 类**

用命令行进入 moon\sun 的父目录 classes，编译 Example6_6_Servlet.java（约定见 6.1.1 节）：

```
classes> javac -cp servlet-api.jar moon\sun\Example6_6_Servlet.java
```

Example6_6_Servlet.java（负责创建 servlet，servlet 运行效果如图 6.8(b)(c)所示）

```java
package moon.sun;
import java.io.*;
import javax.servlet.*;
import javax.servlet.http.*;
public class Example6_6_Servlet extends HttpServlet{
    public void init(ServletConfig config) throws ServletException{
        super.init(config);
    }
    public void doPost(HttpServletRequest request,
            HttpServletResponse response)throws ServletException,IOException{
        request.setCharacterEncoding("utf-8");
        response.setContentType("text/html;charset=utf-8");
        PrintWriter out = response.getWriter();
        out.println("<html><body bgcolor=cyan>");
        String s = request.getParameter("number");
        String []a = s.split("[,,]+");
        double sum = 0;
        for(String item:a) {
            if(item.length()>=1)
              sum += Double.parseDouble(item);
        }
        out.print("<h2>用户的请求方式是" + request.getMethod() + "<br>");
        for(String item:a) {
            if(item.length()>=1)
              out.print(item + " ");
        }
        out.print("<br>的和是" + sum);
        out.println("</h2></body></html>");
    }
    public void doGet(HttpServletRequest request,
            HttpServletResponse response)throws ServletException,IOException{
        request.setCharacterEncoding("utf-8");
        response.setContentType("text/html;charset=utf-8");
        PrintWriter out = response.getWriter();
        out.println("<html><body bgcolor=yellow>");
        String s = request.getParameter("number");
        String []a = s.split("[,,]+");
        double product = 1;
        for(String item:a) {
            if(item.length()>=1)
```

```
                product * = Double.parseDouble(item);
            }
            out.print("<h2>用户的请求方式是" + request.getMethod() + "<br>") ;
            for(String item:a) {
                if(item.length()>=1)
                    out.print(item + " ");
            }
            out.print("<br>的乘积是" + product);
            out.println("<h2></body></html>");
        }
    }
```

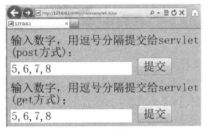
(a) 用get或post方式请求servlet

(b) servlet调用doPost方法

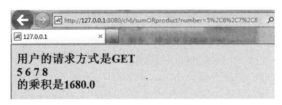
(c) servlet调用doGet方法

图 6.8　doPost 和 doGet 方法

> **web.xml 文件**

向 ch6\WEB\INF\下的部署文件 web.xml 添加如下的 servlet 和 servlet-mapping 标记（知识点见 6.1.2 节），部署的 servlet 的名字是 sumORproduct，访问 servlet 的 url-pattern 是/sumORproduct。

web.xml

```xml
<?xml version="1.0" encoding="utf-8"?>
<web-app>
    <!-- 以下是web.xml文件新添加的内容 -->
    <servlet>
        <servlet-name>sumORproduct</servlet-name>
        <servlet-class>moon.sun.Example6_6_Servlet</servlet-class>
    </servlet>
    <servlet-mapping>
        <servlet-name>sumORproduct</servlet-name>
        <url-pattern>/sumORproduct</url-pattern>
    </servlet-mapping>
</web-app>
```

6.6 重定向与转发

重定向的功能是将用户从当前页面或 servlet 定向到另一个 JSP 页面或 servlet。转发的功能是将用户对当前 JSP 页面或 servlet 的请求转发给另一个 JSP 页面或 servlet。本节学习在 Servlet 类中使用 HttpServletResponse 类的 sendRedirect 重定向方法，以及 RequestDispatcher 类的 forward 转发方法，并指出二者的区别。

❶ sendRedirect 方法

重定向方法 void sendRedirect(String location) 是 HttpServletResponse 类中的方法。当用户请求一个 servlet 时，该 servlet 在处理数据后，可以使用重定向方法将用户重新定向到另一个 JSP 页面或 servlet。重定向方法仅仅是将用户从当前页面或 servlet 定向到另一个 JSP 页面或 servlet，但不能将用户对当前页面或 servlet 的请求（HttpServletRequest 对象）转发给所定向的资源。即重定向的目标页面或 servlet 无法使用 request 获取用户提交的数据。

❷ forward 方法

RequestDispatcher 对象可以把用户对当前 JSP 页面或 servlet 的请求转发给另一个 JSP 页面或 servlet，而且将用户对当前 JSP 页面或 servlet 的请求传递给转发到的 JSP 页面或 servlet。也就是说，当前页面所转发到的 JSP 页面或 servlet 可以使用 request 获取用户提交的数据。下面介绍实现转发的步骤。

（1）得到 RequestDispatcher 对象。用户所请求的当前 JSP 或 servlet 可以让 HttpServletRequest 对象 request 调用

```
public RequestDispatcher getRequestDispatcher(String path)
```

方法返回一个 RequestDispatcher 对象，其中参数 path 是准备要转发到的 JSP 页面的 URL 或 servlet 的 url-pattern。例如：

```
RequestDispatcher dispatcher = request.getRequestDispatcher("target.jsp");
RequestDispatcher dispatcher = request.getRequestDispatcher("targetServlet");
```

（2）转发。在步骤（1）中获取的 RequestDispatcher 对象调用

```
void forward(ServletRequest request,ServletResponse response)
       throws ServletException,ava.io.IOException
```

方法可以将用户对当前 JSP 页面或 servlet 的请求转发给 RequestDispatcher 对象所指定的 JSP 页面或 servlet，例如：

```
dispatcher.forward(request,response);
```

把用户对当前 JSP 页面或 servlet 的请求转变为对转发到的 JSP 页面或 servlet 的请求。

❸ 二者的区别

转发（forward）和重定向方法（sendRedirect）不同的是，用户可以看到转发到的 JSP 页面或 servlet 的运行效果，但是，在浏览器的地址栏中不能看到 forward 方法转发到的 JSP 页面的地址或 servlet 的地址，用户在浏览器的地址栏中所看到的仍然是当前 JSP 页面的 URL 或 servlet 的 url-pattern。如果此时刷新浏览器，那么请求将是当前的 JSP 页面或 servlet。

另外，当 servlet 中执行 forward 方法实施转发操作时，Tomcat 会立刻结束当前 servlet 的执行。而 servlet 中执行 sendRedirect 方法（重定向，见 4.2.3 节）时，Tomcat 服务器还是要把当前 servlet 代码执行完毕后才实施重定向(跳转)操作，但 Tomcat 服务器不再给用户看当前 servlet 代码的执行效果。如果在执行 sendRedirect(URL url)方法后，servlet 紧接着执行了 return 返回语句，那么 Tomcat 服务器会立刻结束当前 servlet 的执行。

❹ 使用转发的好处

使用转发技术可以让 JSP 页面和处理数据的 servlet 解耦，JSP 页面只需和处理转发的 servlet 打交道。例如，在实际问题中，对数据的处理可能有多种需求，那么可以把这些需求分别指派给几个 servlet 来完成，然后编写一个 servlet 负责接收用户的请求，并根据用户的请求信息将用户的请求转发给这些 servlet 中的某个 servlet。当更新某个需求时，只需重新编译相应的 Servlet 类，这有利于 Web 应用程序的维护。

例 6_7 中有 3 个 servlet，其中名字是 sort 的 servlet 负责排序数字，名字是 sum 的 servlet 负责计算数字的代数和，名字是 handleForward 的 servlet 负责转发。用户在 example6_7.jsp 页面提供的 form 表单的 textArea 中输入一组数字（例如 2,-65,1.618）请求 handleForward，如果是单击名字含有 sort 的提交键，handleForward 就将用户的请求转发给 sort，如果单击名字含有 sum 的提交键，handleForward 就将用户的请求转发给 sum。如果用户没有输入任何数字，handleForward 就将用户重定向到 example6_7.jsp。

例 6_7

> **JSP 页面**

example6_7.jsp（效果如图 6.9(a)所示）

```
<%@ page contentType = "text/html" %>
<%@ page pageEncoding = "utf-8" %>
<style>
    #textStyle{
       font-family:宋体;font-size:36;color:blue
    }
</style>
<HTML><body bgcolor = #ffccff>
<p id = textStyle>
<form action = "handleForward" id = textStyle method = post>
   输入数字(用逗号空格或其他非数字字符分隔): <br>
<textArea name = 'digitData' id = textStyle rows = 5 cols = 30>
</textArea>
<br><input type = submit id = textStyle name = submit value = "提交(排序数字 sort)"/>
<br><input type = submit id = textStyle name = submit value = "提交(求数字代数和 sum)"/>
</form>
</p></body></HTML>
```

> **Servlet 类**

用命令行进入 moon\sun 的父目录 classes，编译 3 个 Servlet 源文件(约定见 6.1.1 节)：

```
classes > javac - cp servlet - api.jar moon\sun\Example6_7_Servlet.java
classes > javac - cp servlet - api.jar moon\sun\Example6_7_Servlet_Sum.java
classes > javac - cp servlet - api.jar moon\sun\Example6_7_Servlet_Sort.java
```

Example6_7_Servlet.java（负责创建 servlet，servlet 负责转发，无显式效果）

```java
package moon.sun;
import java.io.*;
import javax.servlet.*;
import javax.servlet.http.*;
public class Example6_7_Servlet extends HttpServlet{
    public void init(ServletConfig config) throws ServletException{
        super.init(config);
    }
    public void doPost(HttpServletRequest request,
            HttpServletResponse response) throws ServletException,IOException{
        RequestDispatcher dispatcher = null;                    //负责转发的对象
        request.setCharacterEncoding("utf-8");
        response.setContentType("text/html;charset=utf-8");
        String mess = request.getParameter("submit");
        String digitData = request.getParameter("digitData");
        if(digitData == null ||digitData.length() == 0) {
            response.sendRedirect("example6_7.jsp");             //重定向到输入数据页面
            return;
        }
        if(mess.contains("sort")) {
            dispatcher = request.getRequestDispatcher("sort");   //转发
            dispatcher.forward(request,response);
        }
        else if(mess.contains("sum")){
            dispatcher = request.getRequestDispatcher("sum");    //转发
            dispatcher.forward(request,response);
        }
    }
    public void doGet(HttpServletRequest request,
        HttpServletResponse response) throws ServletException,IOException{
        doPost(request,response);
    }
}
```

Example6_7_Servlet_Sum.java（负责创建 servlet，servlet 运行效果如图 6.9(b)所示）

```java
package moon.sun;
import java.io.*;
import javax.servlet.*;
import javax.servlet.http.*;
import java.util.regex.Pattern;
import java.util.regex.Matcher;
public class Example6_7_Servlet_Sum extends HttpServlet{
    public void init(ServletConfig config) throws ServletException{
        super.init(config);
    }
    public void doPost(HttpServletRequest request,
            HttpServletResponse response) throws ServletException,IOException{
        request.setCharacterEncoding("utf-8");
        response.setContentType("text/html;charset=utf-8");
```

```
        PrintWriter out = response.getWriter();
        String digitData = request.getParameter("digitData");
        double sum = getPriceSum(digitData);
        out.println("<html><body bgcolor=cyan>");
        out.println("<h1>数字的代数和:" + sum);
        out.println("</h1></body></html>");
    }
    public void doGet(HttpServletRequest request,
        HttpServletResponse response) throws ServletException,IOException{
        doPost(request,response);
    }
    public double getPriceSum(String input){              //定义方法
        Pattern pattern;                                  //模式对象
        Matcher matcher;                                  //匹配对象
        String regex = "-?[0-9][0-9]*[.]?[0-9]*";         //匹配数字的正则表达式
        pattern = Pattern.compile(regex);                 //初始化模式对象
        matcher = pattern.matcher(input);                 //初始化匹配对象,用于检索input
        double sum = 0;
        while(matcher.find()) {
            String str = matcher.group();
            sum += Double.parseDouble(str);
        }
        return sum;
    }
}
```

Example6_7_Servlet_Sort.java(负责创建 servlet, servlet 运行效果如图 6.9(c)所示)

```
package moon.sun;
import java.io.*;
import javax.servlet.*;
import javax.servlet.http.*;
import java.util.regex.Pattern;
import java.util.regex.Matcher;
import java.util.TreeSet;
import java.util.Iterator;
public class Example6_7_Servlet_Sort extends HttpServlet{
    public void init(ServletConfig config) throws ServletException{
        super.init(config);
    }
    public void doPost(HttpServletRequest request,
            HttpServletResponse response) throws ServletException,IOException{
        TreeSet<Double> treeSet = new TreeSet<Double>();      //排序数字
        request.setCharacterEncoding("utf-8");
        response.setContentType("text/html;charset=utf-8");
        PrintWriter out = response.getWriter();
        String digitData = request.getParameter("digitData");
        sort(digitData,treeSet);
        Iterator<Double> iterator = treeSet.iterator();
        out.println("<html><body bgcolor=cyan>");
        out.println("<h1>排序后的数字:<br>");
```

```java
        while(iterator.hasNext()) {
            out.println(iterator.next()+",");
        }
        out.println("</h1></body></html>");
    }
    public void doGet(HttpServletRequest request,
           HttpServletResponse response) throws ServletException,IOException{
        doPost(request,response);
    }
    public void sort(String input,TreeSet<Double> treeSet){
        Pattern pattern;                                //模式对象
        Matcher matcher;                                //匹配对象
        String regex = "-?[0-9][0-9]*[.]?[0-9]*";       //匹配数字的正则表达式
        pattern = Pattern.compile(regex);               //初始化模式对象
        matcher = pattern.matcher(input);               //初始化匹配对象,用于检索input
        double sum = 0;
        while(matcher.find()) {
            String str = matcher.group();
            treeSet.add(Double.parseDouble(str));
        }
    }
}
```

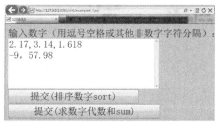

(a) 在JSP页面输入数字　　　　　(b) sum servlet负责求和　　　(c) sort servlet负责排序

图 6.9　重定向与转发

➢ web.xml 文件

向 ch6\WEB\INF\ 下的部署文件 web.xml 添加如下的 servlet 和 servlet-mapping 标记（知识点见 6.1.2 节），部署的 3 个 servlet 的名字分别是 handleForward、sort 和 sum，访问 3 个 servlet 的 url-pattern 分别是 /handleForward、/sort 和 /sum。

web.xml

```xml
<?xml version="1.0" encoding="utf-8"?>
<web-app>
    <!-- 以下是web.xml文件新添加的内容 -->
    <servlet>
        <servlet-name>handleForward</servlet-name>
        <servlet-class>moon.sun.Example6_7_Servlet</servlet-class>
    </servlet>
    <servlet-mapping>
        <servlet-name>handleForward</servlet-name>
        <url-pattern>/handleForward</url-pattern>
```

```xml
        </servlet-mapping>
        <servlet>
            <servlet-name>sort</servlet-name>
            <servlet-class>moon.sun.Example6_7_Servlet_Sort</servlet-class>
        </servlet>
        <servlet-mapping>
            <servlet-name>sort</servlet-name>
            <url-pattern>/sort</url-pattern>
        </servlet-mapping>
        <servlet>
            <servlet-name>sum</servlet-name>
            <servlet-class>moon.sun.Example6_7_Servlet_Sum</servlet-class>
        </servlet>
        <servlet-mapping>
            <servlet-name>sum</servlet-name>
            <url-pattern>/sum</url-pattern>
        </servlet-mapping>
</web-app>
```

6.7 使用 session

视频讲解

HTTP 通信协议是用户与服务器之间一种请求与响应(request/response)的通信协议,属于无状态协议。所谓无状态是指,当用户(浏览器)发送请求给服务器,Tomcat 服务器作出响应后,如果同一个用户再发送请求给 Tomcat 服务器时,Tomcat 服务器并不知道就是刚才的那个用户。简单地说,Tomcat 服务器不会记录用户的信息。

用户在访问一个 Web 服务目录期间,Tomcat 服务器为该用户分配一个 session 对象(称为用户的会话),Tomcat 服务器可以在各个页面以及 servlet 中使用这个 session 记录用户的有关信息,而且 Tomcat 服务器保证不同用户的 session 对象互不相同。有关 session 对象的原理、常用方法可参见 4.3 节。本节学习怎样在 servlet 中使用 session 对象记录有关信息。

HttpServletRequest 对象 request 调用 getSession 方法获取用户的 session 对象:

```
HttpSession session = request.getSession(true);
```

访问某个 Web 服务目录的用户,在不同的 servlet 中获取的 session 对象是完全相同的,不同的用户的 session 对象互不相同。

例 6_8 是一个猜字母游戏,当用户访问或刷新 example6_8.jsp 页面时,随机分配给用户一个英文字母(不区分大小写),并将这个字母存在用户的 session 中。用户链接到 example6_8_input.jsp 页面输入自己的猜测,并将该猜测提交给一个名字为 guess 的 servlet,该 servlet 负责处理用户的猜测。具体处理方式是:如果用户猜小了,将"猜小了"存放到用户的 session 中,然后将用户重新定向到 example6_8_input.jsp。如果猜大了,将"猜大了"存放到用户的 session 中,然后将用户重新定向到 example6_8_input.jsp。如果用户猜成功了,将"猜对了"存放到用户的 session 中,然后将用户重新定向到 example6_8_input.jsp。

例 6_8

➢ JSP 页面

example6_8.jsp(效果如图 6.10(a)所示)

```jsp
<%@ page contentType="text/html" %>
<%@ page pageEncoding="utf-8" %>
<HTML><body bgcolor=#ffccff>
<% session.setAttribute("message","请您猜字母");
   session.setAttribute("count","0");
   char a[] = new char[26];
   int m = 0;
   for(char c = 'a';c<='z';c++) {
       a[m] = c;
       m++;
   }
   int randomIndex = (int)(Math.random()*a.length);
   char ch = a[randomIndex];//获取一个英文字母
   session.setAttribute("savedLetter",new Character(ch)); //字母放入session
%>
<h3>访问或刷新该页面可以随机得到一个英文字母.
<br>单击超链接去猜出这个字母:<br>
<a href="example6_8_input.jsp">去猜字母</a>
</h3></body></HTML>
```

example6_8_input.jsp(效果如图 6.10(b)(c)(d)所示)

```jsp
<%@ page contentType="text/html" %>
<%@ page pageEncoding="utf-8" %>
<style>
   #tom{
       font-family:宋体;font-size:26;color:blue
   }
</style>
<%   String message = (String)session.getAttribute("message");        //会话中的信息
%>
<HTML><body bgcolor=#ffccff>
<table border=1>
<form action="guess" method=post>
<tr><td id=tom>输入您的猜测(a~z之间的字母):</td>
<td><input type=text name=clientGuessLetter id=tom size=10/>
    <input type=submit id=tom value="提交"/></td>
</tr><td id=tom>提示信息:</td>
       <td id=tom><%= message %></td>
</form>
<form action="example6_8.jsp" method=post>
  <tr><td id=tom>单击按钮重新开始:</td>
  <td id=tom><input type=submit id=tom value="随机得到一个字母"/></td>
  </tr>
</form>
</body></HTML>
```

(a) 随机得到一个字母　　　　　　　　　(b) 猜小了

　　　　(c) 猜大了　　　　　　　　　　　　(d) 猜对了

图 6.10　猜字母

> **Servlet 类**

用命令行进入 moon\sun 的父目录 classes，编译 Example6_8_Servlet.java（约定见 6.1.1 节所示）：

```
classes> javac -cp servlet-api.jar moon\sun\Example6_8_Servlet.java
```

Example6_8_Servlet.java（负责创建 servlet，无显式运行效果）

```
package moon.sun;
import java.io.*;
import javax.servlet.*;
import javax.servlet.http.*;
public class Example6_8_Servlet extends HttpServlet {
    public void init(ServletConfig config) throws ServletException {
        super.init(config);
    }
    public void doPost(HttpServletRequest request,
        HttpServletResponse response) throws ServletException,IOException {
        request.setCharacterEncoding("utf-8");
        response.setContentType("text/html;charset=utf-8");
        HttpSession session = request.getSession(true);          //获取客户的会话对象
        String str = request.getParameter("clientGuessLetter");
        Character guessLetter = str.trim().charAt(0);             //获取客户猜测所提交的字母
        //获得曾放入 session 中的字母
        String count = (String)session.getAttribute("count");
        int n = Integer.parseInt(count);
        Character savedLetter =
        (Character)session.getAttribute("savedLetter");
        char realLetter = savedLetter.charValue();
        if(Character.isUpperCase(guessLetter)) {
            guessLetter = Character.toLowerCase(guessLetter);
        }
        if(guessLetter < realLetter) {
            n++;
            session.setAttribute("message","第"+ n +"次,猜小了");
            session.setAttribute("count",""+ n);
```

```java
                response.sendRedirect("example6_8_input.jsp");
            }
            else if(guessLetter > realLetter) {
                n++;
                session.setAttribute("message","第" + n + "次,猜大了");
                session.setAttribute("count",""+ n);
                response.sendRedirect("example6_8_input.jsp");
            }
            else if(guessLetter == realLetter){
                n++;
                session.setAttribute("message","猜对了,猜的次数:" + n);
                session.setAttribute("count","0");
                response.sendRedirect("example6_8_input.jsp");
            }
    }
    public   void   doGet(HttpServletRequest request,
        HttpServletResponse response) throws ServletException,IOException {
        doPost(request,response);
    }
}
```

> **web.xml 文件**

向 ch6\WEB\INF\ 下的部署文件 web.xml 添加如下的 servlet 和 servlet-mapping 标记（知识点见 6.1.2 节），部署的 servlet 的名字是 guess，访问 servlet 的 url-pattern 是 /guess。

web.xml

```xml
<?xml version = "1.0" encoding = "utf-8"?>
<web-app>
  <!-- 以下是 web.xml 文件新添加的内容 -->
    <servlet>
        <servlet-name>guess</servlet-name>
        <servlet-class>moon.sun.Example6_8_Servlet</servlet-class>
    </servlet>
    <servlet-mapping>
        <servlet-name>guess</servlet-name>
        <url-pattern>/guess</url-pattern>
    </servlet-mapping>
</web-app>
```

6.8 上机实验

提供了详细的实验步骤要求,按步骤完成,提升学习效果,积累经验,不断提高 Web 设计能力。

视频讲解

▶ 6.8.1 实验1 绘制多边形数

❶ 实验目的

掌握在 JSP 页面中,使用 form 表单请求 servlet 绘制图形。

❷ 实验要求

（1）编写 inputVertex.jsp，该页面提供一个 form 表单，该 form 表单提供一个 textArea 输入区，用于用户输入多边形的顶点（例如（12,34）(5,10)(100,89)），用户单击 submit 提交键请求名字是 drawPolygon 的 servlet。

（2）编写创建 servlet 的 Servlet 类，该类创建的 servlet 可以绘制多边形。

（3）在 Tomcat 服务器的 webapps 目录下（例如 D:\apache-tomcat-9.0.26\webapps）新建一个名字是 ch6_practice_one 的 Web 服务目录。把 JSP 页面都保存到 ch6_practice_one 目录中。在 ch6_practice_one 目录下建立子目录 WEB-INF（字母大写），然后在 WEB-INF 目录下再建立子目录 classes，将创建 servlet 的类的 Java 源文件按照包名保存在 classes 的相应子目录中（见 6.1.1 节）。

（4）用命令行进入目录 classes，编译 Servlet 源文件（约定见 6.1.1 节）：

classes> javac -cp servlet-api.jar 包名路径\Servlet 源文件

（5）向 ch6_practice_one\WEB\INF\下的部署文件 web.xml 添加 servlet 和 servlet-mapping 标记（知识点见 6.1.2 节），部署的 servlet 的名字是 drawPolygon，访问 servlet 的 url-pattern 是/drawPolygon。

（6）用浏览器访问 JSP 页面 inputVertex.jsp。

❸ 参考代码

参考代码运行效果如图 6.11 所示。

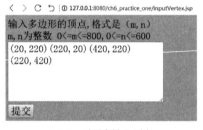

(a) JSP 页面中输入顶点　　　　(b) servlet 绘制多边形

图 6.11　请求 servlet 绘制多边形

➢ **JSP 页面**

inputVertex.jsp（效果如图 6.11(a)所示）

```
<%@ page contentType="text/html" %>
<%@ page pageEncoding="utf-8" %>
<style>
    #tom{
        font-family:宋体;font-size:26;color:blue
    }
</style>
<%
String s = "(20,220)(220,20)(420,220)(220,420)";
%>
<HTML><body bgcolor=#ffccff>
<p id=tom>
```

```
<form action = "drawPolygon" id = tom method = post>
输入多边形的顶点,格式是(m,n)<br>m,n 为整数 0<=m<=800,0<=n<=600<br>
<textArea name = "polygonVertex" id = tom rows = 5 cols = 30>
<%= s %>
</textArea>
<br><input type = submit id = tom value = "提交"/>
</form>
</p></body></HTML>
```

> **Servlet 类**

用命令行进入 moon\sun 的父目录 classes, 编译 DrawPolygon_Servlet.java(约定见 6.1.1 节):

```
classes > javac - cp servlet - api.jar moon\sun\DrawPolygon_Servlet.java
```

DrawPolygon_Servlet.java(负责创建 servlet, 运行效果如图 6.11(b)所示)

```java
package moon.sun;
import java.io.*;
import javax.servlet.*;
import javax.servlet.http.*;
import java.awt.image.BufferedImage;
import java.awt.*;
import java.util.regex.Pattern;
import java.util.regex.Matcher;
import java.awt.geom.*;
import javax.imageio.ImageIO;
public class DrawPolygon_Servlet extends HttpServlet{
    HttpServletRequest request;
    HttpServletResponse response;
    public void init(ServletConfig config) throws ServletException{
        super.init(config);
    }
    public void service(HttpServletRequest request,
                        HttpServletResponse response) throws IOException{
        request.setCharacterEncoding("utf-8");
        String polygonVertex = request.getParameter("polygonVertex");
        if(polygonVertex == null||polygonVertex.length() == 0){
            response.sendRedirect("inputVertex.jsp");      //重定向到输入数据页面
            return;
        }
        response.setContentType("image/jpeg");
        Polygon   polygon = getPolygon(polygonVertex);
        BufferedImage image = getImage(polygon);
        OutputStream outClient =  response.getOutputStream();
        boolean boo =  ImageIO.write(image,"jpeg",outClient);
    }
    Polygon getPolygon(String polygonVertex){            //得到多边形
        Polygon polygon = new Polygon();
        Pattern pattern;                                 //模式对象
        Matcher matcher;                                 //匹配对象
        String regex = "(\\d+[,,]+\\d+)";                //匹配顶点的正则表达式
        pattern = Pattern.compile(regex);                //初始化模式对象
```

```
        matcher = pattern.matcher(polygonVertex);              //用于检索 polygonVertex
        while(matcher.find()) {
            String str = matcher.group();
            String []vertex = str.split("[( ),, ]+");
            polygon.addPoint
            (Integer.parseInt(vertex[0]),Integer.parseInt(vertex[1]));
        }
        return polygon;
    }
    BufferedImage getImage(Shape shape){                       //得到图形的图像
        int width = 1000, height = 800;
        BufferedImage image =
        new BufferedImage(width,height,BufferedImage.TYPE_INT_RGB);   //图像
        Graphics g = image.getGraphics();
        g.fillRect(0, 0, width, height);
        Graphics2D g_2d = (Graphics2D)g;
        g_2d.setColor(Color.blue);
        g_2d.draw(shape);                                      //在图像上绘制图形
        return image;
    }
}
```

> **web.xml 文件**

在 ch6_practice_one\WEB\INF\下新建部署文件 web.xml，添加如下的 servlet 和 servlet-mapping 标记（见 6.1.2 节），部署的 servlet 的名字是 drawPolygon，访问 servlet 的 url-pattern 是/drawPolygon。

web.xml

```xml
<?xml version = "1.0" encoding = "utf-8"?>
<web-app>
    <servlet>
        <servlet-name>drawPolygon</servlet-name>
        <servlet-class>moon.sun.DrawPolygon_Servlet</servlet-class>
    </servlet>
    <servlet-mapping>
        <servlet-name>drawPolygon</servlet-name>
        <url-pattern>/drawPolygon</url-pattern>
    </servlet-mapping>
</web-app>
```

▶ 6.8.2 实验 2 双色球福利彩票

❶ 实验目的

掌握在 JSP 页面中请求 servlet 处理数据。

❷ 实验要求

（1）编写 buyLottery.jsp，用户在该页面输入双色球彩票的 6 个红球和一个蓝球的号码，代表用户购买的彩票，然后单击 form 表单的提交键，请求名字是 lottery 的 servlet。

（2）编写创建 servlet 的 Servlet 类，该类创建的 servlet 可以判断用户的中奖情况。

（3）在 Tomcat 服务器的 webapps 目录下（例如 D:\apache-tomcat-9.0.26\webapps）新

建一个名字是 ch6_practice_two 的 Web 服务目录。把 JSP 页面都保存到 ch6_practice_two 目录中。在 ch6_practice_two 目录下建立子目录 WEB-INF(字母大写),然后在 WEB-INF 目录下再建立子目录 classes。将创建 servlet 的类的 Java 源文件保存在 classes 的相应子目录中(见 6.1.1 节)。

(4) 用命令行进入目录 classes,编译 Servlet 源文件(约定见 6.1.1 节):

classes> javac -cp servlet-api.jar 包名路径\Servlet 源文件

(5) 向 ch6_practice_two\WEB\INF\下的部署文件 web.xml 添加 servlet 和 servlet-mapping 标记(见 6.1.2 节),部署的 servlet 的名字是 lottery,访问 servlet 的 url-pattern 是/lottery。

(6) 用浏览器访问 JSP 页面 buyLottery.jsp。

❸ 参考代码

参考代码运行效果如图 6.12 所示。

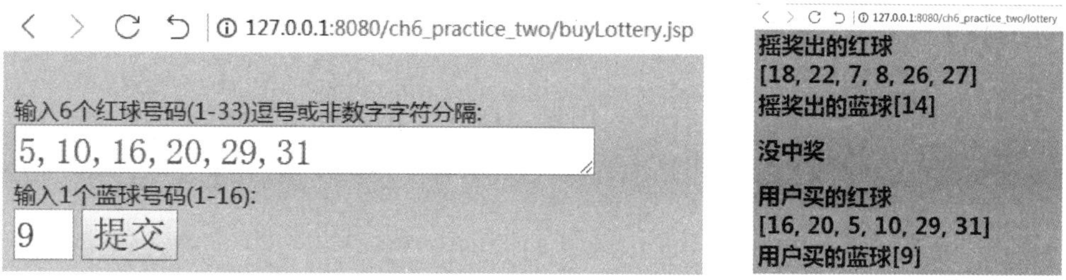

(a) 在JSP页面选择号码　　　　　　　　　(b) servlet判断中奖情况

图 6.12　双色球福利彩票

➢ **JSP 页面**

buyLottery.jsp(效果如图 6.12(a)所示)

```
<%@ page contentType="text/html" %>
<%@ page pageEncoding="utf-8" %>
<style>
   #red{
      font-family:宋体;font-size:26;color:red
   }
</style>
<style>
   #blue{
      font-family:宋体;font-size:26;color:blue
   }
</style>
<HTML><body bgcolor=#ffccff>
<form action="lottery"  method=post>
<br>输入 6 个红球号码(1-33)逗号或非数字字符分隔:<br>
<textArea name="digitRedball" id=red rows=1 cols=30></textArea>
<br>输入 1 个蓝球号码(1-16):<br>
<input type=text name="digitBlueball" id=blue maxlength=2 size=2/>
<input type=submit id=blue value="提交"/>
</form>
</body></HTML>
```

➤ Servlet 类

用命令行进入 moon\sun 的父目录 classes，编译 Lottery_Servlet.java（约定见 6.1.1 节）：

```
classes> javac -cp servlet-api.jar moon\sun\Lottery_Servlet.java
```

Lottery_Servlet.java（负责创建 servlet，运行效果如图 6.12(b) 所示）

```java
package moon.sun;
import java.io.*;
import javax.servlet.*;
import javax.servlet.http.*;
import java.util.regex.Pattern;
import java.util.regex.Matcher;
import java.util.*;
public class Lottery_Servlet extends HttpServlet{
    public void init(ServletConfig config) throws ServletException{
        super.init(config);
    }
    public void service(HttpServletRequest request,
                        HttpServletResponse response) throws IOException{
        request.setCharacterEncoding("utf-8");
        response.setContentType("text/html;charset=utf-8");
        PrintWriter out = response.getWriter();
        out.println("<html><body bgcolor=cyan>");
        String digitRedball = request.getParameter("digitRedball");
        String digitBlueball = request.getParameter("digitBlueball");
        if(digitRedball.length() == 0 ||digitBlueball.length() == 0) {
            response.sendRedirect("buyLottery.jsp");      //重定向到输入数据页面
            return;
        }
        //用户买的全部红、蓝球号码
        HashSet<Integer> userRedball = getUserDigit(digitRedball);
        HashSet<Integer> userBlueball = getUserDigit(digitBlueball);
        if(userRedball.size()!= 6||userBlueball.size()!= 1){    //红球 6 个，蓝球 1 个
            response.sendRedirect("buyLottery.jsp");       //否则重新输入
            return;
        }
        HashSet<Integer> drawLotteryRedball = drawLottery(6,33);    //摇奖红球
        HashSet<Integer> drawLotteryBlueball = drawLottery(1,16);    //摇奖蓝球
        out.print("<h1>摇奖出的红球 br>" + drawLotteryRedball.toString() + "<br>");
        out.print("摇奖出的蓝球" + drawLotteryBlueball.toString() + "</h1>");
        drawLotteryRedball.removeAll(userRedball);        //摇奖红球去除用户红球
        drawLotteryBlueball.removeAll(userBlueball);       //摇奖蓝球去除用户蓝球
        int leftRedball = drawLotteryRedball.size();       //剩余的红球数量
        int leftBlueball = drawLotteryBlueball.size();      //剩余的蓝球数量
        if(leftRedball == 0&&leftBlueball == 0){         //为了减少代码,减少了中奖分类
            out.println("<h1>头奖</h1>");
        }
        else if(leftRedball == 0&&leftBlueball == 1){
            out.println("<h1>二等奖</h1>");
        }
        else if(leftRedball == 1&&leftBlueball == 0){
```

```java
            out.println("<h1>三等奖</h1>");
        }
        else{
            out.println("<h1>没中奖</h1>");
        }
        out.print("<h1>用户买的红球<br>" + userRedball.toString() + "<br>");
        out.print("用户买的蓝球" + userBlueball.toString() + "</h1>");
    }
    public HashSet<Integer> getUserDigit(String input){
        HashSet<Integer> set = new HashSet<Integer>();
        Pattern pattern;                                    //模式对象
        Matcher matcher;                                    //匹配对象
        String regex = "[0-9][0-9]*";                       //匹配正整数的正则表达式
        pattern = Pattern.compile(regex);                   //初始化模式对象
        matcher = pattern.matcher(input);                   //初始化匹配对象,用于检索input
        double sum = 0;
        while(matcher.find()) {
            String str = matcher.group();
            set.add(Integer.parseInt(str));                 //用户买的彩票号码放入集合
        }
        return set;
    }
    public HashSet<Integer> drawLottery(int count, int allNumber){
        LinkedList<Integer> saveNumber = new LinkedList<Integer>();   //存放数
        HashSet<Integer> set = new HashSet<Integer>();
        for(int i = 1; i <= allNumber; i++) {               //1 到 allNumber(球号)
            saveNumber.add(i);                              //顺序存入链表 saveNumber
        }
        Random random = new Random();
        while( count > 0) {
            int index = random.nextInt(saveNumber.size());
            int number = saveNumber.remove(index);          //抽取一个球不放回
            count--;
            set.add(number);                                //开奖号码放入集合
        }
        return  set;
    }
}
```

> **web.xml 文件**

在 ch6_practice_two\WEB\INF\下新建部署文件 web.xml,添加如下的 servlet 和 servlet-mapping 标记(知识点见 6.1.2 节),部署的 servlet 的名字是 lottery,访问 servlet 的 url-pattern 是/lottery。

web.xml

```xml
<?xml version="1.0" encoding="utf-8"?>
<web-app>
    <servlet>
        <servlet-name>lottery</servlet-name>
        <servlet-class>moon.sun.Lottery_Servlet</servlet-class>
    </servlet>
```

```
        <servlet - mapping>
            <servlet - name>lottery</servlet - name>
            <url - pattern>/lottery</url - pattern>
        </servlet - mapping>
</web - app>
```

6.8.3 实验3 分析整数

❶ 实验目的

掌握使用 forward 方法进行转发。

❷ 实验要求

(1) 编写 inputIntegers.jsp，该页面的 form 表单提供一个 textArea 输入区，用户在输入区输入若干个整数，然后单击名字含有 personOne 或 personTwo 的提交键提交给负责转发的名字是 handleForward 的 servlet。

(2) 编写创建 handleForward 的 Servlet 类。如果用户单击的是名字含有 personOne 的提交键，handleForward 就将用户的请求转发到名字是 personOne 的 servlet，如果单击的是名字含有 personTwo 的提交键，handleForward 就将用户的请求转发到名字是 personTwo 的 servlet。如果用户没有输入任何数字，handleForward 就将用户重定向到 inputIntegers.jsp。

(3) 编写创建 personOne 的 Servlet 类和 personTwo 的 Servlet 类。personOne 分析用户提交的整数中哪些是奇数哪些是偶数，personTwo 把用户提交的整数进行归类，比如按 3 求余数进行归类。

(4) 在 Tomcat 服务器的 webapps 目录下（例如 D:\apache-tomcat-9.0.26\webapps）新建一个名字是 ch6_practice_three 的 Web 服务目录。把 JSP 页面都保存到 ch6_practice_three 目录中。在 ch6_practice_three 目录下建立子目录 WEB-INF（字母大写），然后在 WEB-INF 目录下再建立子目录 classes。将创建 servlet 的类的 Java 源文件保存在 classes 的相应子目录中（见 6.1.1 节）。

(5) 用命令行进入目录 classes，编译 Servlet 源文件（约定见 6.1.1 节）：

```
classes> javac - cp servlet - api.jar 包名路径\Servlet 源文件
```

(6) 向 ch6_practice_three\WEB\INF\下的部署文件 web.xml 添加 servlet 和 servlet-mapping 标记（见 6.1.2 节），部署创建的 3 个 servlet 的名字分别是 handleForward、personOne 和 personTwo，访问的 url-pattern 分别是/handleForward、/personOne 和/personTwo。

(7) 用浏览器访问 JSP 页面 inputIntegers.jsp。

❸ 参考代码

参考代码运行效果如图 6.13 所示。

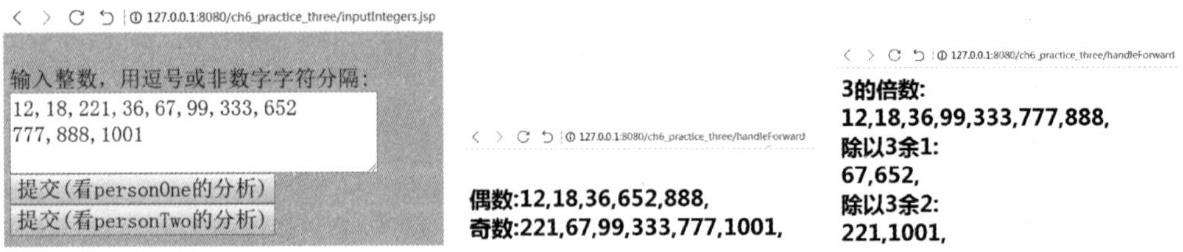

(a) 输入整数　　　(b) personOne 分析整数　　　(c) personTwo 分析整数

图 6.13 使用 forward 实现转发

第6章 Java Servlet基础

➢ JSP 页面

inputIntegers.jsp（效果如图 6.13(a)所示）

```
<%@ page contentType = "text/html" %>
<%@ page pageEncoding = "utf-8" %>
<style>
   #tom{
       font-family:宋体;font-size:26;color:blue
   }
</style>
<HTML><body bgcolor = #ffccff>
<p id = tom>
<form action = "handleForward" id = tom  method = post>
<br>输入整数,用逗号或非数字字符分隔:<br>
<textArea name = "number" id = tom rows = 3 cols = 32></textArea><br>
<input type = submit name = "submit" id = tom  value = "提交(看 personOne 分析)"/><br>
<input type = submit name = "submit" id = tom  value = "提交(看 personTwo 的分析)"/>
</form>
</p></body></HTML>
```

➢ Servlet 类

用命令行进入 moon\sun 的父目录 classes，编译 3 个 Servlet 源文件（约定见 6.1.1 节）：

```
classes> javac -cp servlet-api.jar moon\sun\HandleForward_Servlet.java
classes> javac -cp servlet-api.jar moon\sun\PersonOne_Servlet_Sum.java
classes> javac -cp servlet-api.jar moon\sun\PersonTwo_Servlet_Sort.java
```

HandleForward.java（负责创建 servlet，servlet 负责转发，无显式效果）

```java
package moon.sun;
import java.io.*;
import javax.servlet.*;
import javax.servlet.http.*;
public class HandleForward_Servlet extends HttpServlet{
    public void init(ServletConfig config) throws ServletException{
        super.init(config);
    }
    public void service(HttpServletRequest request,
    HttpServletResponse response) throws ServletException,IOException{
        RequestDispatcher dispatcher = null;                    //负责转发的对象
        request.setCharacterEncoding("utf-8");
        response.setContentType("text/html;charset = utf-8");
        String mess = request.getParameter("submit");
        String integers = request.getParameter("number");
        if(integers == null ||integers.length() == 0) {
            response.sendRedirect("inputIntegers.jsp");         //重定向到输入数据页面
            return;
        }
        if(mess.contains("personOne")) {
            dispatcher = request.getRequestDispatcher("personOne");   //转发
            dispatcher.forward(request,response);
        }
```

```
            else if(mess.contains("personTwo")){
                dispatcher = request.getRequestDispatcher("personTwo");    //转发
                dispatcher.forward(request,response);
            }
        }
}
```

PersonOne_Servlet.java(负责创建 servlet，servlet 运行效果如图 6.13(b)所示)

```
package moon.sun;
import java.io.*;
import javax.servlet.*;
import javax.servlet.http.*;
import java.util.regex.Pattern;
import java.util.regex.Matcher;
public class PersonOne_Servlet extends HttpServlet{
    public void init(ServletConfig config) throws ServletException{
        super.init(config);
    }
    public void service(HttpServletRequest request,
        HttpServletResponse response)throws ServletException,IOException{
        request.setCharacterEncoding("utf-8");
        response.setContentType("text/html;charset=utf-8");
        PrintWriter out = response.getWriter();
        String integers = request.getParameter("number");
        String backMess = analysisIntegers(integers);
        out.print("<html><body><h1>" + backMess + "</body></html>");
    }
    public String analysisIntegers(String input){
        Pattern pattern;                                    //模式对象
        Matcher matcher;                                    //匹配对象
        String regex = "-?[0-9][0-9]*";                     //匹配整数的正则表达式
        pattern = Pattern.compile(regex);                   //初始化模式对象
        matcher = pattern.matcher(input);                   //初始化匹配对象,用于检索 input
        StringBuffer evenNumbers = new StringBuffer("<br>偶数:");
        StringBuffer oddNumbers = new StringBuffer("<br>奇数:");
        while(matcher.find()) {
            String str = matcher.group();
            if(Integer.parseInt(str) % 2 == 0){
                evenNumbers.append(str + ",");
            }
            else {
                oddNumbers.append(str + ",");
            }
        }
        StringBuffer buffer = evenNumbers.append(oddNumbers);
        return new String(buffer);
    }
}
```

PersonTwo_Servlet.java(负责创建 servlet,servlet 运行效果如图 6.13(c)所示)

```java
package moon.sun;
import java.io.*;
import javax.servlet.*;
import javax.servlet.http.*;
import java.util.regex.Pattern;
import java.util.regex.Matcher;
public class PersonTwo_Servlet extends HttpServlet{
    public void init(ServletConfig config) throws ServletException{
        super.init(config);
    }
    public void service(HttpServletRequest request,
        HttpServletResponse response) throws ServletException,IOException{
        request.setCharacterEncoding("utf-8");
        response.setContentType("text/html;charset=utf-8");
        PrintWriter out = response.getWriter();
        String integers = request.getParameter("number");
        String backMess = analysisIntegers(integers);
        out.print("<html><body><h1>" + backMess + "</body></html>");
    }
    public String analysisIntegers(String input){
        Pattern pattern;                              //模式对象
        Matcher matcher;                              //匹配对象
        String regex = "-?[0-9][0-9]*";               //匹配整数的正则表达式
        pattern = Pattern.compile(regex);             //初始化模式对象
        matcher = pattern.matcher(input);             //初始化匹配对象,用于检索 input
        StringBuffer numbersZero = new StringBuffer("3 的倍数:<br>");
        StringBuffer numbersOne = new StringBuffer("<br>除以 3 余 1:<br>");
        StringBuffer numbersTwo = new StringBuffer("<br>除以 3 余 2:<br>");
        while(matcher.find()) {
            String str = matcher.group();
            if(Integer.parseInt(str) % 3 == 0){
                numbersZero.append(str + ",");
            }
            else if(Integer.parseInt(str) % 3 == 1){
                numbersOne.append(str + ",");
            }
            else if(Integer.parseInt(str) % 3 == 2){
                numbersTwo.append(str + ",");
            }
        }
        StringBuffer buffer =
            numbersZero.append(numbersOne.append(numbersTwo));
        return new String(buffer);
    }
}
```

➢ **web.xml 文件**

向 ch6_practice_three\WEB\INF\下的部署文件 web.xml 添加如下的 servlet 和 servlet-mapping 标记(见 6.1.2 节),部署的 3 个 servlet 的名字分别是 handleForward、personOne 和 personTwo,部署的 3 个 servlet 的 url-pattern 分别是/handleForward、/personOne 和/personTwo。

web.xml

```xml
<?xml version="1.0" encoding="utf-8"?>
<web-app>
  <!-- 以下是 web.xml 文件新添加的内容 -->
  <servlet>
      <servlet-name>handleForward</servlet-name>
      <servlet-class>moon.sun.HandleForward_Servlet</servlet-class>
  </servlet>
  <servlet-mapping>
      <servlet-name>handleForward</servlet-name>
      <url-pattern>/handleForward</url-pattern>
  </servlet-mapping>
  <servlet>
      <servlet-name>personOne</servlet-name>
      <servlet-class>moon.sun.PersonOne_Servlet</servlet-class>
  </servlet>
  <servlet-mapping>
      <servlet-name>personOne</servlet-name>
      <url-pattern>/personOne</url-pattern>
  </servlet-mapping>
  <servlet>
      <servlet-name>personTwo</servlet-name>
      <servlet-class>moon.sun.PErsonTwo_Servlet</servlet-class>
  </servlet>
  <servlet-mapping>
      <servlet-name>personTwo</servlet-name>
      <url-pattern>/personTwo</url-pattern>
  </servlet-mapping>
</web-app>
```

6.9 小结

- Java Servlet 的核心是在服务器端创建响应用户请求的对象，即创建 servlet 对象。
- servlet 对象第一次被请求加载时，服务器创建一个 servlet 对象，这个对象调用 init 方法完成必要的初始化工作。init 方法只被 servlet 对象调用一次，当后续的客户请求该 servlet 对象服务时，服务器将启动一个新的线程，在该线程中，servlet 对象调用 service 方法响应客户的请求。每个客户的每次请求都导致 service 方法被调用执行，调用过程运行在不同的线程中，互不干扰。
- Servlet 类继承的 service 方法检查 HTTP 的请求类型（get、post 等），并在 service 方法中根据用户的请求方式，对应地调用 doGet 或 doPost 方法。因此，Servlet 类不必重写 service 方法，直接继承该方法即可。可以在 Servlet 类中重写 doPost 或 doGet 方法来响应用户的请求。
- RequestDispatcher 对象可以把用户对当前 JSP 页面或 servlet 的请求转发给另一个 JSP 页面或 servlet，而且将用户对当前 JSP 页面或 servlet 的请求和响应传递给转发到的 JSP 页面或 servlet。也就是说，当前页面所要转发到的目标页面或 servlet 对象可以使用 request 获取用户提交的数据。

习题 6

1. 假设 Web 服务目录 mymoon 中的 JSP 页面要使用一个 servlet，该 servlet 的包名为 blue.sky。请说明，应当怎样保存 servlet 的字节码文件。

2. 假设 Web 服务目录是 mymoon，star 是 mymoon 的一个子目录，JSP 页面 a.jsp 保存在 star 中，a.jsp 准备请求一个 servlet，该 servlet 的包名为 tom.jiafei。下列哪个叙述是正确的？

    A. 创建 servlet 的字节码文件保存在 \mymoon\WEB-INF\classes\tom\jiafei 中

    B. 创建 servlet 的字节码文件保存在 \mymoon\star\WEB-INF\classes\tom\jiafei 中

    C. 创建 servlet 的字节码文件保存在 \mymoon\WEB-INF\star\classes\tom\jiafei 中

    D. 创建 servlet 的字节码文件保存在 \mymoon\WEB-INF\classes\start\tom\jiafei 中

3. 假设 Web 服务目录是 mymoon，star 是 mymoon 的一个子目录，JSP 页面 a.jsp 保存在 star 中，a.jsp 准备请求一个 servlet，该 servlet 的包名为 tom.jiafei。下列哪个叙述是正确的？

    A. web.xml 文件保存在 \mymoon\WEB-INF\classes 中

    B. web.xml 文件保存在 \mymoon\WEB-INF\ 中

    C. web.xml 文件保存在 \mymoon\WEB-INF\star\ 中

    D. web.xml 文件保存在 \mymoon\star\WEB-INF\ 中

4. servlet 对象是驻留在服务器端，还是在客户端被创建？

5. servlet 对象被创建后将首先调用 init 方法还是 service 方法？

6. "servlet 第一次被请求加载时调用 init 方法。当后续的客户请求 servlet 对象时，servlet 对象不再调用 init 方法"，这样的说法是否正确？

7. servlet 第一次被请求加载后，当后续的客户请求 servlet 对象时，下列哪个叙述是正确的？

    A. servlet 调用 service 方法   B. servlet 调用 init 方法

    C. servlet 调用 doPost 方法   D. servlet 调用 doGet 方法

8. 假设创建 servlet 的类是 tom.jiafei.Dalian，创建的 servlet 对象的名字是 myservlet，应当怎样配置 web.xml 文件？

9. 如果 Servlet 类不重写 service 方法，那么应当重写哪两个方法？

10. HttpServletResponse 类的 sendRedirect 方法和 RequestDispatcher 类的 forward 方法有何不同？

11. servlet 对象怎样获得用户的会话对象？

12. 编写 inputCircle.jsp，页面提供 form 表单，该 form 表单提供两个 text 文本框，用于用户输入圆的圆心（例如（12,34））和圆的半径，用户单击 submit 提交键请求名字是 drawCircle 的 servlet。编写创建 servlet 的 Servlet 类，该类创建的 servlet 可以绘制圆。

第 7 章 MVC 模式

本章导读

    主要内容
- MVC 介绍
- JSP 中的 MVC 模式
- 模型的生命周期与视图更新
- MVC 模式的简单实例

    难点
- 模型的生命周期与视图更新

    关键实践
- 等差、等比级数和
- 点餐

本章将介绍 MVC 模式，MVC 模式的核心思想是将"模型""视图"和"控制器"进行有效组合。掌握该模式对于设计合理的 Web 应用以及学习使用某些流行的 Web 框架，如 Spring、Struts 等，有着十分重要的意义。

本章在 Tomcat 安装目录的 webapps 目录下建立 Web 服务目录 ch7。另外，须在 Web 服务目录 ch7 下建立目录结构 ch7\WEB-INF\classes(WEB-INF 字母大写)。

本章使用的 bean 的包名(除非特别说明)均为 save.data。在 classes 下建立目录结构 \save\data，创建 bean 的类的字节码文件均保存在\WEB-INF\classes\save\data 中，如图 7.1 所示(知识点见 5.1.2 节)。本章的 servlet 的包名(除非特别说明)均为 handle.data。在 classes 下建立目录结构\handle\data，创建 servlet 的类的字节码文件均保存在\WEB-INF\classes\handle\data 中，如图 7.1 所示。知识点见 6.1.1 节。

图 7.1 bean 和 servlet 的存放位置

本章使用的 javax.servlet 和 javax.servlet.http 包中的类不在 JDK 提供的核心类库中，为了方便编译 Java 源文件，请事先将 Tomcat 安装目录 lib 子目录中的 servlet-api.jar 文件复制(不要剪贴)到\ch7\WEB-INF\classes\中(见 6.1.1 节)。另外，保存 Java 源文件时，"保存类型"选择为"所有文件"，将"编码"选择为"ANSI"。保存 JSP 文件和部署文件 web.xml 时，"保存类型"选择为"所有文件"，将"编码"选择为"UTF-8"。

第 7 章 MVC 模式

7.1 MVC 模式介绍

视频讲解

模型-视图-控制器(Model-View-Controller),简称为 MVC。MVC 已经成为软件设计者必须熟练使用的开发模式。本章必须理解、掌握在 JSP 程序设计中怎样具体体现 MVC 开发模式(其他语言的程序设计是非常类似的,仅仅是具体使用的 API 不同而已)。

MVC 是一种通过三部分构造一个软件或组件的理想办法。
- 模型(model):用于存储数据的对象。
- 视图(view):向控制器提交所需数据、显示模型中的数据。
- 控制器(controller):负责具体的业务逻辑操作,即控制器根据视图提出的要求对数据做出(商业)处理,将有关结果存储到模型中,并负责让模型和视图进行必要的交互,当模型中的数据变化时,让视图更新显示。

从面向对象的角度看,MVC 开发模式可以使程序容易维护,也更容易扩展。在设计程序时,可以将某个对象看作"模型",然后为"模型"提供恰当的显示组件,即"视图"。在 MVC 模式中,"视图""模型"和"控制器"之间是松耦合结构,便于系统的维护和扩展。

7.2 JSP 中的 MVC 模式

视频讲解

目前,随着软件规模的扩大,MVC 模式正在被运用到各种应用程序的设计中。那么在 JSP Web 程序设计中,MVC 模式是怎样具体体现的呢?

我们已经知道,JSP 页面擅长数据的显示,即适合作为用户的视图,例如,JSP 页面里可以有 HTML 标记、JavaScript、CSS、Java 表达式等。本教材侧重 JSP 本身的重点内容,即服务器端,让 JSP 页面代码尽量简明,所以没有在 JSP 页面(客户端)使用 JavaScript、CSS(熟悉这些内容,可以美化客户端,见 2.1 节的说明)。在 JSP 页面中可以使用 HTML 标记、JSP 指令(例如 getProperty 指令)或 Java 程序片、Java 表达式来为用户显示数据,避免使用大量的 Java 程序片来进行数据的逻辑处理(简明扼要的除外)。servlet 擅长数据的处理,应当尽量避免在 servlet 中使用 out 流输出大量的 HTML 标记来显示数据,否则一旦要修改显示外观就要重新编译 servlet。

通过前面的学习,特别是在学习了第 5 章后,已经体会到一些小型的 Web 应用可以使用 JSP 页面调用 JavaBean 完成数据的处理,实现代码复用。在 JSP+JavaBean 模式中,JavaBean 不仅要提供修改和返回数据的方法,而且要经常参与数据的处理。当 Web 应用变得复杂时,我们希望 JavaBean 仅仅负责提供修改和返回数据的方法即可,不必参与数据的具体处理,而是把数据的处理交给称作控制器的 servlet 对象去完成,即 servlet 控制器负责处理数据,并将有关的结果存储到 JavaBean 中,实现存储与处理的分离。负责视图功能的 JSP 页面可以使用 Java 程序片或用 JavaBean 标记显示 JavaBean 中的数据。

在 JSP 中,MVC 模式的实现如图 7.2 所示,具体实现如下:
- 模型(Model):一个或多个 JavaBean 对象,用于存储数据。JavaBean 主要提供简单的 setXxx 方法和 getXxx 方法,在这些方法中不涉及对数据的具体处理细节,以便增强模型的通用性。

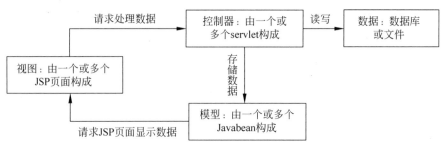

图 7.2 JSP 中的 MVC 模式

- 视图(View)：一个或多个 JSP 页面，其作用是向控制器提交必要的数据和显示数据。JSP 页面可以使用 HTML 标记、JavaBean 标记以及 Java 程序片或 Java 表达式来显示数据。视图的主要工作就是显示数据，对数据的逻辑操作由控制器负责。
- 控制器(Controller)：一个或多个 servlet 对象，根据视图提交的要求进行数据处理操作，并将有关的结果存储到 JavaBean 中，然后 servlet 使用转发或重定向的方式请求视图中的某个 JSP 页面显示数据。例如让某个 JSP 页面通过使用 JavaBean 标记、Java 程序片或 Java 表达式显示控制器存储在 JavaBean 中的数据。

7.3 模型的生命周期与视图更新

视频讲解

使用 MVC 模式和前面学习的 JSP＋JavaBean 模式有很大的不同。在 JSP＋JavaBean 模式中，由 JSP 页面通过使用 useBean 标记：

`< jsp:useBean id = "名字" class = "创建 bean 的类" scope = "生命周期"/>`

创建 bean。而在 MVC 模式中，由控制器 servlet 创建 bean，并将有关数据存储到所创建的 bean 中，然后 servlet 请求某个 JSP 页面使用 getProperty 动作标记：

`< jsp:getProperty name = "名字" property = "bean 的属性"/>`

显示 bean 中的数据。

在 MVC 模式中，当用控制器 servlet 创建 bean 时，就可以使用 bean 类的带参数的构造方法。类中的方法的命名继续保留 getXxx 规则，但可以不遵守 setXxx 规则(有关规则细节见 5.1.1 节)。其理由是：我们不希望 JSP 页面修改 JavaBean 中的数据，只需要它显示 bean 中的数据。

在 MVC 模式中，servlet 创建的 bean 也涉及生命周期(有关 bean 的生命周期见 5.1.3 节)。生命周期分为 request bean、sessionbean 和 application bean。

▶ 7.3.1 request bean

❶ bean 的创建

servlet 创建 request bean 的步骤如下：

(1) 用 BeanClass 类的某个构造方法创建 bean 对象，例如：

`BeanClass bean = new BeanClass();`

(2) 将所创建的 bean 对象存放到 HttpServletRequest 对象 request 中,并指定查找该 bean 的 id。该步骤决定了 bean 为 request bean。例如:

```
request.setAttribute("keyWord",bean);
```

执行上述操作,就会把 bean 存放到 Tomcat 服务器管理的内置对象 pageContext 中,该 bean 被指定的 id 是 keyWord,生命周期是 PageContext.REQUEST_SCOPE(request)。

❷ 视图更新

在 MVC 模式中,由 servlet(控制器)负责根据模型中数据的变化通知 JSP 页面(视图)更新,其手段是使用转发,即使用 RequestDispatcher 对象向某个 JSP 页面发出请求,让所请求的 JSP 页面显示 bean(模型)中的数据(不能使用重定向,即不能用 sendRedirect 方法)。

因为 servlet 创建 bean 的步骤(2)决定了 bean 为 request bean,因此,当 servlet 使用 RequestDispatcher 对象向某个 JSP 页面发出请求时(进行转发操作),该 request bean 只对 servlet 所请求的 JSP 页面有效,该 JSP 页面对请求作出响应之后,request bean 所占有的内存被释放,结束自己的生命。

servlet 请求一个 JSP 页面,例如 show.jsp 的代码如下:

```
RequestDispatcher dispatcher = request.getRequestDispatcher("show.jsp");
dispatcher.forward(request,response);
```

servlet 所请求的 JSP 页面,例如 show.jsp 页面可以使用如下标记获得 servlet 所创建的 request bean:

```
<jsp:useBean id = "keyWord" class = "save.data.BeanClass" scope = "request"/>
```

id 的值是 servlet 创建 request bean 时,为 bean 指定的关键字。然后 JSP 页面可以使用相应的标记或 Java 程序片显示该 request bean 中的数据,例如使用:

```
<jsp:getProperty name = "keyWord" property = "bean 的变量"/>
```

标记显示 request bean 中的数据。如果上述代码执行成功,用户就看到了 show.jsp 页面显示 request bean 中的数据的效果。

▶ 7.3.2 session bean

❶ bean 的创建

servlet 创建 session bean 的步骤如下:

(1) 用 BeanClass 类的某个构造方法创建 bean 对象,例如:

```
BeanClass bean = new BeanClass();
```

(2) 将所创建的 bean 对象存放到 HttpServletSession 对象 session 中,并指定查找该 bean 的 id。该步骤决定了 bean 为 session bean。例如:

```
HttpSession session = request.getSession(true);
session.setAttribute("keyWord",bean);
```

内置对象执行上述操作,就会把 bean 存放到 Tomcat 服务器管理的内置对象 pageContext

中,该 bean 被指定的 id 是 keyWord,生命周期是 PageContext.SESSION_SCOPE(session)。

❷ 视图更新

servlet 创建 bean 的步骤(2)决定了 bean 为 session bean,只要用户的 session 没有消失,该 session bean 就一直存在。Web 服务目录的各个 JSP 都可以使用

```
<jsp:useBean id="keyWord" class="save.data.BeanClass" scope="session"/>
```

标记获得 servlet 所创建的 session bean(id 的值是 servlet 创建 session bean 时,为 bean 指定的关键字),然后使用相应的标记或程序片显示该 session bean 中的数据,例如使用

```
<jsp:getProperty name="keyWord" property="bean 的变量"/>
```

标记显示该 session bean 中的数据。

对于 session bean,如果 servlet 希望某个 JSP 显示其中的数据,可以使用 RequestDispatcher 对象转发到该页面,也可以使用 HttpServletResponse 类中的重定向方法 (sendRedirect)定向到该页面。

需要注意的是,不同用户的 session bean 是互不相同的,即占有不同的内存空间。

▶ 7.3.3 application bean

❶ bean 的创建

servlet 创建 application bean 的步骤如下:

(1) 用 BeanClass 类的某个构造方法创建 bean 对象,例如:

```
BeanClass bean = new BeanClass();
```

(2) servlet 使用 getServletContext()方法返回服务器的 ServletContext 内置对象的引用,将所创建的 bean 对象存放到服务器这个 ServletContext 内置对象中,并指定查找该 bean 的关键字。该步骤决定了 bean 的生命周期为 application。例如:

```
getServletContext().setAttribute("keyWord", bean);
```

这样就会把 bean 存放到 Tomcat 服务器管理的内置对象 pageContext 中,该 bean 被指定的 id 是 keyWord,生命周期是 PageContext.APPLICATION_SCOPE(application)。

❷ 视图更新

servlet 创建 bean 的步骤(2)决定了 bean 为 application bean。当 servlet 创建 application bean 后,只要 Tomcat 服务器不关闭,该 bean 就一直存在。一个用户在访问 Web 服务目录的各个 JSP 中都可以使用

```
<jsp:useBean id="keyWord" class="save.data.BeanClass" scope="application"/>
```

标记获得 servlet 所创建的 application bean(id 的值是 servlet 创建 application bean 时为 bean 指定的关键字),然后使用相应的标记或程序片显示该 application bean 中的数据,例如使用

```
<jsp:getProperty name="keyWord" property="bean 的变量"/>
```

标记显示该 application bean 中的数据。

对于 application bean,如果 servlet 希望某个 JSP 显示其中的数据,可以使用

RequestDispatcher 对象向该 JSP 页面发出请求,也可以使用 HttpServletResponse 类中的重定向方法(sendRedirect)。

需要注意的是,所有用户在同一个 Web 服务目录中的 application bean 是相同的,即占有相同的内存空间。

7.4 MVC 模式的简单实例

视频讲解

本节结合几个简单的实例体现 MVC 三个部分的设计与实现。

▶ 7.4.1 简单的计算器

本节的例 7_1 设计一个 Web 应用,只有一个 JSP 页面 example7_1.jsp、一个 request bean 和一个 servlet。JSP 页面 example7_1.jsp 提供一个表单,用户可以通过表单输入两个数,选择运算符号提交给 servlet 控制器。bean 负责存储运算数、运算符号和运算结果,servlet 控制器负责运算,将结果存储在 request bean 中,并负责请求 JSP 页面 example7_1.jsp 显示 request bean 中的数据。

例 7_1

▶ **bean**(模型)

Example7_1_Bean.java 中的 getXxx 和 setXxx 方法可以返回和设置模型中的数据,但不参与数据的处理。

用命令行进入 save\data 的父目录 classes,编译 Example7_1_Bean.java(见本章开始的约定):

classes > javac save\data\Example7_1_Bean.java

Example7_1_Bean.java

```java
packagesave.data;
public class Example7_1_Bean {
    double numberOne,numberTwo,result;
    String operator = " + ";
    public void setNumberOne(double n){
        numberOne = n;
    }
    public double getNumberOne(){
        return numberOne;
    }
    public void setNumberTwo(double n){
        numberTwo = n;
    }
    public double getNumberTwo(){
        return numberTwo;
    }
    public void setOperator(String s){
        operator = s.trim();;
    }
    public String getOperator(){
        return operator;
    }
```

```
        public void setResult(double r){
            result = r;
        }
        public double getResult(){
            return result;
        }
    }
```

注：请读者比较 Example7_1_Bean.java 和 5.4.2 节中 ComputerBean 的不同之处,这里的 Example7_1_Bean 没有参与计算,仅仅负责存储数据。

➢ **JSP 页面(视图)**

视图部分由一个 example7_1.jsp 页面构成,该页面负责提供输入和显示数据的视图。用户可以在该页面输入参与运算的数据,然后将数据提交到名字是 computer 的 request servlet。computer 负责计算四则运算的结果,并将结果存储到 id 为 digitBean 的 bean 中,然后请求视图 example6_2.jsp 显示数据模型 bean 中的数据。

example7_1.jsp(效果如图 7.3 所示)

```
<%@ page contentType = "text/html" %>
<%@ page pageEncoding = "utf-8" %>
<jsp:useBean id = "digitBean" class = "save.data.Example7_1_Bean" scope = "request"/>
<style>
    #tom{
        font-family:宋体;font-size:26;color:blue
    }
</style>
<HTML><body bgcolor = #ffccff>
<form action = "computer" id = tom method = post>
<table>
<tr><td id = tom>输入两个数:</td>
<td id = tom>
<input type = text name = "numberOne"
        value = <% = digitBean.getNumberOne() %> id = tom size = 6/></td>
<td><input type = text name = "numberTwo"
        value = <% = digitBean.getNumberTwo() %> id = tom size = 6/></td>
</tr>
<tr><td id = tom>选择运算符号:</td>
<td id = tom>
<select id = tom name = "operator">
    <option value = " + "> + (加)
    <option value = " - "> - (减)
    <option value = " * "> * (乘)
    <option value = "/">/(除)
</select>
</td>
<td><input type = "submit" id = tom value = "提交" name = "sub"/></td>
</tr>
</table></form>
<p id = tom>
```

第 7 章　MVC模式

运算结果：
<jsp:getProperty name = "digitBean" property = "numberOne"/>
<jsp:getProperty name = "digitBean" property = "operator"/>
<jsp:getProperty name = "digitBean" property = "numberTwo"/> =
<jsp:getProperty name = "digitBean" property = "result"/>
</p></body></HTML>

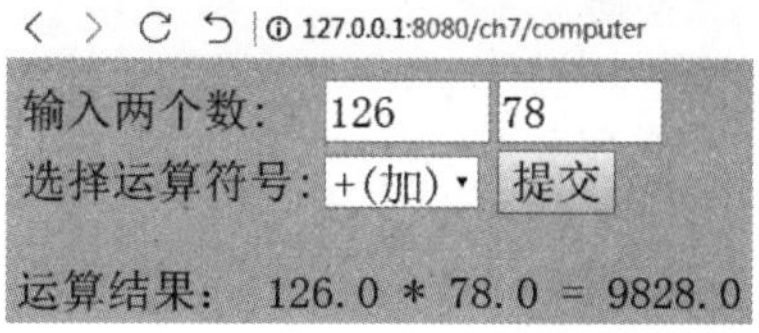

图 7.3　输入并显示有关结果

> **servlet（控制器）**

Example7_1_Sevlet 负责创建名字是 compute 的 servlet。名字为 compute 的 servlet 控制器负责计算四则运算的结果，并将结果存放在 id 是 digitBean 的 bean 中，然后用重定向的方法，请求 example6_2.jsp 显示 bean 中的数据。

用命令行进入 handle\data 的父目录 classes，编译 Example7_1_Servlet.java（见本章开始的约定）：

```
classes> javac - cp .;servlet - api.jar  handle/data/Example7_1_Servlet.java
```

注意"．；"和"servlet-api.jar"之间不要有空格，"．；"的作用是保证 Java 源文件能使用 import 语句引入当前 classes 目录中其他自定义包中的类，例如 save.data 包中的 bean 类"．；"是 javac 默认具有的功能，在使用-cp 参数时，尽量保留这一功能）。

Example7_1_Servlet.java

```java
package handle.data;
import save.data.*;
import java.io.*;
import javax.servlet.*;
import javax.servlet.http.*;
public class Example7_1_Servlet extends HttpServlet{
   public void init(ServletConfig config) throws ServletException{
      super.init(config);
   }
   public void doPost(HttpServletRequest request,
      HttpServletResponse response) throws ServletException,IOException{
      Example7_1_Bean digitBean = null;
      digitBean = new Example7_1_Bean();            //创建 JavaBean 对象
      //digitBean 是 request bean:
      request.setAttribute("digitBean",digitBean); //
      String str1 = request.getParameter("numberOne");
      String str2 = request.getParameter("numberTwo");
      if(str1 == null||str2 == null)
         return;
```

```java
        if(str1.length() == 0||str2.length() == 0)
            return;
        double numberOne = Double.parseDouble(str1);
        double numberTwo = Double.parseDouble(str2);
        String operator = request.getParameter("operator");
        double result = 0;
        if(operator.equals("+"))
            result = numberOne + numberTwo;
        else if(operator.equals("-"))
            result = numberOne - numberTwo;
        else if(operator.equals("*"))
            result = numberOne * numberTwo;
        else if(operator.equals("/"))
            result = numberOne/numberTwo;
        digitBean.setNumberOne(numberOne);           //数据存储在digitBean中
        digitBean.setNumberTwo(numberTwo);
        digitBean.setOperator(operator);
        digitBean.setResult(result);
        //请求example7_1.jsp显示digitBean中的数据
        RequestDispatcher dispatcher =
        request.getRequestDispatcher("example7_1.jsp");
        dispatcher.forward(request,response);
    }
    public void  doGet(HttpServletRequest request,
        HttpServletResponse response)throws ServletException,IOException{
        doPost(request,response);
    }
}
```

➢ **web.xml**(部署文件)

向 ch7\WEB\INF\下的部署文件 web.xml 添加如下的 servlet 和 servlet-mapping 标记(见 6.1.2 节),部署的 servlet 的名字是 computer,访问 servlet 的 url-pattern 是/computer。

web.xml

```xml
<?xml version = "1.0" encoding = "utf-8"?>
<web-app>
    <!-- 以下是 web.xml 文件新添加的内容 -->
    <servlet>
        <servlet-name>computer</servlet-name>
        <servlet-class>handle.data.Example7_1_Servlet</servlet-class>
    </servlet>
    <servlet-mapping>
        <servlet-name>computer</servlet-name>
        <url-pattern>/computer</url-pattern>
    </servlet-mapping>
</web-app>
```

▶ **7.4.2 表白墙**

本节的例 7_2 设计一个 Web 应用,有 3 个 JSP 页面 example7_2.jsp、example7_2_show.jsp 和 example7_3_delete.jsp。一个 ExpressWish 类,ExpressWish 对象负责存放用户(表白人)

的表白信息,例如表白人的 id、昵称、标题、内容、时间等数据,一个 application bean(模拟表白墙)负责存放 ExpressWish 对象。例 7_2 中有一个 servlet,负责处理数据。JSP 页面 example7_2.jsp 提供一个 form 表单,用户使用 form 表单将表白信息(昵称、标题、内容)提交给 servlet 控制器。servlet 给表白人提交的数据分配一个 id 和时间,然后将表白人的 id、昵称、标题、内容、时间存储到 application bean 中,然后定向到 example7_2_show.jsp 页面显示 application bean 中的数据。example7_3_delete.jsp 页面负责删除留言(删除留言所需密码 123456 由管理员掌控)。

例 7_2

➢ **bean**(模型)

创建 application bean 需要 ExpressWish_Bean 类,ExpressWish 类是 ExpressWish_Bean 需要的一个类(ExpressWish_Bean 类需要组合 ExpressWish 类的实例)。

用命令行进入 save\data 的父目录 classes,编译 ExpressWish.java 和 ExpressWish_Bean.java(见本章开始的约定):

```
classes > javac save\data\ExpressWish.java
classes > javac save\data\ExpressWish_Bean.java
```

ExpressWish.java

```java
package save.data;
public class ExpressWish {
    String contents ;                    //表白内容
    String title;                        //标题
    String dateTime;                     //时间
    String peopleName;                   //表白人
    String id;
    public void setId(String id){
        this.id = id;
    }
    public String getId(){
        return id;
    }
    public void setPeopleName(String s){
        peopleName = s;
    }
    public String getPeopleName(){
        return peopleName;
    }
    public void setContent(String s){
        contents = s;
    }
    public String getContent(){
        return contents;
    }
    public void setTitle(String s){
        title = s;
    }
    public String getTitle(){
```

```
        return title;
    }
    public void setDateTime(String s){
        dateTime = s;
    }
    public String getDateTime(){
        return dateTime ;
    }
}
```

ExpressWish_Bean.java

```
package save.data;
import java.util.HashMap;
import java.util.ArrayList;
import java.util.Iterator;
public class ExpressWish_Bean {
    public HashMap<String,ExpressWish> wishList;
    ArrayList<ExpressWish> wishes;                 //存放 wishList 中的表白信息的 ArrayList
    public ExpressWish_Bean(){
        wishList = new HashMap<String,ExpressWish>();
        wishes = new ArrayList<ExpressWish>();
    }
    public void addExpressWish(String id,ExpressWish expressWish){
        wishList.put(id,expressWish);
        putToArrays(wishList);                     //再把全部表白放到 ArrayList wishes
    }
    public void removeExpressWish(String id){
        wishList.remove(id);
        putToArrays(wishList);
    }
    public String getId(int index) {               //返回某个表白者
        return wishes.get(index).getId();
    }
    public String getPeopleName(int index) {       //返回某个表白者
        return wishes.get(index).getPeopleName();
    }
    public String getTitle(int index){
        return wishes.get(index).getTitle();
    }
    public String getContent(int index){
        return wishes.get(index).getContent();
    }
    public String getDateTime(int index){
        return wishes.get(index).getDateTime();
    }
    public int size() {
        return wishes.size();
    }
    void putToArrays(HashMap<String,ExpressWish> list){    //把表白放到 wishes
        wishes.clear();
```

第7章 MVC模式

```
            Iterator<ExpressWish> iterator = list.values().iterator();
            while(iterator.hasNext()){
                ExpressWish wish = iterator.next();
                wishes.add(wish);
            }
        }
    }
```

> **JSP 页面（视图）**

视图部分由 example7_2.jsp、example_7_2_show.jsp 和 example_7_2_delete.jsp 3 个页面构成，JSP 页面 example7_2.jsp 提供一个 form 表单，用户使用 form 表单将表白信息（昵称、标题、内容）提交给 servlet 控制器。example_7_2_show.jsp 负责显示存储在 bean 中的全部表白内容，example7_3_delete.jsp 页面负责删除 bean 中的某个表白内容。

example7_2.jsp（效果如图 7.4(a)所示）

```
<%@ page contentType = "text/html" %>
<%@ page pageEncoding = "utf-8" %>
<HTML>
<style>
    #tom{
        font-family:宋体;font-size:18;color:blue
    }
</style>
<body bgcolor = #ffccff>
<form action = "handleExpress" id = "tom" method = "post">
表白者：<input type = "text" id = "tom" name = "peopleName" size = 28/>
<br>标题：<input type = "text" id = "tom" name = "title" size = 30/>
<br>内容：<br>
<textArea name = "contents" id = "tom" rows = "10" cols = 36>
</textArea>
<br><input type = "submit" id = "tom" value = "提交表白" name = "submit"/>
</form>
<p id = "tom">
<a href = "example7_2_show.jsp">查看表白墙</a>
</p></body></HTML>
```

example7_2_show.jsp（效果如图 7.4(b)所示）

```
<%@ page contentType = "text/html" %>
<%@ page pageEncoding = "utf-8" %>
<jsp:useBean id = "wishWallBean"
class = "save.data.ExpressWish_Bean" scope = "application"/>
<style>
    #tom{
        font-family:宋体;font-size:26;color:blue
    }
</style>
<HTML><body bgcolor = white>
<table border = 1>
    <tr><th id = tom> id </th><th id = tom>表白人</th><th id = tom>标题</th>
```

```
            < th id = tom >时间</th><th id = tom >表白内容</th>
<%    for(int i = 0;i < wishWallBean.size();i++){
        out.print("<tr>");
        out.print("<td id = tom >" + wishWallBean.getId(i) + "</td>");
        out.print("<td id = tom >" + wishWallBean.getPeopleName(i) + "</td>");
        out.print("<td id = tom >" + wishWallBean.getTitle(i) + "</td>");
        out.print("<td id = tom >" + wishWallBean.getDateTime(i) + "</td>");
        out.print("<td><textArea rows = 5 cols = 20
                    id = tom >" + wishWallBean.getContent(i) +
                "</textArea></td>");
        out.print("</tr>");
    }
%></table>
<a id = tom href = "example7_2.jsp">去表白</a>
</body></HTML>
```

example7_2_delete.jsp(效果如图 7.4(c)所示)

```
<%@ page contentType = "text/html" %>
<%@ page pageEncoding = "utf - 8" %>
<jsp:useBean id = "wishWallBean" class =
             "save.data.ExpressWish_Bean" scope = "application"/>
<HTML><body bgcolor = pink >
<p style = "font - family:宋体;font - size:18;color:blue">
管理员删除表白的页面.
<form action = "" method = post >
输入密码:<input type = "password" name = "password"size = 12 /><br>
输入表白 id:<input type = "text" name = "peopleId" size = 6 />
<br><input type = "submit" name = "submit"  value = "删除"/>
</form>
<% request.setCharacterEncoding("utf - 8");
   String password = request.getParameter("password");
   String id = request.getParameter("peopleId");
   if(password == null ) password = "";
   if(id == null ) id = "";
   if(password.equals("123456")){
       wishWallBean.removeExpressWish(id);
   }
%>
<a href = "example7_2_show.jsp">查看表白墙</a>
</p></body></HTML>
```

> **servlet**(控制器)

ExpressWish_Servlet 负责创建名字是 handleExpress 的 servlet。名字为 handleExpress 的 servlet 控制器负责给提交的数据分配一个 id 和时间,然后将表白人的 id、昵称、标题、内容、时间存储到 application bean 中。

用命令行进入 handle\data 的父目录 classes,编译 ExpressWish_Servlet.java(见本章开始的约定,如图 7.1 所示):

```
javac - cp .;servlet - api.jar   handle/data/ExpressWish_Servlet_Servlet.java
```

第7章 MVC模式

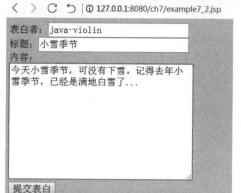

(a) 输入表白提交给servlet　　　　　(b) 删除某个表白

(c) 显示表白墙

图 7.4　表白墙

注意". ;"和"servlet-api. jar"之间不要有空格，". ;"的作用是保证 Java 源文件能使用 import 语句引入当前 classes 目录中其他自定义包中的类，例如 save. data 包中的类。

ExpressWish_Servlet. java

```java
package handle.data;
import save.data.ExpressWish;
import save.data.ExpressWish_Bean;
import java.util.*;
import java.io.*;
import java.time.LocalDateTime;
import javax.servlet.*;
import javax.servlet.http.*;
public class ExpressWish_Servlet extends HttpServlet{
    int index;                                               //id
    public void init(ServletConfig config) throws ServletException{
        super.init(config);
    }
```

```java
        synchronized long getIndex() {              //synchronized 修饰的方法
            index = index + 1;
            return index;
        }
        public void service(HttpServletRequest request,
            HttpServletResponse response) throws ServletException,IOException{
            request.setCharacterEncoding("utf-8");
            ExpressWish_Bean wishWallBean = null;        //wishWallBean 存放表白墙内容
            ServletContext application = getServletContext();
            wishWallBean =
            (ExpressWish_Bean)application.getAttribute("wishWallBean");
            if(wishWallBean == null ){                   //wishWallBean 不存在就创建 wishWallBean
                wishWallBean = new ExpressWish_Bean();
                application.setAttribute("wishWallBean",wishWallBean);
            }
            String peopleName = request.getParameter("peopleName");   //表白者
            String title = request.getParameter("title");             //标题
            String content = request.getParameter("contents");        //表白内容
            ExpressWish wish = new ExpressWish();
            if(peopleName.length() == 0||title.length() == 0||content.length() == 0){
                response.sendRedirect("example7_2.jsp");
                return;
            }
            wish.setPeopleName(peopleName);
            wish.setTitle(title);
            wish.setContent(content);
            LocalDateTime dateTime = LocalDateTime.now();
            String str = dateTime.toString();
            String time = str.substring(0,str.lastIndexOf("."));      //不要纳秒
            wish.setDateTime(time);
            long number = getIndex();
            wish.setId("" + number);
            wishWallBean.addExpressWish("" + number,wish);            //添加一条表白
            response.sendRedirect("example7_2_show.jsp");             //显示表白墙
        }
    }
```

> web.xml（部署文件）

向 ch7\WEB\INF\下的 web.xml 添加如下的 servlet 和 servlet-mapping 标记（见 6.1.2 节），部署的 servlet 的名字是 handleExpress，访问 servlet 的 url-pattern 是/handleExpress。

web.xml

```xml
<?xml version = "1.0" encoding = "utf-8"?>
<web-app>
    <!-- 以下是 web.xml 文件新添加的内容 -->
    <servlet>
        <servlet-name>handleExpress</servlet-name>
        <servlet-class>handle.data.ExpressWish_Servlet</servlet-class>
    </servlet>
    <servlet-mapping>
        <servlet-name>handleExpress</servlet-name>
```

```
            <url-pattern>/handleExpress</url-pattern>
        </servlet-mapping>
</web-app>
```

注:建议读者将这里的表白墙和4.4.2节的留言板进行比较,进一步体会MVC模式的优点。

7.5 上机实验

提供了详细的实验步骤要求,按步骤完成,提升学习效果,积累经验,不断提高Web设计能力。

▶ 7.5.1 实验1 等差、等比级数和

❶ 实验目的

掌握MVC模式三部分的设计。使用request bean存储数据,使用servlet处理数据,并将有关数据存储到request bean,使用JSP页面显示request bean数据。

❷ 实验要求

(1)编写一个创建bean的类,该类可以储存等差(等比)级数的首项、公差(公比)和等差(等比)级数的和。

(2)编写一个JSP页面inputNumber.jsp。该页面提供form表单,该form表单提供用户输入等差(等比)级数的首和公差(公比)。用户单击"求等差级数和"提交键或"求等比级数和"提交键请求名字是computeSum的servlet。

(3)编写创建servlet的Servlet类,该类创建的servlet根据有户提交的数据,负责创建一个request bean。如果用户单击的是"求等差级数和"("求等比级数和")提交键,servlet计算等差(等比)级数的和,并将结果存储到request bean,并请求inputNumber.jsp页面显示等差(等比)级数的和。

(4)在Tomcat服务器的webapps目录下新建一个名字是ch7_practice_one的Web服务目录。把JSP页面保存到ch7_practice_one目录中。在ch7_practice_one下建立子目录WEB-INF(字母大写),然后在WEB-INF下再建立子目录classes,将创建servlet和bean的类的Java源文件按照包名保存在classes的相应子目录中。

(5)向ch7_practice_one\WEB\INF\下的部署文件web.xml添加servlet和servlet-mapping标记(知识点见6.1.2节),部署servlet的名字和访问servlet的url-pattern。

(6)用浏览器访问JSP页面inputNumber.jsp。

❸ 参考代码

参考代码运行效果如图7.5所示。

图7.5 求等差(等比)级数和

➢ bean(模型)

用命令行进入 save\data 的父目录 classes,编译 Series_Bean.java(参考本章开始的约定):

```
classes> javac save\data\Series_Bean.java
```

Series_Bean.java

```java
package save.data;
public class Series_Bean{
    double firstItem;                      //级数首项
    double var;                            //公差或公比
    int number;                            //求和项数
    double sum;                            //求和结果
    String name = "";                      //级数类别
    public void setFirstItem(double a){
        firstItem = a;
    }
    public double getFirstItem(){
        return firstItem;
    }
    public void setVar(double b){
        var = b;
    }
    public double getVar(){
        return var;
    }
    public void setNumber(int n){
        number = n;
    }
    public double getNumber(){
        return number;
    }
    public void setSum(double s){
        sum = s;
    }
    public double getSum(){
        return sum;
    }
    public void setName(String na){
        name = na;
    }
    public String getName(){
        return name;
    }
}
```

➢ JSP 页面(视图)

inputNumber.jsp 提供 form 表单,该 form 表单提供用户输入等差(等比)级数的首和公差(公比)。用户单击"求等差级数和"提交键或"求等比级数和"提交键请求名字是 computeSum 的 servlet。

inputNumber.jsp(效果如图 7.5 所示)

```jsp
<%@ page contentType="text/html" %>
<%@ page pageEncoding="utf-8" %>
<jsp:useBean id="seriesData" class="save.data.Series_Bean"
             scope="request"/>
<style>
   #tom{
      font-family:宋体;font-size:28;color:blue
   }
</style>
<HTML><body bgcolor=#FFBBFF>
<form action="computeSum" id=tom method="post">
等差级数求和：
   <br>输入首项:<input type=text id=tom name="firstItem" size=4>
       输入公差(公比):<input type=text id=tom name="var" size=4>
       求和项数:<input type=text id=tom name="number" size=4>
   <input type=submit name="submit" id=tom value="提交(求等差级数和)" />
   <input type=submit name="submit" id=tom value="提交(求等比级数和)">
</form>
<table border=1 id=tom>
   <tr>
     <th>级数的首项</th>
     <th><jsp:getProperty name="seriesData" property="name"/></th>
     <th>所求项数</th>
     <th>求和结果</th>
   </tr>
     <td><jsp:getProperty name="seriesData" property="firstItem"/></td>
     <td><jsp:getProperty name="seriesData" property="var"/></td>
     <td><jsp:getProperty name="seriesData" property="number"/></td>
     <td><jsp:getProperty name="seriesData" property="sum"/></td>
   </tr>
</table>
</body></HTML>
```

> ➢ **servlet**(控制器)

ComputeSum_Servlet 负责创建名字是 computeSum 的 servlet。名字为 computeSum 的 servlet 控制器负责计算等差或等比级数的和，并将结果存放在 id 是 seriesData 的 request bean 中，然后用重定向的方法，请求 inputNumber.jsp 显示 request bean 中的数据。

用命令行进入 handle\data 的父目录 classes，编译 ComputeSum_Servlet.java(参考本章开始的约定)：

```
classes> javac -cp .;servlet-api.jar  handle/data/ComputeSum_Servlet.java
```

事先将 Tomcat 安装目录 lib 子目录中的 servlet-api.jar 文件复制(不要剪贴)到\ch7_practice_one\WEB-INF\classes 中(知识点见 6.1.1 节)。

注意".;"和"servlet-api.jar"之间不要有空格，".;"的作用是保证 Java 源文件能使用 import 语句引入当前 classes 目录中其他自定义包中的类，例如 save.data 包中的 bean 类。

ComputeSum_Servlet.java

```java
package handle.data;
import save.data.*;
import java.io.*;
import javax.servlet.*;
import javax.servlet.http.*;
public class ComputeSum_Servlet extends HttpServlet{
    public void init(ServletConfig config) throws ServletException{
        super.init(config);
    }
    public void service(HttpServletRequest request,
        HttpServletResponse response) throws ServletException,IOException{
        Series_Bean seriesData = new Series_Bean();            //创建 bean
        request.setCharacterEncoding("utf-8");
        request.setAttribute("seriesData",seriesData);         //request bean
        String mess = request.getParameter("submit");
        String firstItem = request.getParameter("firstItem");  //首项
        String var = request.getParameter("var");              //或公比公差
        String number = request.getParameter("number");        //求和项数
        if(firstItem.length() == 0||var.length() == 0||number.length() == 0){
            response.sendRedirect("inputNumber.jsp");
            return;
        }
        if(mess.contains("等差")) {
            compute(firstItem,var,number,seriesData,0);
        }
        else if(mess.contains("等比")) {
            compute(firstItem,var,number,seriesData,1);
        }
        //请求 inputNumber.jsp 显示 seriesData 中的数据
        RequestDispatcher dispatcher =
        request.getRequestDispatcher("inputNumber.jsp");
        dispatcher.forward(request,response);
    }
    void compute(String firstItem,String var,String number,
                Series_Bean seriesData,int type){
        double a = Double.parseDouble(firstItem);
        double d = Double.parseDouble(var);
        int n = Integer.parseInt(number);
        seriesData.setFirstItem(a);                //将数据存储在数据模型 seriesData 中
        seriesData.setVar(d);
        seriesData.setNumber(n);
        double sum = 0, item = a;
        int i = 1;
        if(type == 0) {
            seriesData.setName("等差级数的公差");
            while(i<=n){                           //计算等差数列的和
                sum = sum + item;
                i++;
                item = item + d;
            }
```

```
            seriesData.setSum(sum);
        }
        else if(type == 1){
            seriesData.setName("等比级数的公比");
            while(i<=n){           //计算等比数列的和
                sum = sum + item;
                i++;
                item = item * d;
            }
            seriesData.setSum(sum);
        }
    }
}
```

➤ **web.xml**（部署文件）

向 ch7_practice_one\WEB\INF\下的 web.xml 添加如下的 servlet 和 servlet-mapping 标记（见 6.1.2 节），部署的 servlet 的名字是 computeSum，访问 servlet 的 url-pattern 是/computeSum。

web.xml

```
<?xml version = "1.0" encoding = "utf-8"?>
<web-app>
    <!-- 以下是 web.xml 文件新添加的内容 -->
    <servlet>
        <servlet-name>computeSum</servlet-name>
        <servlet-class>handle.data.ComputeSum_Servlet</servlet-class>
    </servlet>
    <servlet-mapping>
        <servlet-name>computeSum</servlet-name>
        <url-pattern>/computeSum</url-pattern>
    </servlet-mapping>
</web-app>
```

▶ 7.5.2 实验2 点餐

❶ 实验目的

掌握 MVC 模式三部分的设计。使用 session bean 存储数据，使用 servlet 处理数据，并将有关数据存储到 session bean，使用 JSP 页面显示 session bean 数据。

❷ 实验要求

(1) 编写一个创建 bean 的类，该类可以储存餐单相关的数据，例如菜名、价格等信息。

(2) 编写一个 JSP 页面 inputMenu.jsp。该页面提供一个 form 表单，该表单提供用户输入餐单信息。用户单击"添加到餐单"提交键请求名字是 addMenu 的 servlet。

(3) 编写创建 servlet 的 Servlet 类，该类创建的 servlet 根据用户提交的餐单数据，解析出餐单的消费总额，并将餐单中消费分项按价格排序，并将这些信息存放到 session bean 中。servlet 请求 showMenu.jsp 页面显示 session bean 中的数据。

(4) 编写 showMenu.jsp 页面，该页面显示或删除 session bean 中的数据（显示餐单，同时也可以删除餐单中的某道菜）。

(5) 在 Tomcat 服务器的 webapps 目录下（例如 D:\apache-tomcat-9.0.26\webapps）新

建名字是 ch7_practice_two 的 Web 服务目录。把 JSP 页面保存到 ch7_practice_two 目录中。在 ch7_practice_two 下建立子目录 WEB-INF(字母大写)，然后在 WEB-INF 下再建立子目录 classes,将创建 servlet 和 bean 的类的 Java 源文件按照包名保存在 classes 的相应子目录中。

(6) 向 ch7_practice_two\WEB\INF\下的部署文件 web.xml 添加 servlet 和 servlet-mapping 标记(见 6.1.2 节),部署 servlet 的名字和访问 servlet 的 url-pattern。

(7) 用浏览器访问 JSP 页面 inputMenu.jsp。

❸ 参考代码

参考代码运行效果如图 7.6 所示。

(a) 输入名称和价格　　　　　　　　　(b) 查看、删除餐单

图 7.6　点餐

➢ **bean**(模型)

创建 session bean 需要使用 MenuBean 类,Food 类是 MenuBean 类需要的一个类。用命令行进入 save\data 的父目录 classes,编译 Food.java 和 MenuBean.java(参考本章开始的约定,如图 7.1 所示):

```
classes > javac save\data\Food.java
classes > javac save\data\MenuBean.java
```

Food.java

```java
package save.data;
public class Food implements Comparable<Food>{
    String foodName ;                           //食物名称
    double price;                               //价格
    public void setFoodName(String name){
        foodName = name;
    }
    public String getFoodName(){
        return foodName;
    }
    public void setPrice(double d){
        price = d;
    }
    public double getPrice(){
        return price;
    }
    public int compareTo(Food food){            //Food 对象按价格比较大小
        return (int)(food.getPrice() * 1000 - price * 1000);
    }
}
```

MenuBean.java

```java
package save.data;
import java.util.ArrayList;
import java.util.Collections;
public class MenuBean {
    String time ;                              //点餐时间
    String totalPrice;                         //餐单总额
    ArrayList<Food> foodList;                  //存放 Food 对象的 ArrayList
    public MenuBean(){
        foodList = new ArrayList<Food>();
    }
    public void addFood(Food food){
        foodList.add(food);
        Collections.sort(foodList);            //排序 foodList
    }
    public void removeFood(int index){
        if(index >= 0){
            foodList.remove(index);
            Collections.sort(foodList);        //排序 foodList
        }
    }
    public String getFoodName(int index) {     //返回某个 Food 的名字
        return foodList.get(index).getFoodName();
    }
    public double getPrice(int index) {        //返回某个 Food 的价格.
        return foodList.get(index).getPrice();
    }
    public int size() {
        return foodList.size();
    }
    public void setTime(String time){
        this.time = time;
    }
    public String getTime(){
        return time;
    }
    public String getTotalPrice(){
        double sum = 0;
        for(Food food:foodList){
           sum += food.getPrice();
        }
        totalPrice = String.format("%.2f",sum);   //保留 2 位小数
        return totalPrice;
    }
}
```

➢ **JSP 页面（视图）**

inputMenu.jsp 提供 form 表单，该 form 表单提供用户输入某道菜的名字和价格。用户单击"添加到餐单"提交键请求名字是 addMenu 的 servlet。showMenu.jsp 负责显示 session bean 中的数据，以及删除 session bean 中的某个数据（即删除某道菜）。

inputMenu.jsp(效果如图 7.6(a)所示)

```jsp
<%@ page contentType="text/html" %>
<%@ page pageEncoding="utf-8" %>
<jsp:useBean id="menu" class="save.data.MenuBean" scope="session"/>
<style>
    #textStyle{
        font-family:宋体;font-size:36;color:blue
    }
</style>
<HTML><body bgcolor=#ffccff>
<form action="addMenu" id=textStyle method=post>
输入餐单(每次输入一个消费项目):<br>
名称:<input type=text name='foodName' id=textStyle value='剁椒鱼头' size=8 />
价格:<input type=text name='price' id=textStyle value='26.9' size=8 />
<br><input type=submit id=textStyle value="添加到餐单">
</form>
</body></HTML>
```

showMenu.jsp(效果如图 7.6(b)所示)

```jsp
<%@ page contentType="text/html" %>
<%@ page pageEncoding="utf-8" %>
<jsp:useBean id="menu" class="save.data.MenuBean" scope="session"/>
<style>
    #tom{
        font-family:宋体;font-size:26;color:blue
    }
</style>
<HTML><body bgcolor=pink>
<table border=1>
    <tr><th id=tom>序号</th><th id=tom>食物名称</th><th id=tom>价格</th>
<% request.setCharacterEncoding("utf-8");
    String index = request.getParameter("删除");
    if(index!=null){
        menu.removeFood(Integer.parseInt(index));
    }
    for(int i=0;i<menu.size();i++){
        out.print("<tr>");
        out.print("<td id=tom>"+(i+1)+"</td>");
        out.print("<td id=tom>"+menu.getFoodName(i)+"</td>");
        out.print("<td id=tom>"+menu.getPrice(i)+"</td>");
        out.print("<td><form action = "+"showMenu.jsp"+" method=post>");
        out.print("<input type=hidden name = 删除 value = "+i+" />");
        out.print("<input type=submit  value = 删除该食物 />");
        out.print("</form></td>");
        out.print("</tr>");
    }
%></table>
<p id=tom>
餐单总额(共有<%= menu.size() %>道食物):
```

```
<jsp:getProperty name = "menu" property = "totalPrice" /><br>
点餐时间:
<jsp:getProperty  name = "menu" property = "time" /></p>
<a id = tom href = "inputMenu.jsp">继续点餐</a>
</body></HTML>
```

> **servlet（控制器）**

HandleMenu_Servlet 负责创建名字是 addMenu 的 servlet。名字为 addMenu 的 servlet 控制器负责创建 id 是 menu 的 session bean，并将 Food 对象存储到 session bean 中，然后将用户重定向到 showMenu.jsp 页面查看餐单。

用命令行进入 handle\data 的父目录 classes，编译 HandleMenu_Servlet.java（参考本章开始的约定）:

```
classes> javac -cp .;servlet-api.jar  handle/data/HandleMenu_Servlet.java
```

事先将 Tomcat 安装目录 lib 子目录中的 servlet-api.jar 文件复制（不要剪贴）到\ch7_practice_one\WEB-INF\classes 中（知识点见 6.1.1 节）。

注意".;"和"servlet-api.jar"之间不要有空格，".;"的作用是保证 Java 源文件能使用 import 语句引入当前 classes 目录中其他自定义包中的类，例如 save.data 包中的 bean 类。

HandleMenu_Servlet.java

```java
HandleMenu_Servlet.java
package handle.data;
import save.data.Food;
import save.data.MenuBean;
import java.util.*;
import java.io.*;
import java.time.LocalTime;
import javax.servlet.*;
import javax.servlet.http.*;
public class HandleMenu_Servlet extends HttpServlet{
    public void init(ServletConfig config) throws ServletException{
        super.init(config);
    }
    public void service(HttpServletRequest request,
        HttpServletResponse response)throws ServletException,IOException{
        request.setCharacterEncoding("utf-8");
        MenuBean menu = null;                        //餐单
        HttpSession session = request.getSession(true);
        menu = (MenuBean)session.getAttribute("menu");
        if(menu == null ){                           //menu 不存在就创建 menu
            menu = new MenuBean();
            session.setAttribute("menu",menu);       //session bean
        }
        String foodName = request.getParameter("foodName");//食物名称
        String price = request.getParameter("price");
        Food food = new Food();
        if(foodName.length() == 0||price.length() == 0){
            response.sendRedirect("inputMenu.jsp");
```

```
                return;
            }
            food.setFoodName(foodName);
            food.setPrice(Double.parseDouble(price));
            LocalTime dateTime = LocalTime.now();
            String str = dateTime.toString();
            String time = str.substring(0,str.lastIndexOf("."));    //不要纳秒
            menu.setTime(time);
            menu.addFood(food);                                     //添加一道食物
            response.sendRedirect("showMenu.jsp");                  //显示餐单
        }
    }
```

➢ **web.xml（部署文件）**

向 ch7_practice_two\WEB\INF\下的 web.xml 添加如下的 servlet 和 servlet-mapping 标记（见 6.1.2 节），部署的 servlet 的名字是 addMenu，访问 servlet 的 url-pattern 是 /addMenu。

web.xml

```
<?xml version = "1.0" encoding = "utf - 8"?>
<web - app>
    <!-- 以下是 web.xml 文件新添加的内容 -->
    <servlet>
        <servlet - name>addMenu</servlet - name>
        <servlet - class>handle.data.HandleMenu_Servlet</servlet - class>
    </servlet>
    <servlet - mapping>
        <servlet - name>addMenu</servlet - name>
        <url - pattern>/addMenu</url - pattern>
    </servlet - mapping>
</web - app>
```

7.6 小结

➢ MVC 模式的核心思想是有效地组合"视图""模型"和"控制器"。在 JSP 技术中，视图是一个或多个 JSP 页面，其作用主要是向控制器提交必要的数据和为模型提供数据显示；模型是一个或多个 JavaBean 对象，用于存储数据；控制器是一个或多个 servlet 对象，根据视图提交的要求进行数据处理操作，并将有关的结果存储到 JavaBean 中，然后 servlet 使用重定向方式请求视图中的某个 JSP 页面更新显示。

➢ 在 MVC 模式中，模型也可以由控制器负责创建和初始化。

习题 7

1. 在 JSP 中，MVC 模式中的数据模型的角色由谁担当？
2. 在 JSP 中，MVC 模式中的控制器的角色由谁担当？

3. 在 JSP 中，MVC 模式中的视图的角色由谁担当？
4. MVC 的好处是什么？
5. MVC 模式中用到的 JavaBean 是由 JSP 页面还是 servlet 负责创建？
6. 设计一个 Web 应用。用户可以通过一个 JSP 页面输入三角形的三边或梯形的上底、下底和高给一个 servlet 控制器，控制器负责计算三角形和梯形的面积，并将结果存储到数据模型中，然后请求另一个 JSP 页面显示数据模型中的数据。

第 8 章　JSP中使用数据库

本章导读

　　主要内容
* MySQL 数据库管理系统
* 连接 MySQL 数据库
* 查询记录
* 更新、添加与删除记录
* 用结果集操作数据库中的表
* 预处理语句
* 事务
* 分页显示记录
* 连接 SQL Server 与 Access 数据库
* 使用连接池
* 标准化考试训练

　　难点
* 分页显示记录
* 使用连接池
* 标准化考试训练

　　关键实践
* 小星星广告网

　　在许多 Web 应用中，服务器需要和用户进行必要的数据交互。例如，服务器需要将用户提供的数据永久、安全地保存在服务器端，需要为用户提供数据查询等，此时，Web 应用就可能需要和数据库打交道，其原因是数据库在数据查询、修改、保存、安全等方面有着其他数据处理手段无法替代的地位。许多优秀的数据库管理系统在数据管理，特别是在基于 Web 的数据管理方面在扮演着重要的角色。

　　本章并非讲解数据库原理，而是讲解如何在 JSP 中使用 JDBC 提供的 API 和数据库进行信息交互。而且只要掌握与某种数据库管理系统所管理的数据库交互信息方法，就会很容易地掌握和其他数据库管理系统所管理的数据库交互信息方法。所以，为了便于教学，本书使用的数据库管理系统是 MySQL 数据库管理系统（读者可以选择任何熟悉的数据库管理系统学习本章的内容，见 8.9 节）。

　　为了理解数据库操作的基本步骤，大部分例子在 JSP 页面中使用 Java 程序片，学习使用数据库（便于理解、掌握数据库操作的基本步骤）。为了体现一个 Web 应用将数据的处理和显示相分离，8.8 节的例子采用了 MVC 模式（有关 MVC 的知识请参见第 7 章）。

　　本章在 Tomcat 安装目录的 webapps 目录下建立 Web 服务目录 ch8。另外，需要在 Web 服务目录 ch8 下建立目录结构 ch8\WEB-INF\classes（WEB-INF 字母大写）。

　　本章使用的 bean 的包名（除非特别说明），均为 save.data。在 classes 下建立目录结构

\save\data，创建 bean 的类的字节码文件保存在\WEB-INF\classes\save\data 中（知识点见 5.1.2 节）。本章的 servlet 的包名（除非特别说明），均为 handle.data。在 classes 下建立目录结构\handle\data，创建 servlet 的类的字节码文件保存在\WEB-INF\classes\handle\data 中（知识点见 6.1.1 节）。

本章使用的 javax.servlet 和 javax.servlet.http 包中的类不在 JDK 提供的核心类库中，为了方便编译 Java 源文件，请事先将 Tomcat 安装目录 lib 子目录中的 servlet-api.jar 文件复制（不要剪贴）到\ch8\WEB-INF\classes 中（知识点见 6.1.1 节）。另外，保存 Java 源文件时，"保存类型"选择为"所有文件"，将"编码"选择为"ANSI"。保存 JSP 文件和部署文件 web.xml 时，"保存类型"选择为"所有文件"，将"编码"选择为"UTF-8"。

8.1 MySQL 数据库管理系统

MySQL 数据库管理系统，简称 MySQL，是目前流行的开源数据库管理系统，其社区版（MySQL Community Edition）是可免费下载的开源数据库管理系统。MySQL 最初由瑞典 MySQL AB 公司开发，目前由 Oracle 公司负责源代码的维护和升级。Oracle 将 MySQL 分为社区版和商业版，并保留 MySQL 开放源码这一特点。目前许多 Web 开发项目都选用社区版 MySQL，其主要原因是社区版 MySQL 的性能卓越，满足许多 Web 应用已经绰绰有余，而且社区版 MySQL 是开源数据库管理系统，可以降低软件的开发和使用成本。

▶ 8.1.1 下载、安装 MySQL

❶ 下载

MySQL 是开源项目，很多网站都提供免费下载。可以使用任何搜索引擎搜索关键字："MySQL 社区版下载"获得有关的下载地址。直接输入 Oracle 的官方网址 https://dev.mysql.com/downloads/mysql/请求下载页，然后在出现的页面中（在页面的下部）选择 Windows（x86,64-bit），ZIP Archive 8.0.18（272.3M），然后单击 Download（下载）按钮，如图 8.1 所示。

图 8.1 下载 MySQL 社区版 MySQL

在出现的新页面中忽略页面上的注册 Sign up，直接单击超链接"No thanks, just start my download"即可。这里我们下载的是 mysql-8.0.18-winx64.zip（适合 64 位机器的 Windows 版）。

❷ 安装

将下载的 mysql-8.0.18-winx64.zip 解压缩到本地计算机，例如解压缩到 D:\。本教材将下载的 mysql-8.0.18-winx64.zip 解压缩到 D:\，形成的安装目录结构如图 8.2 所示。

图 8.2　MySQL 的安装目录

▶ 8.1.2　启动 MySQL 数据库服务器

视频讲解

如果使用 Windows 10 操作系统(不含 Windows 7),需要将 Windows 10 操作系统缺少的 vcruntime140_1.dll 存放到 C:\Windows\System32 目录中(如果不缺少 vcruntime140_1.dll,可以忽略这部分内容)。可以在搜索引擎中搜索 vcruntime140_1.dll,然后选择一个下载地址。

❶ 初始化

首次启动 MySQL 数据库需要进行一些必要的初始化工作(不要进行两次初始化,除非重新安装了 MySQL)。用管理员身份启动命令行窗口,可以右击任何已有的"命令提示符"快捷图标,选择"以管理员身份运行";或使用文件资源管理器在 C:\Windows\System32 下找到 cmd.exe,右击 cmd.exe,选择"以管理员身份运行"。然后在命令行窗口进入 MySQL 安装目录的 bin 子目录,输入"mysqld --initialize"命令,按 Enter 键确认,如图 8.3 所示。

```
D:\mysql-8.0.18-winx64\bin> mysqld -- initialize
```

```
D:\>cd   D:\mysql-8.0.18-winx64\bin
D:\mysql-8.0.18-winx64\bin> mysqld --initialize
D:\mysql-8.0.18-winx64\bin>
```

图 8.3　进行必要的初始化

执行成功后,MySQL 安装目录下多出一个 data 子目录(用于存放数据库,对于早期的 5.6 版本,安装后就有该目录)。初始化的目的是在 MySQL 安装目录下初始化 data 子目录,并授权一个 root 用户。对于 Windows 10 系统,root 用户的初始默认密码可以在 data 目录的 DESKTOP-4DGOGO5.err 文件中找到(选择用记事本打开该文件):A temporary password is generated for root@localhost:drH&&1svhvoa,可以看出 root 用户的临时密码是 drH&&1svhvoa。对于 Windows 7 系统,可以在 data 目录的 Pc2015224130.err 文件中发现 root 用户默认是无密码的(注意,对于 Windows 10 系统,root 的初始密码是随机的;对于 Windows 7 系统,root 初始无密码)。

❷ 启动

MySQL 是一个网络数据库管理系统,可以使远程的计算机访问它所管理的数据库。安装好 MySQL 后，需要启动 MySQL 提供的数据库服务器（数据库引擎），以便使远程的计算机访问它所管理的数据库。用管理员身份启动命令行窗口，然后进入 MySQL 安装目录的 bin 子目录，输入"net start mysql"（Windows 7 输入"mysqld"），按 Enter 键确认启动 MySQL 数据库服务器（以后再启动 MySQL 就不需要初始化了），如图 8.4 所示。MySQL 服务器占用的端口是 3306（3306 是 MySQL 服务器使用的端口号）。

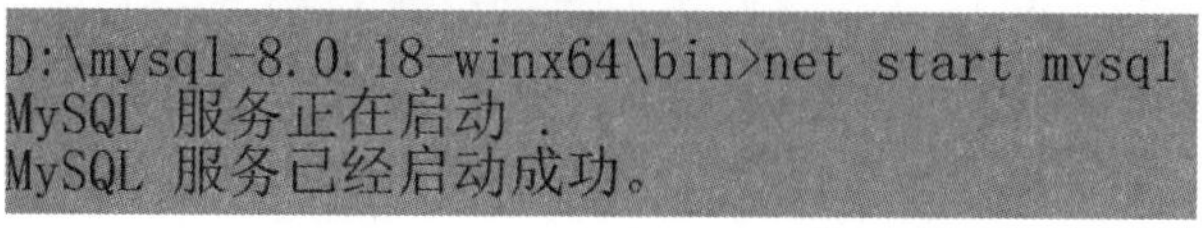

```
D:\mysql-8.0.18-winx64\bin>net start mysql
MySQL 服务正在启动.
MySQL 服务已经启动成功。
```

图 8.4　启动 MySQL 服务器

注：对于 Windows 7 系统，输入"mysqld"启动 MySQL 数据库服务器，成功后 MySQL 数据库服务器将占用当前的 MS-DOS 窗口。

❸ 停止

进入 MySQL 安装目录的 bin 子目录，输入"net stop mysql"，按 Enter 键确认停止 MySQL 数据库服务器。

注：对于 Windows 7 系统，关闭 MySQL 数据库服务器占用的当前 MS-DOS 窗口，将关闭 MySQL 数据库服务器。

❹ root 用户

MySQL 数据库服务器启动后，MySQL 授权可以访问该服务器的用户只有一个，名字是 root，临时密码是 drH&.&.1svhvoa（如果是 Windows 7，root 初始无密码）。应用程序以及 MySQL 客户端管理工具软件都必须借助 MySQL 授权的"用户"来访问数据库服务器。如果没有任何"用户"可以访问启动的 MySQL 数据库服务器，那么这个服务器就如同虚设、没有意义了。MySQL 数据库服务器启动后，不仅可以用 root 用户访问数据库服务器，而且可以再授权能访问数据库服务器的新用户（只有 root 用户有权建立新的用户）。关于建立新的用户的命令见 8.1.3 节。

MySQL 8.0 版本必须对 root 用户进行身份确认，否则将导致其他 mysql 客户户程序，如 Navicat for MySQL 等，无法访问 MySQL 8.0.21 数据库服务器。因此，MySQL 数据库服务器启动后，再用管理员身份打开另一个命令行窗口，使用 mysqladmin 命令确认 root 用户和 root 用户的密码，或确认 root 用户，并修改 root 用户的密码。在新的命令行窗口进入 MySQL 的安装目录 D:\mysql-8.0.18-winx64\bin，使用 mysqladmin 命令：

```
mysqladmin -u root -p password
```

按 Enter 键确认后，将提示输入 root 的当前密码（无密码就直接按 Enter 键确认），Windows 10 安装的 MySQL 的 root 用户的初始密码是 drH&.&.1svhvoa，如果输入正确，将继续提示输入 root 的新密码，以及确认新密码，如图 14.5 所示。

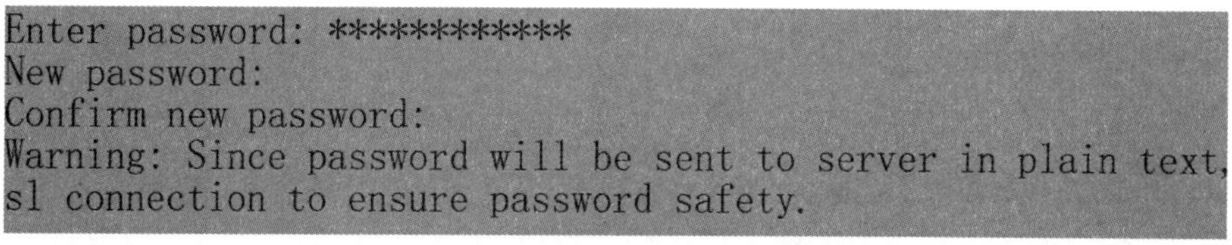

```
Enter password: ***********
New password:
Confirm new password:
Warning: Since password will be sent to server in plain text,
sl connection to ensure password safety.
```

图 8.5　确认 root 用户和密码

注：本书始终让root用户的密码是无密码。

8.1.3 MySQL客户端管理工具

视频讲解

所谓MySQL客户端管理工具，就是专门让客户端在MySQL服务器上建立数据库的软件。可以下载图形用户界面(GUI)的MySQL管理工具，并使用该工具在MySQL服务器上进行创建数据库、在数据库中创建表等操作，MySQL管理工具有免费的也有需要购买的。读者可以在搜索引擎中搜索MySQL客户端管理工具，选择一款MySQL客户端管理工具，例如Navicat for MySQL(目前Navicat for MySQL的最新版本，连接MySQL8.0或之前的版本MySQL没有任何问题)，可以在搜索引擎搜索Navicat for MySQL或登录http://www.navicat.com.cn/download下载试用版或购买商业版，例如下载navicat121_mysql_cs_x64.exe，安装即可。

MySQL管理工具必须和数据库服务器建立连接后，才可以建立数据库及相关操作。因此，在使用客户端管理工具之前须启动MySQL数据库服务器(见前面的8.1.2节)。

本书为了加强训练在命令行使用SQL语句，首先讲解MySQL自带的命令行客户端管理工具，然后再简单介绍Navicat for MySQL。如果会使用命令行客户端管理工具建立数据库等操作，就很容易掌握任何用户界面(GUI)的MySQL管理工具。

❶ 命令行客户端

启动MySQL数据库服务器后，也可以用命令行方式创建数据库(要求有比较好的SQL语句基础)。如果读者有比较好的数据库知识基础，特别是掌握SQL语句的知识，那么使用命令行方式管理MySQL数据库也是很方便的，本节介绍几个简单的命令，以满足本书应用的需求。

注：可以在网络上搜索到MySQL命令详解，详细讲解MySQL本身的知识内容不属于本书的范畴。

为了启动命令行客户端(即和MySQL数据库服务器建立连接)，须打开一个新的命令行窗口(不必管理员身份)，进入MySQL安装目录下的bin子目录。执行mysql.exe，即启动命令行客户端。执行格式为：

```
mysql -h ip -u root -p
```

对于本机调试(即客户端和数据库服务器同机)，执行格式为：

```
mysql -u root -p
```

然后按要求输入密码即可(如果密码是空，可以不输入密码，如图8.6所示)。

如果在远程的数据库服务器(假设ip是192.168.0.1)，建立数据库或管理数据库，执行格式为：

```
mysql -h 192.168.0.1 -u root -p
```

然后按要求输入密码即可。

成功启动命令行客户端后，MS-DOS窗口出现"mysql>"字样效果，如图8.6所示。如果想关闭命令行客户端，输入exit即可。

第8章　JSP中使用数据库

```
D:\mysql-8.0.18-winx64\bin>mysql -u root -p
Enter password:
Welcome to the MySQL monitor. Commands end with ; or \g.
Your MySQL connection id is 10
Server version: 8.0.18 MySQL Community Server - GPL

Copyright (c) 2000, 2019, Oracle and/or its affiliates. All rights reserved.

Oracle is a registered trademark of Oracle Corporation and/or its
affiliates. Other names may be trademarks of their respective
owners.

Type 'help;' or '\h' for help. Type '\c' to clear the current input statement.

mysql>
```

图 8.6　启动 MySQL 命令行客户端

❷ 创建数据库

启动命令行客户端后就可以使用 SQL 语句进行创建数据库、建表等操作。在 MS-DOS 命令行窗口输入 SQL 语句需要用";"号结束,在编辑 SQL 语句的过程中可以使用\c 终止当前 SQL 语句的编辑。需要提醒的是,可以把一个完整的 SQL 语句命令分成几行来输入,最后用分号作结束标志即可。

注:建议用记事本编辑相关的 SQL 语句,然后复制、粘贴到命令行窗口。

下面使用命令行客户端创建一个名字为 Book 的数据库。在当前命令行客户端的命令行窗口输入创建数据库的 SQL 语句:

create database bookDatabase;

如果数据库已经存在,将提示数据库已经存在,不再创建数据库,否则将创建数据库(如图 8.7 所示)。如果删除已有数据库,比如数据库 bookDatabase,执行:

drop database bookDatabase;

```
mysql> create database bookDatabase;
Query OK, 1 row affected (0.13 sec)
```

图 8.7　创建数据库

❸ 建表

创建数据库后就可以使用 SQL 语句在该库中创建表。为了在数据库中创建表,必须首先进入该数据库(即使用数据库),命令格式是:"user 数据库名;"或"user 数据库名"。在当前命令行客户端管理工具占用的命令行窗口输入:

use bookDatabase

回车确认(进入数据库也可以没有分号)进入数据库 bookDatabase,操作如图 8.8 所示。

```
mysql> use bookDatabase
Database changed
mysql>
```

图 8.8　进入 Book 数据库

下面在数据库 bookDatabase 建立一个名字为 bookList 表,该表的字段为:

ISBN(varchar) name(varchar) price(float) publishDate(date)

输入创建 bookList 表的 SQL 语句(建议用记事本编辑相关的 SQL 语句,然后右击,复制、粘贴到命令行窗口):

```
create table bookList(
ISBN varchar(100) not null,
name varchar(100) character set gb2312,
price float,
publishDate date,
primary key(ISBN)
);
```

创建表的 SQL 语句操作效果如图 8.9 所示。

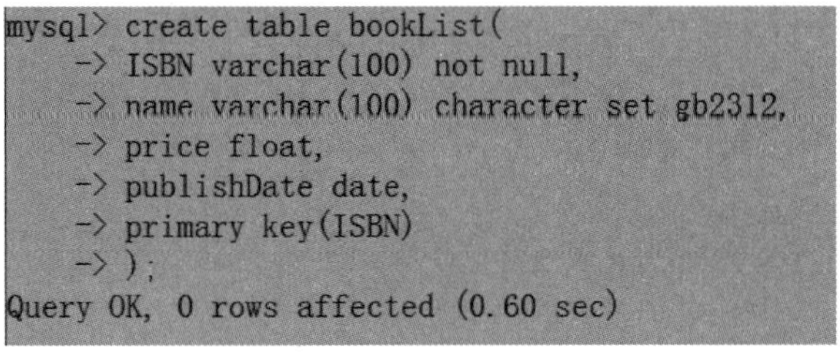

```
mysql> create table bookList(
    -> ISBN varchar(100) not null,
    -> name varchar(100) character set gb2312,
    -> price float,
    -> publishDate date,
    -> primary key(ISBN)
    -> );
Query OK, 0 rows affected (0.60 sec)
```

图 8.9 创建 bookList 表

创建 bookList 表之后就可以使用 SQL 语句对 bookList 表进行添加、更新和查询等操作(如果已经退出数据库,需要再次进入数据库)。在当前命令行客户端占用的窗口输入插入记录的 SQL 语句(如图 8.10 所示),记录之间用逗号分隔:

```
insert into bookList values('7302014655','高等数学',28.67,'2020-12-10'),
('7352014658','大学英语',58.5,'1999-9-10'),
('7987302464259','Java2 实用教程第 5 版',59.5,'2017-5-1');
```

```
mysql> insert into bookList values('7302014655','高等数学',28.67,'2020-12-10'),
    -> ('7352014658','大学英语',58.5,'1999-9-10'),
    -> ('7987302464259','Java2实用教程第5版',59.5,'2017-5-1');
Query OK, 3 rows affected (1.19 sec)
Records: 3  Duplicates: 0  Warnings: 0
```

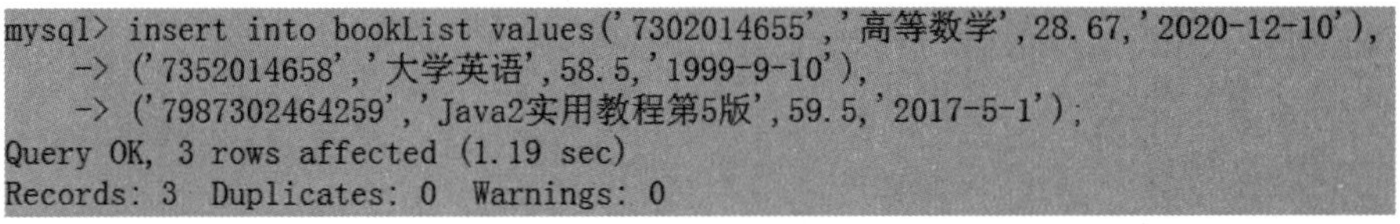

图 8.10 向 bookList 表添加记录

在当前命令行客户端占用的窗口输入查询记录的 SQL 语句:

```
select * from bookList;
```

查询 bookList 表中的全部记录,效果如图 8.11 所示。

❹ 导入.sql 文件中的 SQL 语句

在使用命令行客户端时,如果觉得在命令行输入 SQL 语句不方便,那么可以事先将需要的 SQL 语句保存在一个扩展名是.sql 的文本文件中,然后在命令行客户端占用的命令行窗口使用 source 命令导入.sql 文件中的 SQL 语句。

```
mysql> select * from bookList;
+--------------+--------------------+--------+--------------+
| ISBN         | name               | price  | publishDate  |
+--------------+--------------------+--------+--------------+
| 7302014655   | 高等数学            | 28.67  | 2020-12-10   |
| 7352014658   | 大学英语            | 58.5   | 1999-09-10   |
| 7987302464259| Java2实用教程第5版  | 59.5   | 2017-05-01   |
+--------------+--------------------+--------+--------------+
3 rows in set (0.00 sec)
```

图 8.11　查询 bookList 表中的记录

在数据库建立之后,使用这样的方式操作数据库也很方便。例如,插入记录 SQL 语句和查询 SQL 语句存放在一个 a.sql 文本文件中(a.sql 按 ANSI 编码保存在 D:\myFile),a.sql 如下:

a.sql

```
insert into bookList values('8302084658','月亮湾',38.67,'2021-12-10'),
('9352914657','雨后',78,'1998-5-19'),
('9787302198048','Java 设计模式',29,'1999-5-16'),
('97873902488644','Java 课程设计第 3 版',32,'2018-1-10');
select * from bookList;
```

在当前命令行客户端占用的窗口输入如下命令:

```
source D:/myFile/a.sql
```

回车确认,导入 sql 文件中的 SQL 语句。如果 a.sql 文件中存在错误的 SQL 语句,将提示错误信息,否则将成功执行这些 SQL 语句(效果如图 8.12 所示)。

```
Query OK, 4 rows affected (0.28 sec)
Records: 4  Duplicates: 0  Warnings: 0

+----------------+--------------------+--------+--------------+
| ISBN           | name               | price  | publishDate  |
+----------------+--------------------+--------+--------------+
| 7302014655     | 高等数学            | 28.67  | 2020-12-10   |
| 7352014658     | 大学英语            | 58.5   | 1999-09-10   |
| 7987302464259  | Java2实用教程第5版  | 59.5   | 2017-05-01   |
| 8302084658     | 月亮湾              | 38.67  | 2021-12-10   |
| 9352914657     | 雨后                |   78   | 1998-05-19   |
| 9787302198048  | Java设计模式        |   29   | 1999-05-16   |
| 97873902488644 | Java课程设计第3版    |   32   | 2018-01-10   |
+----------------+--------------------+--------+--------------+
```

图 8.12　导入 .sql 文件中的 SQL 语句

❺ **删除数据库或表**

删除数据库的命令:drop database <数据库名>,例如删除名为 tiger 的数据库:

```
drop database tiger;
```

删除表的命令:drop table <表名>,例如使用 bookDatabase 数据库后,执行:

```
drop table bookList;
```

将删除 bookDatabase 数据库中的 bookList 表。

❻ 使用 Navicat for MySQL

使用图形用户界面(GUI)的 MySQL 客户端管理工具,可以更加方便地创建数据库、在数据库中创建表等。登录 http://www.navicat.com.cn/download 下载试用版或购买商业版,例如下载试用版 navicat121_mysql_cs_x64.exe,安装后运行 Navicat.exe 启动 Navicat,然后单击"连接"按钮,建立一个连接,比如名字是 geng 的连接。成功建立连接后,在连接名字上右击,打开连接就可以在 MySQL 数据库服务器上建立数据库了。在连接名字上右击,选择建立数据库菜单,或查看数据库菜单,在数据库下的"表"上右击,在数据库中建表,如图 8.13 所示。使用 Navicat(见图 8.13)可以看到前面我们用命令行客户端建立的 bookDatabase 数据库,以及数据库中的 bookList 表。单击图 8.13 界面上的"用户"按钮,可以新增访问 MySQL 数据库服务器的用户。

图 8.13 Navicat for MySQL

8.2 连接 MySQL 数据库

视频讲解

为了使 Java 编写的程序不依赖于具体的数据库,Java 提供了专门用于操作数据库的 API,即 JDBC(Java DataBase Connectivity)。JDBC 操作不同的数据库时,仅仅是加载的数据库连接器不同以及和数据库建立连接的方式不同而已。使用 JDBC 的应用程序和数据库建立连接之后,就可以使用 JDBC 提供的 API 操作数据库(如图 8.14 所示)。

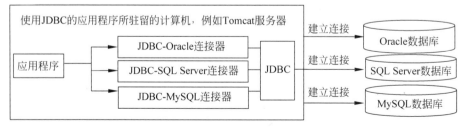

图 8.14 使用 JDBC 操作数据库

MySQL 数据库服务器启动后(见 8.1.2 节),应用程序为了能和数据库建立连接、交互信息,需要获得数据库驱动,即连接器,如图 8.14 所示。

注:对于准备采用 Microsoft Access 数据库管理系统学习本章内容的读者,建议首先阅读 8.9 节,然后将所有例子中的数据库连接器和连接方式更换成适合 Access 数据库的即可。

❶ 下载 JDBC-MySQL 数据库连接器

应用程序为了能访问 MySQL 数据库服务器上的数据库,必须要保证应用程序所驻留的计算机上安装有相应 JDBC-MySQL 数据库连接器。直接在浏览器的地址栏中直接输入:https://dev.mysql.com/downloads/connector/j/。然后在给出的下拉列表 Select Operating System 中选择 Platform Independent(即 Java 平台),然后选择 Platform Independent (Architecture Independent),ZIP 格式,单击 download 下载按钮即可。本书下载的是 mysql-connector-java-8.0.18.zip(Linux 系统可下载 tar.gz 格式文件),将该 zip 文件解压至硬盘,在解压目录下的 mysql-connector-java-8.0.18.jar 文件就是连接 MySQL 数据库的 JDBC-MySQL 数据库连接器(作者也将该文件放在了教学资源的源代码文件夹中)。

将 MySQL 数据库的 JDBC-MySQL 数据库连接器 mysql-connector-java-8.0.18.jar 保存到 Tomcat 安装目录下的 lib 文件夹中(例如 D:\apache-tomcat-9.0.26\lib),并重新启动 Tomcat 服务器。

❷ 加载 JDBC-MySQL 数据库连接器

应用程序负责加载的 JDBC-MySQL 连接器,代码如下(注意字符序列和 8.0 版本之前的 com.mysql.jdbc.Driver 不同):

```
try{   Class.forName("com.mysql.cj.jdbc.Driver ");
}
catch(Exception e){}
```

MySQL 数据库驱动是 mysql-connector-java-8.0.18.jar 文件中的 Driver 类,该类的包名是 com.mysql.cj.jdbc.Driver(包名和以前的版本不同)。Driver 类不是 Java 运行环境类库中的类,是连接器 mysql-connector-java-8.0.18.jar 中的类。

注:不要忘记将 mysql-connector-java-8.0.18.jar 保存到 Tomcat 安装目录下的 lib 文件夹中,并重新启动 Tomcat 服务器。

❸ 连接数据库

java.sql 包中的 DriverManager 类有两个用于建立连接的类方法(static 方法):

```
(1) Connection getConnection
            (java.lang.String, java.lang.String, java.lang.String)
(2) Connection getConnection(java.lang.String)
```

上述两个方法都可能抛出 SQLException 异常,DriverManager 类调用上述方法可以和数据库建立连接,即可返回一个 Connection 对象。

为了能和 MySQL 数据库服务器管理的数据库建立连接,必须保证该 MySQL 数据库服务器已经启动,如果没有更改过 MySQL 数据库服务器的配置,那么该数据库服务器占用的端口是 3306。假设 MySQL 数据库服务器所驻留的计算机的 IP 地址是 192.168.100.1(命令行运行 ipconfig 可以得到当前计算机的 IP 地址)。

应用程序要和 MySQL 数据库服务器管理的数据库 Book(在 8.1.3 节建立的数据库)建立连接,而有权访问数据库 Book 的用户的 id 和密码分别是 root 和空。建立连接的代码如下:

```
Connection con;
Stringurl =
"jdbc:mysql://192.168.100.1:3306/Book?useSSL = false&serverTimezone = GMT";
String user = "root";
String password = "";
try{
        con = DriverManager.getConnection(url,user,password);        //连接代码
}
catch(SQLException e){
        System.out.println(e);
}
```

对丁 MySQL 8.0 版本,必须设置 serverTimezone 参数的值(值可以是 MySQL 8.0 支持的时区之一即可,例如 EST、CST、GMT 等),例如:serverTimezone = CST 或 serverTimezone = GMT(CST 是 Eastern Standard Time 的缩写,CST 是 China Standard Time 的缩写,GMT 是 Greenwich Mean Tim 缩写)。如果 root 用户密码是 99,将 password = ""更改为 password = "99"即可。

MySQL 5.7 以及之后的版本建议应用程序和数据库服务器建立连接时明确设置 SSL (Secure Sockets Layer),即在连接信息里明确使用 useSSL 参数,并设置值是 true 或 false,如果不设置 useSSL 参数,程序运行时总会提示用户程序进行明确设置(但不影响程序的运行)。对于早期的 MySQL 版本,用户程序不必设置该项。

应用程序和某个数据库建立连接之后,就可以通过 SQL 语句和该数据库中的表交互信息,比如查询、修改、更新表中的记录。

注:如果用户要和连接 MySQL 驻留在同一计算机上,使用的 IP 地址可以是 127.0.0.1 或 localhost。另外,由于 3306 是 MySQL 数据库服务器的默认端口号,连接数据库时允许应用程序省略默认的 3306。

❹ 注意汉字问题

需要特别注意的是,如果数据库的表中的记录有汉字,那么在建立连接时需要额外多传递一个参数 characterEncoding,并取值 GB2312 或 UTF-8:

```
String url =
"jdbc:mysql://localhost/bookDatabase?" +
"useSSL = false&serverTimezone = GMT&characterEncoding = utf - 8";
con = DriverManager.getConnection(url, "root","");        //连接代码
```

例 8_1 是一个简单的 JSP 页面,该页面中的 Java 程序片代码负责加载 JDBC-驱动程序,并连接到数据库 bookDatabase,查询 bookList 表中的全部记录(见 8.1.3 节曾建立的 bookDatabase 数据库)。

第8章 JSP中使用数据库

例 8_1

example8_1.jsp（效果如图 8.15 所示）

ISBN	名称	价格	日期
7302014655	高等数学	28.67	2020-12-10
7352014658	大学英语	58.5	1999-09-10
7987302464259	Java2实用教程第5版	59.5	2017-05-01
8302084658	月亮湾	38.67	2021-12-10
9352914657	雨后	78.0	1998-05-19
9787302198048	Java设计模式	29.0	1999-05-16
97873902488644	Java课程设计第3版	32.0	2018-01-10

图 8.15　连接 bookDatabase 数据库，查询 bookList 表

```jsp
<%@ page contentType="text/html" %>
<%@ page pageEncoding="utf-8" %>
<%@ page import="java.sql.*" %>
<style>
    #tom{
        font-family:宋体;font-size:18;color:blue
    }
</style>
<HTML><body bgcolor=#EEDDFF>
<% Connection con = null;
    Statement sql;
    ResultSet rs;
    try{   //加载JDBC-MySQL 8.0 连接器
        Class.forName("com.mysql.cj.jdbc.Driver");
    }
    catch(Exception e){
        out.print("<h1>" + e);
    }
    String url = "jdbc:mysql://localhost:3306/bookDatabase?" +
    "useSSL = false&serverTimezone = CST&characterEncoding = utf-8";
    String user = "root";
    String password = "";
    out.print("<table border = 1>");
    out.print("<tr>");
    out.print("<th id = tom width = 100>" + "ISBN");
    out.print("<th id = tom width = 100>" + "名称");
    out.print("<th id = tom width = 50>" + "价格");
    out.print("<th id = tom width = 50>" + "日期");
    out.print("</tr>");
    try{
        con = DriverManager.getConnection(url,user,password);   //连接数据库
        sql = con.createStatement();
        String SQL = "SELECT * FROM bookList";                  //SQL 语句
        //String SQL = "select * from bookList " +
        //"where year(publishDate) between 1999 and 2021 and price >= 30 " +
        //" order by publishDate"
        rs = sql.executeQuery(SQL);                             //查表
        while(rs.next()) {
            out.print("<tr>");
```

```
                    out.print("<td id=tom>" + rs.getString(1) + "</td>");
                    out.print("<td id=tom>" + rs.getString(2) + "</td>");
                    out.print("<td id=tom>" + rs.getFloat(3) + "</td>");
                    out.print("<td id=tom>" + rs.getDate(4) + "</td>");
                    out.print("</tr>");
                }
                out.print("</table>");
                con.close();
            }
            catch(SQLException e) {
                out.print("<h1>" + e);
            }
    %>
    </body></HTML>
```

注：如果 JSP 页面无法正确运行，并且 Tomcat 服务器窗口提示 No suitable driver found for jdbc:mysql…，请检查是否已经将 mysql-connector-java-8.0.18.jar 数据库连接器保存到 Tomcat 安装目录下的 lib 文件夹中，并重新启动了 Tomcat 服务器。

8.3 查询记录

和数据库建立连接后，就可以使用 JDBC 提供的 API 与数据库交互信息，例如查询、修改和更新数据库中的表等。JDBC 与数据库表进行交互的主要方式是使用 SQL 语句（其他方式见 8.5 节）。JDBC 提供的 API 可以将标准的 SQL 语句发送给数据库，实现和数据库的交互。

▶ 8.3.1 结果集与查询

视频讲解

对一个数据库中的表进行查询，然后将查询结果返回到一个 ResultSet 对象中，习惯称 ResultSet 对象为结果集对象。

使用 Statement 声明一个 SQL 语句对象，然后让已创建的连接对象 con 调用方法 createStatement() 返回 SQL 语句对象，代码如下：

```
try{   Statement sql = con.createStatement();
    }
catch(SQLException e ){
        System.out.println(e);
}
```

有了 SQL 语句对象后，这个对象就可以调用相应的方法查询数据库中的表，并将查询结果存放在一个 ResultSet 结果集中。也就是说 SQL 查询语句对数据库的查询操作将返回一个 ResultSet 结果集，ResultSet 结果集由以列（也称字段）为结构的数据行组成。例如，对于

```
ResultSet rs = sql.executeQuery("SELECT * FROM bookList");
```

内存中的结果集对象 rs 的列数是 4 列，刚好和 bookList 表的列数相同，第 1 列至第 4 列分别是 ISBN、name、price 和 publishDate 列；而对于

```
ResultSet rs = sql.executeQuery("SELECT name,price FROM bookList");
```

结果集对象 rs 只有两列,第 1 列是 name 列,第 2 列是 price 列。

ResultSet 结果集一次只能看到一个数据行,使用 next()方法可走到下一数据行。获得一行数据后,ResultSet 结果集可以使用 getXxx 方法获得字段值(列值),将位置索引(第一列使用 1,第二列使用 2 等)或列名传递给 getXxx 方法的参数即可。表 8.1 给出了 ResultSet 结果集的若干方法。

表 8.1　ResultSet 结果集的若干方法

返回类型	方法名称	返回类型	方法名称
boolean	next()	byte	getByte(String columnName)
byte	getByte(int columnIndex)	Date	getDate(String columnName)
Date	getDate(int columnIndex)	double	getDouble(String columnName)
double	getDouble(int columnIndex)	float	getFloat(String columnName)
float	getFloat(int columnIndex)	int	getInt(String columnName)
int	getInt(int columnIndex)	long	getLong(String columnName)
long	getLong(int columnIndex)	String	getString(String columnName)
String	getString(int columnIndex)		

注:无论列(字段)是何种属性,总可以使用 getString(int columnIndex)或 getString(String columnName)方法返回列(字段)值的串表示。

▶ 8.3.2　随机查询

视频讲解

前面 8.4.1 节给出顺序查询的步骤,如例 8_1 所示。ResultSet 结果集使用 next()方法顺序地查询记录,但有时候要在结果集中前后移动、显示结果集指定的一条(一行)记录或随机显示若干条记录等。这时,必须返回一个可滚动的结果集(结果集的游标可以上下移动)。为了得到一个可滚动的结果集,须使用下述方法先获得一个 Statement 对象:

```
Statement stmt = con.createStatement(int type, int concurrency);
```

然后,根据参数的 type、concurrency 的取值情况,stmt 返回相应类型的结果集:

```
ResultSet re = stmt.executeQuery(SQL 语句);
```

(1) type 的取值决定滚动方式,取值可以是:

- ResultSet.TYPE_FORWORD_ONLY:结果集的游标只能向下滚动。
- ResultSet.TYPE_SCROLL_INSENSITIVE:结果集的游标可以上下移动,当数据库变化时,当前结果集不变。
- ResultSet.TYPE_SCROLL_SENSITIVE:结果集的游标可以上下移动,当数据库变化时,当前结果集同步改变。

(2) Concurrency 取值决定是否可以用结果集更新数据库,Concurrency 取值:

- ResultSet.CONCUR_READ_ONLY:不能用结果集更新数据库中的表。
- ResultSet.CONCUR_UPDATABLE:能用结果集更新数据库中的表。

(3) 滚动查询经常用到 ResultSet 的下述方法:

- public boolean previous()：将游标向上移动,当移到结果集第一行前面时,返回 false。
- public void beforeFirst()：将游标移动到结果集的初始位置,即在第一行之前。
- public void afterLast()：将游标移到结果集最后一行之后。
- public void first()：将游标移到结果集的第一行。
- public void last()：将游标移到结果集的最后一行。
- public boolean isAfterLast()：判断游标是否在最后一行之后。
- public boolean isBeforeFirst()：判断游标是否在第一行之前。
- public boolean isFirst()：判断游标是否指向结果集的第一行。
- public boolean isLast()：判断游标是否指向结果集的最后一行。
- public int getRow()：得到当前游标所指行的行号,行号从 1 开始。如果结果集没有行,返回 0。
- public boolean absolute(int row)：将游标移到参数 row 指定的行号。

注：如果 row 取负值,就是倒数的行数。absolute(-1)表示移到最后一行,absolute(-2)表示移到倒数第 2 行。当移动到第一行前面或最后一行的后面时,该方法返回 false。

可以看出,用滚动集的 absolute(int row)方法就可以随机查询某条记录。

例 8_2 随机查询 bookDatabase 数据库中 bookList 表的 3 条记录(见 8.1.3 节建立的数据库)。例子中首先将游标移动到最后一行,然后再获取最后一行的行号,以便获得 bookList 表中的记录数目。例子中的 int [] getRandomNumber(int max,int amount)方法返回 1 至 max 之间的 amount 个不同的随机数。

例 8_2

example8_2.jsp(效果如图 8.16 所示)

```jsp
<%@ page contentType="text/html" %>
<%@ page pageEncoding="utf-8" %>
<%@ page import="java.sql.*" %>
<%@ page import="java.util.LinkedList" %>
<%@ page import="java.util.Random" %>
<%! public int [] getRandomNumber(int max,int amount){
      LinkedList<Integer> list = new LinkedList<Integer>();
      for(int i=1;i<=max;i++){
          list.add(i);
      }
      int result[] = new int[amount];
      while(amount>0){
          int index = new Random().nextInt(list.size());
          int m = list.remove(index);
          result[amount-1] = m;
          amount--;
      }
      return result;
    }
%>
<style>
    #tom{
```

```jsp
         font-family:宋体;font-size:18;color:blue
      }
</style>
<HTML><body bgcolor=#EEDDFF>
<% Connection con=null;
   Statement sql;
   ResultSet rs;
   try{    //加载 JDBC-MySQL 8.0 连接器
      Class.forName("com.mysql.cj.jdbc.Driver");
   }
   catch(Exception e){
      out.print("<h1>"+e);
   }
   String url = "jdbc:mysql://localhost:3306/bookDatabase?"+
    "useSSL=false&serverTimezone=CST&characterEncoding=utf-8";
   String user  = "root";
   String password  = "";
   out.print("<table border=1>");
   out.print("<tr>");
   out.print("<th id=tom width=100>"+"ISBN");
   out.print("<th id=tom width=100>"+"名称");
   out.print("<th id=tom width=50>"+"价格");
   out.print("<th id=tom width=50>"+"日期");
   out.print("</tr>");
   int count=3;
   try{
     con=DriverManager.getConnection(url,user,password);       //连接数据库
     sql=con.createStatement(ResultSet.TYPE_SCROLL_SENSITIVE,
                             ResultSet.CONCUR_READ_ONLY);
     rs=sql.executeQuery("SELECT * FROM bookList ");
     rs.last();
     int max=rs.getRow();
     out.print("表共有"+max+"条记录,随机抽取"+count+"条记录: ");
     int [] a =getRandomNumber(max,count);
     for(int i:a){   // i 依次取数组每个单元的值
        rs.absolute(i);                                     //游标移动到第 i 行
        out.print("<tr>");
        out.print("<td id=tom>"+rs.getString(1)+"</td>");
        out.print("<td id=tom>"+rs.getString(2)+"</td>");
        out.print("<td id=tom>"+rs.getFloat(3)+"</td>");
        out.print("<td id=tom>"+rs.getDate(4)+"</td>");
        out.print("</tr>") ;
     }
     out.print("</table>");
     con.close();
   }
   catch(SQLException e) {
      out.print("<h1>"+e);
   }
%>
</body></HTML>
```

图 8.16　随机查询

8.3.3　条件查询

❶ where 子语句

一般格式：

视频讲解

```
select 字段 from 表名 where 条件
```

(1) 字段值和固定值比较，例如：

```
select name,price from bookList where name = '高等数学'
```

(2) 字段值在某个区间范围，例如：

```
select * from bookList where price > 28.68 and price <= 87.7
select * from bookList where price > 56 and name != '月亮湾'
```

使用某些特殊的日期函数，如 year、month、day：

```
select * from bookList where year(publishDay) < 1980
select * from bookList where year(publishDay) between 2015 and 2020
```

(3) 用操作符 like 进行模式般配，使用％代替 0 个或多个字符，用一个下画线代替一个字符。例如查询 name 含有"程序"两个字的记录：

```
select * from bookLidt where name like '%程序%'
```

❷ 排序

用 order by 子语句对记录排序，

```
select * from 表名 order by 字段名(列名)
select * from 表名 where 条件 order by 字段名(列名)
```

例如：

```
select * from bookList order by price
select * from bookList where name like '%编程%' order by name
```

将例 8_1 中的查询 bookList 表中 SQL 语句：

```
String SQL = "SELECT * FROM bookList";
```

替换成：

```
String SQL = "select * from bookList " +
    "where year(publishDate) between 1999 and 2021 and price >= 30 " +
    " order by publishDate";
```

即查询价格大于或等于30，出版日期在2017年至2021年的图书，并按出版时间publishData排序。效果如图8.17所示。

ISBN	名称	价格	日期
7352014658	大学英语	58.5	1999-09-10
7987302464259	Java2实用教程第5版	59.5	2017-05-01
97873902488644	Java课程设计第3版	32.0	2018-01-10
8302084658	月亮湾	38.67	2021-12-10

图8.17　条件查询

8.4　更新、添加与删除记录

视频讲解

Statement对象调用方法：

```
public int executeUpdate(String sqlStatement);
```

通过参数sqlStatement指定的方式实现对数据库表中记录的更新、添加和删除操作。方法执行成功（成功更新、添加或删除），将返回一个正整数，否则返回0。

❶ 更新

```
update 表 set 字段 = 新值 where <条件子句>
```

下述SQL语句将bookList表中name值为"大学英语"的记录的publishDate字段的值更新为2019-12-26：

```
update bookList set publishDate = '2019 - 12 - 26' where name = '大学英语'
```

❷ 添加

```
insert into 表(字段列表) values (对应的具体的记录)
```

或

```
insert into 表 values (对应的具体的记录)
```

下述SQL语句将向bookList表中添加2条新的记录（记录之间用逗号分隔）：

```
insert into bookList values
            ('2306084657', '春天', 35.8, '2020 - 3 - 20'),
            ('5777564629', '冬日', 29.9, '2019 - 12 - 23')
```

❸ 删除

```
delete from 表名 where <条件子句>
```

下述SQL语句将删除bookList表中的ISBN字段值为9352914657的记录：

```
delete from bookList where ISBN = '9352914657'
```

注：需要注意的是，当返回结果集后，没有立即输出结果集的记录，而接着执行了更新语句，那么结果集就不能输出记录了。要想输出记录就必须重新返回结果集。

例 8_3 将 bookList 表中 name 值为"大学英语"的记录的 publishDate 字段的值更新为 '2019-12-26',向 bookList 表中添加 2 条新的记录:('2306084657','春天',35.8,'2020-3-20')和('5777564629','冬日',29.9,'2019-12-23'),删除了 ISBN 字段值为 9352914657 的记录。

例 8_3

example8_3.jsp(效果如图 8.18 所示)

```jsp
<%@ page contentType="text/html" %>
<%@ page pageEncoding="utf-8" %>
<%@ page import="java.sql.*" %>
<style>
    #tom{
        font-family:宋体;font-size:18;color:blue
    }
</style>
<HTML><body bgcolor=#EEDFF>
<% Connection con = null;
    Statement sql;
    ResultSet rs;
    try{//加载 JDBC-MySQL 8.0 连接器
        Class.forName("com.mysql.cj.jdbc.Driver");
    }
    catch(Exception e){
        out.print("<h1>" + e);
    }
    String url = "jdbc:mysql://localhost:3306/bookDatabase?" +
    "useSSL=false&serverTimezone=CST&characterEncoding=utf-8";
    String user = "root";
    String password = "";
    out.print("<table border=1>");
    out.print("<tr>");
    out.print("<th id=tom width=100>" + "ISBN");
    out.print("<th id=tom width=100>" + "名称");
    out.print("<th id=tom width=50>" + "价格");
    out.print("<th id=tom width=50>" + "日期");
    out.print("</tr>");
    try{
        con = DriverManager.getConnection(url,user,password);      //连接数据库
        sql = con.createStatement();
        String updateSQL =
        "update bookList set publishDate='2019-12-26' where name='大学英语'";
        String indsertSQL =
        "insert into bookList values('2306084657','春天',35.8,'2020-3-20')," +
        "('5777564629','冬日', 29.9,'2019-12-23')";                //插入语句
        String deleteSQL =
        "delete from bookList where ISBN='9352914657'";            //删除语句
        int ok = sql.executeUpdate(updateSQL);                     //更新
        ok = sql.executeUpdate(indsertSQL);                        //插入
        ok = sql.executeUpdate(deleteSQL);                         //删除
        out.print("<h1>" + ok);
        rs = sql.executeQuery("SELECT * FROM bookList");           //查表
```

```
            while(rs.next()) {
                out.print("<tr>");
                out.print("<td id = tom>" + rs.getString(1) + "</td>");
                out.print("<td id = tom>" + rs.getString(2) + "</td>");
                out.print("<td id = tom>" + rs.getFloat(3) + "</td>");
                out.print("<td id = tom>" + rs.getDate(4) + "</td>");
                out.print("</tr>") ;
            }
            out.print("</table>");
            con.close();
        }
        catch(SQLException e) {
            out.print("<h1>" + e);
        }
    %>
</body></HTML>
```

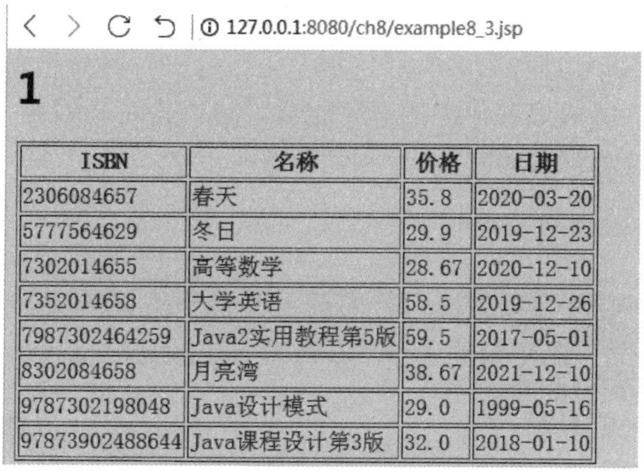

图 8.18　更新、添加与删除记录

8.5　用结果集操作数据库中的表

尽管可以用 SQL 语句对数据库中表进行更新、插入操作，但也可以使用内存中 ResultSet 结果集对底层数据库表进行更新和插入操作(这些操作由系统自动转化为相应的 SQL 语句)。优点是不必熟悉有关更新、插入的 SQL 语句，而且方便编写代码。缺点是，必须得到一个可滚动的、可以用结果集更新数据库的 ResultSet 结果集 rs(见 8.3.2 节)。

▶ 8.5.1　更新记录

使用结果集更新数据库表中第 n 行记录中某列的值的步骤如下。

(1) 游标移动到第 n 行。结果集 rs 调用 absolute()方法将游标移到第 n 行：

```
rs.absolute(n);
```

(2) 结果集 rs 将第 n 行的 column 列的列值更新。结果集可以使用下列方法更新列值：

```
updateInt(String column, int x),updateInt(int column int x)
```

```
updateLong(String column, long x), updateLong(int column, long x)
updateDouble(String column, double x), updateDouble(int column, double x)
updateString(String column, String x), updateString(int column, String x)
updateDate(String column, Date x), updateDate(int column, Date x)
```

(3) 更新数据库中的表。最后,结果集 rs 调用 updateRow()方法用结果集中的第 n 行更新数据库表中的第 n 行记录。

以下代码片段按照上述步骤,更新 bookList 表中的第 3 行记录的 name 列(字段)的值。

```
rs.absolute(3);
rs.updateString(2, "操作系统");        //也可以写成 rs.updateString("name","操作系统");
rs.updateRow();
```

▶ 8.5.2 插入记录

使用结果集向数据库表中插入(添加)一行记录步骤如下:

(1) 将结果集 rs 的游标移动到插入行。结果集中有一个特殊区域,用作构建要插入的行的暂存区域,习惯上将该区域位置称作结果集的插入行。为了向数据库表中插入一行新的记录,必须首先将结果集的游标移动到插入行,代码如下:

```
rs.moveToInsertRow();
```

(2) 更新插入行的列值,即设置插入行的列值。结果集可以用 updateXxx()方法更新插入行的列值,例如,准备插入的记录是('7307014659 ','数据结构', 58 , '2020-08-10'),那么执行下列操作:

```
rs.updateString(1, "7307014659");
rs.updateString(2, "数据结构");
rs.updateFloat(3,58);
rs.updateDate(4, '2020 - 08 - 10');
```

(3) 插入记录。结果集调用 insertRow()方法用结果集中的插入行向数据库表中插入一行新记录。

例 8_4 将 bookList 表中 ISBN 值为 8302084658 的 price 的值更新为 56,向 bookList 表中添加 2 条新的记录:('8306084656','四月天',39.2,'2020-5-20')和('6777564622','Java 面向对象程序设计第 3 版(微课版)', 59.9,'2020-1-23')。

例 8_4

example8_4.jsp(效果如图 8.19 所示)

```
<%@ page contentType = "text/html" %>
<%@ page pageEncoding = "utf - 8" %>
<%@ page import = "java.sql.*" %>
<style>
    #tom{
        font - family:宋体;font - size:18;color:blue
    }
</style>
```

```jsp
<HTML><body bgcolor=#EEDDFF>
<% Connection con=null;
    Statement sql;
    ResultSet rs;
    try{    //加载JDBC-MySQL 8.0连接器
        Class.forName("com.mysql.cj.jdbc.Driver");
    }
    catch(Exception e){
        out.print("<h1>"+e);
    }
    String url="jdbc:mysql://localhost:3306/bookDatabase?"+
    "useSSL=false&serverTimezone=CST&characterEncoding=utf-8";
    String user = "root";
    String password = "";
    out.print("<table border=1>");
    out.print("<tr>");
    out.print("<th id=tom width=100>"+"ISBN");
    out.print("<th id=tom width=100>"+"名称");
    out.print("<th id=tom width=50>"+"价格");
    out.print("<th id=tom width=50>"+"日期");
    out.print("</tr>");
    try{
        con=DriverManager.getConnection(url,user,password);       //连接数据库
        sql=con.createStatement(ResultSet.TYPE_SCROLL_SENSITIVE,
                    ResultSet.CONCUR_UPDATABLE);     //可更新数据库的结果集
        rs=sql.executeQuery("SELECT * FROM bookList");
        rs.absolute(6);                //游标移到第6行,第6行ISBN的值是8302084658
        rs.updateFloat("price",56);    //结果集第6行的price更新为56
        rs.updateRow();                //bookList表中第6行的price更新为56
        rs.moveToInsertRow();          //游标移到第插入行
        rs.updateString(1,"8306084656");
        rs.updateString(2,"四月天");
        rs.updateFloat(3,39.2f);
        rs.updateString(4,"2020-5-20");         //向插入行插入记录
        rs.insertRow();                         //向bookList表中插入一条记录
        rs.moveToInsertRow();                   //游标移到第插入行
        rs.updateString(1,"6777564622");
        rs.updateString(2,"Java面向对象程序设计第3版(微课版)");
        rs.updateFloat(3,59.9f);
        rs.updateString(4,"2020-1-23");         //向插入行插入记录
        rs.insertRow();                         //向bookList表中插入一条记录
        rs=sql.executeQuery("SELECT * FROM bookList");  //查表
        while(rs.next()) {
            out.print("<tr>");
            out.print("<td id=tom>"+rs.getString(1)+"</td>");
            out.print("<td id=tom>"+rs.getString(2)+"</td>");
            out.print("<td id=tom>"+rs.getFloat(3)+"</td>");
            out.print("<td id=tom>"+rs.getDate(4)+"</td>");
            out.print("</tr>") ;
        }
        out.print("</table>");
        con.close();
```

```
            }
        catch(SQLException e) {
            out.print("< h1 >" + e);
            out.print("< h1 > ISBN 是主键,主键值不能有相同的");
            }
    %>
</body></HTML>
```

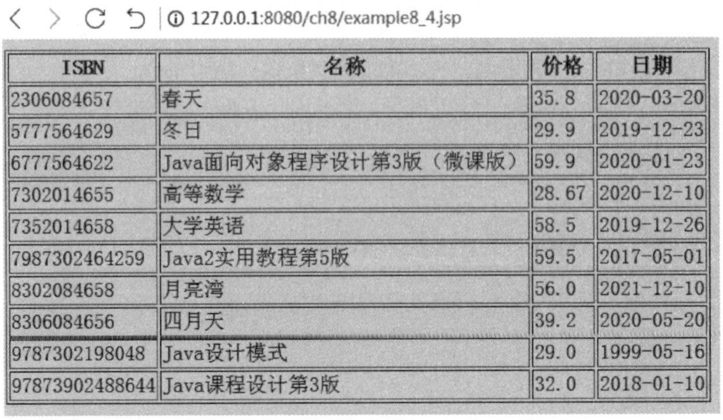

图 8.19　使用结果集操作数据库中的表

8.6　预处理语句

Java 提供了更高效率的数据库操作机制,就是 PreparedStatement 对象,该对象被习惯地称作预处理语句对象。本节学习怎样使用预处理语句对象操作数据库中的表。

▶ 8.6.1　预处理语句的优点

视频讲解

当向数据库发送一个 SQL 语句,例如"select * from bookList",数据库中的 SQL 解释器负责将把 SQL 语句生成底层的内部命令,然后执行该命令,完成有关的数据操作。如果不断地向数据库提交 SQL 语句势必增加数据库中 SQL 解释器的负担,影响执行的速度。如果应用程序能针对连接的数据库,事先就将 SQL 语句解释为数据库底层的内部命令,然后直接让数据库去执行这个命令,显然不仅减轻了数据库的负担,而且也提高了访问数据库的速度。

Connection 连接对象 con 调用 prepareStatement(String sql)方法:

```
PreparedStatement pre = con.prepareStatement(String sql);
```

对参数 sql 指定的 SQL 语句进行预编译处理,生成该数据库底层的内部命令,并将该命令封装在 PreparedStatement 对象 pre 中,那么该对象调用下列方法都可以使得该底层内部命令被数据库执行:

```
ResultSet executeQuery()
boolean execute()    (执行成功返回 false)
int executeUpdate()   (执行成功返回 1)
```

第8章　JSP中使用数据库

只要编译好了PreparedStatement对象pre,那么pre可以随时执行上述方法,显然提高了访问数据库的速度。

下面的例8_5使用预处理语句来查询bookDatabase数据库中bookList表的全部记录(有关bookList表见8.1.3节),请读者比较例8_5和例8_1的不同之处。

例8_5

example8_5.jsp

```jsp
<%@ page contentType="text/html" %>
<%@ page pageEncoding="utf-8" %>
<%@ page import="java.sql.*" %>
<style>
   #tom{
      font-family:宋体;font-size:18;color:blue
   }
</style>
<HTML><body bgcolor=#EEDDFF>
<% Connection con;
   PreparedStatement pre = null;        //预处理语句,和例8_1不同
   ResultSet rs;
   try{                                 //加载JDBC-MySQL 8.0连接器
      Class.forName("com.mysql.cj.jdbc.Driver");
   }
   catch(Exception e){
      out.print("<h1>" + e);
   }
   String url = "jdbc:mysql://localhost:3306/bookDatabase?" +
   "useSSL=false&serverTimezone=CST&characterEncoding=utf-8";
   String user = "root";
   String password = "";
   out.print("<table border=1>");
   out.print("<tr>");
   out.print("<th id=tom width=100>" + "ISBN");
   out.print("<th id=tom width=100>" + "名称");
   out.print("<th id=tom width=50>" + "价格");
   out.print("<th id=tom width=50>" + "日期");
   out.print("</tr>");
   try{
      con = DriverManager.getConnection(url,user,password);  //连接数据库
      String SQL = "SELECT * FROM bookList";                 //SQL语句
      pre = con.prepareStatement(SQL);                       //进行预处理,和例8_1不同
      rs = pre.executeQuery();                               //和例8_1不同
      while(rs.next()) {
         out.print("<tr>");
         out.print("<td id=tom>" + rs.getString(1) + "</td>");
         out.print("<td id=tom>" + rs.getString(2) + "</td>");
         out.print("<td id=tom>" + rs.getFloat(3) + "</td>");
         out.print("<td id=tom>" + rs.getDate(4) + "</td>");
         out.print("</tr>");
      }
      out.print("</table>");
```

```
            con.close();
        }
        catch(SQLException e) {
            out.print("< h1 >" + e);
        }
%>
</body></HTML>
```

8.6.2 使用通配符

视频讲解

在对 SQL 进行预处理时可以使用通配符？（英文问号）来代替字段的值,只要在预处理语句执行之前再设置通配符所表示的具体值即可。例如：

```
pre = con.prepareStatement("SELECT * FROMbookList WHERE price < ? ");
```

那么在 pre 对象执行之前,必须调用相应的方法设置通配符？代表的具体值,例如：

```
pre.setFloat(1,65);
```

指定上述预处理语句 pre 中通配符？代表的值是 65。通配符按照它们在预处理的"SQL 语句"中从左至右依次出现的顺序分别被称作第 1 个、第 2 个……第 m 个通配符。例如,下列方法：

```
void setFloat(int parameterIndex, int x);
```

用来设置通配符的值,其中参数 parameterIndex 用来表示 SQL 语句中从左到右的第 parameterIndex 个通配符号,x 是该通配符所代表的具体值。

尽管

```
pre = con.prepareStatement("SELECT * FROMbookList WHERE price < ? ");
pre.setDouble(1,65);
```

的功能等同于

```
pre = con.prepareStatement("SELECT * FROM message WHEREprice < 65 ");
```

但是,使用通配符可以使得应用程序更容易动态地改变 SQL 语句中关于字段值的条件。

预处理语句设置通配符？的值的常用方法有：

```
void setDate(int parameterIndex, Date x)
void setDouble(int parameterIndex, double x)
void setFloat(int parameterIndex, float x)
void setInt(int parameterIndex, int x)
void setLong(int parameterIndex, long x)
void setString(int parameterIndex, String x)
```

例 8_6 使用预处理语句(注意通配符？的用法)将 bookList 表中 name 值为"大学英语"和"高等数学"的 publishDate 的值分别更新为 2021-2-10 和 2021-6-20,向 bookList 表中添加 2 条新的记录：('92306084659','数据库原理',56.9,'2019-2-10')和('82306884758','计算机组成原理',66.7,'2020-8-10')。

例 8_6
example8_6.jsp（效果如图 8.20 所示）

```jsp
<%@ page contentType="text/html" %>
<%@ page pageEncoding="utf-8" %>
<%@ page import="java.sql.*" %>
<style>
   #tom{
      font-family:宋体;font-size:18;color:blue
   }
</style>
<HTML><body bgcolor=#EEDDFF>
<% Connection con;
   PreparedStatement pre = null;                    //预处理语句.
   ResultSet rs;
   try{  //加载 JDBC-MySQL 8.0 连接器
      Class.forName("com.mysql.cj.jdbc.Driver");
   }
   catch(Exception e){
      out.print("<h1>" + e);
   }
   String url = "jdbc:mysql://localhost:3306/bookDatabase?" +
   "useSSL=false&serverTimezone=CST&characterEncoding=utf-8";
   String user = "root";
   String password = "";
   out.print("<table border=1>");
   out.print("<tr>");
   out.print("<th id=tom width=100>" + "ISBN");
   out.print("<th id=tom width=100>" + "名称");
   out.print("<th id=tom width=50>" + "价格");
   out.print("<th id=tom width=50>" + "日期");
   out.print("</tr>");
   try{
      con = DriverManager.getConnection(url,user,password);    //连接数据库
      String querySQL = "select * from booklist";
      //带?的 SQL 语句
      String updateSQL = "update bookList set publishDate = ?where name = ?";
      String insertSQL = "insert into bookList values(?,?,?,?)";
      pre = con.prepareStatement(updateSQL);       //进行预处理返回预处理语句 SQL
      pre.setString(1,"2021-2-10");
      pre.setString(2,"大学英语");
      boolean boo = pre.execute();                 //执行更新
      pre.setString(1,"2021-6-20");
      pre.setString(2,"高等数学");
      boo = pre.execute();                         //执行更新
      pre = con.prepareStatement(insertSQL);       //进行预处理返回预处理语句 SQL
      pre.setString(1,"92306084659");
      pre.setString(2,"数据库原理");
      pre.setFloat(3,56.9f);
      pre.setString(4,"2019-2-10");
      int ok = pre.executeUpdate();;               //插入新记录
```

```
        pre.setString(1,"82306884758");
        pre.setString(2,"计算机组成原理");
        pre.setFloat(3,66.7f);
        pre.setString(4,"2020-8-10");
        ok = pre.executeUpdate();                              //插入新记录
        out.print("<h1>"+ok+" "+boo);
        pre = con.prepareStatement(querySQL);
        rs = pre.executeQuery();
        while(rs.next()) {
           out.print("<tr>");
           out.print("<td id=tom>"+rs.getString(1)+"</td>");
           out.print("<td id=tom>"+rs.getString(2)+"</td>");
           out.print("<td id=tom>"+rs.getFloat(3)+"</td>");
           out.print("<td id=tom>"+rs.getDate(4)+"</td>");
           out.print("</tr>") ;
        }
        out.print("</table>");
        con.close();
     }
     catch(SQLException e) {
        out.print("<h1>"+e);
        out.print("<h1>ISBN是主键,主键值不能有相同的");
     }
  %>
</body></HTML>
```

ISBN	名称	价格	日期
2306084657	春天	35.8	2020-03-20
5777564629	冬日	29.9	2019-12-23
6777564622	Java面向对象程序设计第3版（微课版）	59.9	2020-01-23
7302014655	高等数学	28.67	2021-06-20
7352014658	大学英语	58.5	2021-02-10
7987302464259	Java2实用教程第5版	56.0	2017-05-01
82306884758	计算机组成原理	66.7	2020-08-10
8302084658	月亮湾	56.0	2021-12-10
8306084656	四月天	39.2	2020-05-20
92306084659	数据库原理	56.9	2019-02-10
9787302198048	Java设计模式	29.0	1999-05-16
97873902488644	Java课程设计第3版	32.0	2018-01-10

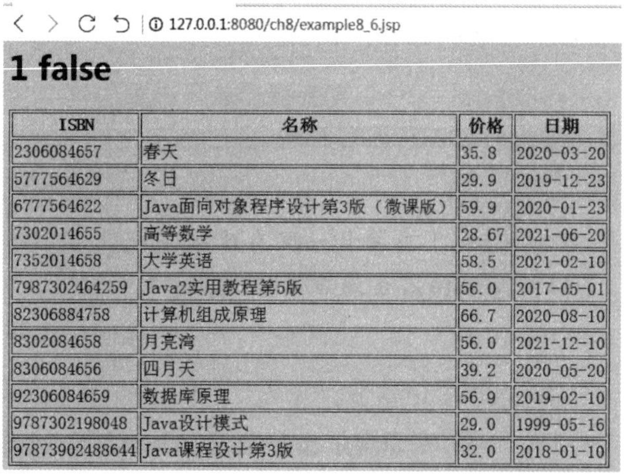

图 8.20　使用预处理语句

8.7　事务

视频讲解

事务由一组 SQL 语句组成。所谓“事务处理”是指应用程序保证事务中的 SQL 语句要么全部都执行,要么一个都不执行。

第8章　JSP中使用数据库

事务处理是保证数据库中数据完整性与一致性的重要机制。应用程序和数据库建立连接之后，可能使用多条 SQL 语句操作数据库中的一个表或多个表。一个管理资金转账的应用程序为了完成一个简单的转账业务可能需要两条 SQL 语句，例如，用户 geng 给另一个用户 zhang 转账 50 元，那么需要一条 SQL 语句完成将用户 geng 的 userMoney 的值由原来的 100 更改为 50(减去 50 的操作)，另一条 SQL 语句完成将 zhang 的用户的 userMoney 的值由原来的 20 更新为 70(增加 50 的操作)。应用程序必须保证这两条 SQL 语句要么全都执行，要么全都不执行。

JDBC 事务处理步骤如下：

❶ 用 setAutoCommit(boolean b)方法关闭自动提交模式

事务处理的第一步骤是使用 setAutoCommit(boolean autoCommit)方法关闭自动提交模式。这样做的理由是，和数据库建立连接的对象 con 的提交模式是自动提交模式，即该连接 con 产生的 Statement 或 PreparedStatement 对象对数据库提交任何一个 SQL 语句操作都会立刻生效，使得数据库中的数据发生变化，这显然不能满足事物处理的要求。例如，在转账操作时，将用户 geng 的 userMoney 的值由原来的 100 更改为 50(减去 50 的操作)的操作不应当立刻生效，而应等到 zhang 的用户的 userMoney 的值由原来的 20 更新为 70(增加 50 的操作)后一起生效。如果第 2 个语句 SQL 语句操作未能成功，那么第一个 SQL 语句操作就不应当生效。因此，为了能进行事务处理，必须关闭 con 的自动提交模式(自动提交模式是连接 con 的默认设置)。连接 con 首先调用 setAutoCommit(boolean autoCommit)方法，将参数 autoCommit 取值为 false 来关闭自动提交模式：

```
con.setAutoCommit(false);
```

❷ 用 commit()方法处理事务

连接 con 调用 setAutoCommit(false)后，产生的 Statement 对象对数据库提交任何一个 SQL 语句操作都不会立刻生效，这样一来，就有机会让 Statement 对象(PreparedStatement 对象)提交多个 SQL 语句，这些 SQL 语句就是一个事务。事务中的 SQL 语句不会立刻生效，直到连接 con 调用 commit()方法。con 调用 commit()方法就是让事务中的 SQL 语句全部生效。

❸ 用 rollback()方法处理事务失败

连接 con 调用 commit()方法进行事务处理时，只要事务中任何一个 SQL 语句没有生效，就抛出 SQLException 异常。在处理 SQLException 异常时，必须让 con 调用 rollback()方法，其作用是：撤销事务中成功执行过的 SQL 语句对数据库数据所做的更新、插入或删除操作，即撤销引起数据发生变化的 SQL 语句操作，将数据库中的数据恢复到 commit()方法执行之前的状态。

为了例 8_7 的需要，我们创建了 bank 数据库，并在 bank 数据库中创建了 user 表(有关创建数据库和表的操作见 8.1.3 节)，user 表的字段及属性如下：

```
(id char(20),name varchar(100),userMony double),
```

我们将 bank.sql 文件保存在 D:/myfile 目录中。

bank.sql

```
create database bank;
use   bank;
```

```
create table user(
    id char(20),
    name varchar(100),
    userMoney double,
    primary key(id)
    );
insert into user values('0001','geng',950),
('0002','zhang',1000);
select * from user;
```

启动命令行客户端(见 8.1.3 节),执行:

```
source d:/myfile/bank.sql;
```

完成创建数据 bank 以及在数据库 bank 中创建 user 表的操作。

下面的例 8_7 使用了事务处理,将 user 表中 name 字段是 geng 的 userMoney 的值减少 50,并将减少的 50 增加到 name 字段是 zhang 的 userMoney 属性值上。

例 8_7

example8_7.jsp(效果如图 8.21 所示)

```
<%@ page contentType="text/html" %>
<%@ page pageEncoding="utf-8" %>
<%@ page import="java.sql.*" %>
<HTML><body bgcolor=#EEDFF>
<% Connection con;
    PreparedStatement pre = null;                           //预处理语句
    ResultSet rs;
    try{  //加载 JDBC-MySQL 8.0 连接器
        Class.forName("com.mysql.cj.jdbc.Driver");
    }
    catch(Exception e){
        out.print("<h1>" + e);
    }
    String url = "jdbc:mysql://localhost:3306/bank?" +
    "useSSL = false&serverTimezone = CST&characterEncoding = utf-8";
    String user = "root";
    String password = "";
    try {
        con = DriverManager.getConnection(url,user,password);   //连接数据库
    }
    catch(SQLException exp){
        return;
    }
    try{
        String querySQL = "select * from user where id = ?";
        pre = con.prepareStatement(querySQL);
        pre.setString(1,"0001");
        rs = pre.executeQuery();
        rs.next();
        double userOne = rs.getDouble(3);
        out.print("转账:" + rs.getString(1) + "," + rs.getString(2) + "," + userOne);
```

```
        pre.setString(1,"0002");
        rs = pre.executeQuery();
        rs.next();
        double userTwo = rs.getDouble(3);
        out.print("< br >转账
前:" + rs.getString(1) + "," + rs.getString(2) + "," + userTwo);
        String updateSQL = "update user set userMoney = ? where id = ?";
        con.setAutoCommit(false);              //关闭自动提交模式
        pre = con.prepareStatement(updateSQL);
        int n = 50;
        pre.setDouble(1,userOne - n);          //减少 50
        pre.setString(2,"0001");
        pre.execute();                         //执行更新(但不立刻生效)
        pre.setDouble(1,userTwo + n);
        pre.setString(2,"0002");
        pre.execute();                         //执行更新(但不立刻生效)
        con.commit();                          //开始事务处理,如果发生异常直接执行 catch 块
        con.setAutoCommit(true);               //恢复自动提交模式
        querySQL = "select * from user where id = ? or id = ?";
        pre = con.prepareStatement(querySQL);
        pre.setString(1,"0001");
        pre.setString(2,"0002");
        rs = pre.executeQuery();
        while(rs.next()) {
            out.print("< br >");
            out.print("转账后:" + rs.getString(1) + "," + rs.getString(2) + "," +
                      rs.getString(3));
        }
        con.close();
    }
    catch(SQLException e) {
        try{  con.rollback();                  //撤销事务所做的操作
        }
        catch(SQLException exp){}
    }
%>
</body></HTML>
```

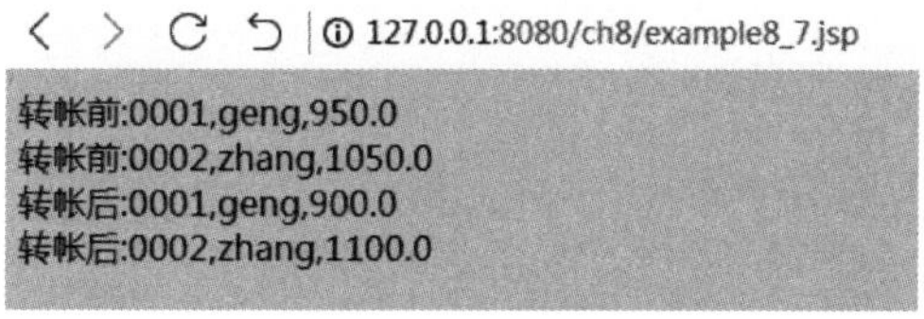

转帐前:0001,geng,950.0
转帐前:0002,zhang,1050.0
转帐后:0001,geng,900.0
转帐后:0002,zhang,1100.0

图 8.21 事务处理

8.8 分页显示记录

视频讲解

本节学习怎样查询数据库中某个表的全部记录,并用分页的方式显示记录。
程序在查询的时候,为了代码更加容易维护,希望知道数据库表的字段(列)

的名字以及表的字段的个数,一个办法是使用返回到程序中的结果集来获取相关的信息。例如,对于 bookDatabae 数据库的 bookList 表(见 8.1.3 节),如果执行下列查询返回结果集 rs:

```
ResultSet rs = sql.executeQuery("SELECT * FROM bookList");
```

首先让 ResultSet 对象 rs 调用 getMetaData()方法返回一个 ResultSetMetaData 对象:

```
ResultSetMetaData metaData = rs.getMetaData();
```

然后 ResultSetMetaData 对象(例如 metaData)调用 getColumnCount()方法就可以返回结果集 rs 中的列(字段)的数目:

```
int columnCount = metaData.getColumnCount();
```

ResultSetMetaData 对象(例如 metaData)调用 getColumnName(int i)方法就可以返回结果集 rs 中的第 i 列(字段)的名字:

```
String columnName = metaData.getColumnName(i);
```

对于 bookDatabae 数据库的 bookList 表,metaData.getColumnCount()的值就是 4,metaData.getColumnName(1)、metaData.getColumnName(2)、metaData.getColumnName(3)、metaData.getColumnName(4)的值依次是 ISBN、name、price 和 publishDate。

如果执行下列查询返回结果集 rs:

```
ResultSet rs = sql.executeQuery("SELECT name,price FROM bookList");
```

metaData.getColumnCount()是 2,metaData.getColumnName(1)、metaData.getColumnName(2)的值依次是 name 和 price。

可以使用二维数组 table 存放表的记录,即用二维数组 table 中的行(一维数组 table[i])存放一条记录。如果一个表中有许多记录,那么二维数组 table 就有多行。为避免长时间占用数据库的连接,应当将全部记录存放到二维数组中,然后关闭数据库连接。假设 table 存放了 m 行记录,准备每页显示 n 行,那么,总页数的计算公式是:

- 如果 m 除以 n 的余数大于 0,总页数等于 m 除以 n 的商加 1;
- 如果 m 除以 n 的余数等于 0,总页数等于 m 除以 n 的商。

即

```
总页数 = (m%n) = = 0?(m/n):(m/n+1);
```

如果准备显示第 p 页的内容,应当从 table 第(p-1)*n 行开始,连续输出 n 行(最后一页可能不足 n 行)。

本节例 8_8 使用 MVC 模式(有关知识见第 7 章)用分页的方式显示记录。

例 8_8

➢ **bean(模型)**

bean 的 id 是 recordList,是 session bean,用于存储从数据库的表中查询到的记录。用命令行进入 save\data 的父目录 classes,编译 Record_Bean.java(见本章开始的约定):

```
classes> javac save\data\BookList_Bean.java
```

Record_Bean.java（负责创建 session bean）

```java
package save.data;
public class Record_Bean{
    String []columnName ;                               //存放列名
    String [][] tableRecord = null;                     //存放查询到的记录
    int pageSize = 3;                                   //每页显示的记录数
    int totalPages = 1;                                 //分页后的总页数
    int currentPage = 1    ;                            //当前显示页
    int totalRecords ;                                  //全部记录
    public void setTableRecord(String [][] s){
        tableRecord = s;
    }
    public String [][] getTableRecord(){
        return tableRecord;
    }
    public void setColumnName(String [] s) {
        columnName = s;
    }
    public String [] getColumnName() {
        return columnName;
    }
    public void setPageSize(int size){
        pageSize = size;
    }
    public int getPageSize(){
        return pageSize;
    }
    public int getTotalPages(){
        return totalPages;
    }
    public void setTotalPages(int n){
        totalPages = n;
    }
    public void setCurrentPage(int n){
        currentPage = n;
    }
    public int getCurrentPage(){
        return currentPage ;
    }
     public int getTotalRecords(){
        totalRecords = tableRecord.length;
        return totalRecords ;
    }
}
```

> **JSP 页面（视图）**

视图部分由两个 JSP 页面构成，其中 example8_8_input.jsp 页面负责提供输入数据的视图，即用户可以在该页面输入数据库的名、表名、密码等信息，然后提交给名字是 query 的 servlet。query 负责查询数据库，并将结果存储到 id 为 recordBean 的 session bean 中，然后请求视图中的 example8_8_show.jsp 页面显示 recordBean 的数据（分页显示）。

example8_8_input.jsp（效果如图8.22(a)所示）

```jsp
<%@ page contentType="text/html" %>
<%@ page pageEncoding="utf-8" %>
<style>
    #tom{
        font-family:宋体;font-size:26;color:blue
    }
</style>
<HTML><body bgcolor=#ffccff>
<form action="queryAllServlet" id=tom method=post>
<b>数据库<input type='text' id=tom name='dataBase' value='bookDatabase'/>
<br>表名<input type='text' id=tom name='tableName' value='bookList'/>
<br>用户名(默认root)<input type='text' id=tom name='user' value='root'/>
<br>用户密码(默认空)<input type='text' id=tom name='password'/>
<br><input type='submit' name='submit' id=tom value='提交'>
</form>
</body></HTML>
```

example8_8_show.jsp（效果如图8.22(b)所示）

```jsp
<%@ page contentType="text/html" %>
<%@ page pageEncoding="utf-8" %>
<style>
    #tom{
        font-family:宋体;font-size:26;color:blue
    }
</style>
<jsp:useBean id="recordBean" class="save.data.Record_Bean"
                    scope="session"/>
<HTML><body bgcolor=#9FEFDF><center>
<jsp:setProperty name="recordBean" property="pageSize" param="pageSize"/>
<jsp:setProperty name="recordBean" property="currentPage"
                    param="currentPage"/>
</p>
<table id=tom border=1>
<%
    String [][] table = recordBean.getTableRecord();
    if(table == null) {
        out.print("没有记录");
        return;
    }
    String []columnName = recordBean.getColumnName();
    if(columnName!= null) {
        out.print("<tr>");
        for(int i=0;i<columnName.length;i++){
            out.print("<th>"+columnName[i]+"</th>");
        }
        out.print("</tr>");
    }
    int totalRecord = table.length;
    int pageSize = recordBean.getPageSize();         //每页显示的记录数
```

```
        int totalPages = recordBean.getTotalPages();
        if(totalRecord % pageSize == 0)
            totalPages = totalRecord/pageSize;              //总页数
        else
            totalPages = totalRecord/pageSize + 1;
    recordBean.setPageSize(pageSize);
    recordBean.setTotalPages(totalPages);
    if(totalPages >= 1) {
        if(recordBean.getCurrentPage()< 1)
            recordBean.setCurrentPage(recordBean.getTotalPages());
        if(recordBean.getCurrentPage()> recordBean.getTotalPages())
            recordBean.setCurrentPage(1);
        int index = (recordBean.getCurrentPage() - 1) * pageSize;
        int start = index;                              //table 的 currentPage 页起始位置
        for(int i = index;i < pageSize + index;i++) {
            if(i == totalRecord)
                break;
            out.print("<tr>");
            for(int j = 0;j < columnName.length;j++) {
                out.print("<td>" + table[i][j] + "</td>");
            }
            out.print("</tr>");
        }
    }
%>
</table>
<p id = tom>全部记录数:<jsp:getProperty name = "recordBean"
                        property = "totalRecords"/>。
<br>每页最多显示
<jsp:getProperty name = "recordBean" property = "pageSize"/>条记录.
<br>当前显示第<jsp:getProperty name = "recordBean" property = "currentPage"/>页
(共有<jsp:getProperty name = "recordBean" property = "totalPages"/>页).</p>
<table id = tom>
<tr>
  <td><form action = "" method = post>
    <input type = hidden name = "currentPage"
                  value = "<% = recordBean.getCurrentPage() - 1 %>"/>
    <input type = submit id = tom value = "上一页"/></form>
  </td>
  <td><form action = "" method = post>
    <input type = hidden name = "currentPage"
                  value = "<% = recordBean.getCurrentPage() + 1 %>"/>
    <input type = submit id = tom name = "g" value = "下一页"></form>
  </td>
  <td><form action = "" id = tom method = post>
    输入页码:<input type = textid = tom   name = "currentPage" size = 2>
    <input type = submit id = tom value = "提交"></form>
  </td>
</tr>
<tr><td></td><td></td>
  <td><form action = "" id = tom method = post>
```

```
        每页显示
        <input type = text id = tom name = "pageSize"
         value = <% = recordBean.getPageSize()%> size = 1>条记录
        <input type = submit id = tom value = "确定"></form></td>
</tr>
</table>
<a href = "example8_8_input.jsp">重新输入数据库和表名</a>
</center></body></HTML>
```

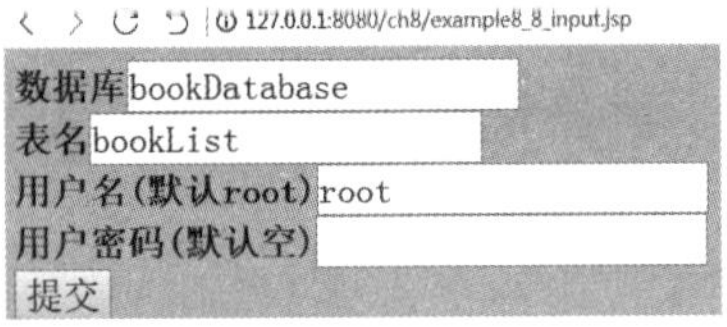

(a) 输入数据库和表名 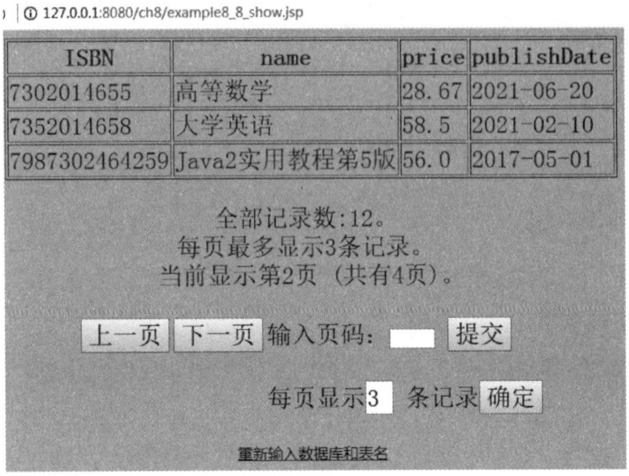　　　　　　　　　(b) 显示记录

图 8.22　分页显示记录

> **servlet（控制器）**

Query_Servlet 负责创建名字是 query 的 servlet。query 查询数据库表中的全部记录,将结果存放到 id 是 recordBean 的 session bean 中,然后请求 example8_8_show.jsp 显示 recordBean 中的数据。

用命令行进入 handle\data 的父目录 classes,编译 Query_Servlet.java(本章开始的约定):

```
classes> javac - cp .;servlet - api.jar   handle/data/Query_Servlet.java
```

注意".;"和"servlet-api.jar"之间不要有空格,".;"的作用是保证 Java 源文件能使用 import 语句引入当前 classes 目录中其他自定义包中的类,例如 save.data 包中的 bean 类 "(.;"是 javac 默认具有的功能,在使用-cp 参数时,尽量保留这一功能)。

Query_Servlet.java

```
package handle.data;
import save.data.Record_Bean;
import java.sql.*;
import java.io.*;
import javax.servlet.*;
import javax.servlet.http.*;
public class Query_Servlet extends HttpServlet{
    public void init(ServletConfig config) throws ServletException{
        super.init(config);
        try{  Class.forName("com.mysql.cj.jdbc.Driver");       //加载连接器
```

```java
    }
    catch(Exception e){}
}
public void service(HttpServletRequest request,HttpServletResponse
    response) throws ServletException,IOException{
    request.setCharacterEncoding("utf-8");
    String dataBase = request.getParameter("dataBase");
    String tableName = request.getParameter("tableName");
    String user = request.getParameter("user");
    String password = request.getParameter("password");
    boolean boo = ( dataBase == null||dataBase.length() == 0);
    boo = boo||( tableName == null||tableName.length() == 0);
    boo = boo||( user == null||user.length() == 0);
    if(boo) {
        response.sendRedirect("example8_8_input.jsp");
        return;
    }
    HttpSession session = request.getSession(true);
    Connection con = null;
    Record_Bean recordBean = null;
    try{
        recordBean = (Record_Bean)session.getAttribute("recordBean");
        if(recordBean == null){
            recordBean = new Record_Bean();                    //创建 bean
            session.setAttribute("recordBean",recordBean);     //是 session bean
        }
    }
    catch(Exception exp){}
    String url = "jdbc:mysql://127.0.0.1:3306/" + dataBase + "?" +
    "useSSL = false&serverTimezone = CST&characterEncoding = utf-8";
    try{
        con = DriverManager.getConnection(url,user,password);
        Statement sql = con.createStatement(ResultSet.TYPE_SCROLL_SENSITIVE,
                            ResultSet.CONCUR_READ_ONLY);
        ResultSet rs = sql.executeQuery("SELECT * FROM " + tableName);
        ResultSetMetaData metaData = rs.getMetaData();
        int columnCount = metaData.getColumnCount();           //得到结果集的列数
        String []columnName = new String[columnCount];
        for( int i = 0;i < columnName.length;i++) {
            columnName[i] = metaData.getColumnName(i + 1);     //得到列名
        }
        recordBean.setColumnName(columnName);                  //更新 bean
        rs.last();
        int rows = rs.getRow();                                //得到记录数
        String [][] tableRecord = recordBean.getTableRecord();
        tableRecord = new String[rows][columnCount];
        rs.beforeFirst();
        int i = 0;
        while(rs.next()){
            for( int k = 0;k < columnCount;k++)
                tableRecord[i][k] = rs.getString(k + 1);
```

```
                i++;
            }
            recordBean.setTableRecord(tableRecord);    //更新 bean
            con.close();
            response.sendRedirect("example8_8_show.jsp");    //重定向
        }
        catch(SQLException e){
            response.getWriter().print("<h2>" + e);
            System.out.println(e);
        }
    }
}
```

> **web.xml**（部署文件）

向 ch8\WEB\INF\下的部署文件 web.xml 添加如下的 servlet 和 servlet-mapping 标记（知识点见 6.1.2 节），部署的 servlet 的名字是 query，访问 servlet 的 url-pattern 是/query。

web.xml

```
<?xml version = "1.0" encoding = "utf - 8"?>
<web - app>
    <!-- 以下是 web.xml 文件新添加的内容 -->
    <servlet>
        <servlet - name>query</servlet - name>
        <servlet - class>handle.data.Query_Servlet</servlet - class>
    </servlet>
    <servlet - mapping>
        <servlet - name>query</servlet - name>
        <url - pattern>/query</url - pattern>
    </servlet - mapping>
</web - app>
```

8.9 连接 SQL Server 与 Access

视频讲解

▶ 8.9.1 连接 Microsoft SQL Server 数据库

本节介绍怎样连接 SQL Server 数据库。

❶ Microsoft SQL Server 2012

SQL Server 2012 是一个功能强大且可靠的免费数据库管理系统，它为轻量级（lightweight）网站和桌面应用程序提供丰富和可靠的数据存储。登录 http://www.microsoft.com，例如，登录 http://www.microsoft.com/zh-cn/download/default.aspx（微软的下载中心），然后在热门下载里选择"服务器"选项，然后选择下载 Microsoft SQL Server 2012 Express 以及相应的管理工具 Microsoft SQL Server 2008 Management Studio Express 或 Microsoft SQL Server Management Studio Express。对于 64 位操作系统可下载 SQLEXPR_x64_CHS.exe，对于 32 位操作系统可下载 SQLEXPR32_x86_CHS.exe。

Microsoft SQL Server Management Studio Express(SSMS)是微软提供的数据库管理软件，用于访问、配置、管理和开发 SQL Server 的所有组件，同时它还合并了多种图形工具和丰富的脚本编辑器。利用它们，技术水平各不相同的开发人员和管理员都可以使用 SQL Server。对于 64 位操作系统可下载 CHS\x64\SQLManagementStudio_x64_CHS.exe，对于 32 位操作系统可下载 CHS\x86\SQLManagementStudio_x86_CHS.exe。

安装好 SQL Server 2012 后，须启动 SQL Server 2012 提供的数据库服务器（数据库引擎），以便使远程的计算机访问它所管理的数据库。在安装 SQL Server 2012 时如果选择的是自动启数据库服务器，数据库服务器会在开机后自动启动，否则须手动启动 SQL Server 2012 服务器。可以单击"开始"|"程序"|Microsoft SQL Server，启动 SQL Server 2012 服务器。

为了便于调试程序，我们在同一台计算机同时安装了 Tomcat 服务器和 SQL Server 2012 数据库服务器，即使这样，为了能让 Tomcat 服务器管理的 Web 应用程序访问 SQL Server 2012 管理的数据库，也必须启动 SQL Server 2012 提供的数据库服务器。

❷ 建立数据库

打开 SSMS 提供的"对象资源管理器"，将出现如图 8.23 所示的操作界面。

图 8.23 所示的界面上的"数据库"目录下是已有的数据库的名称。在"数据库"目录上右击可以建立新的数据库，例如建立名称是 bookDatabase 的数据库。

创建好数据库后，就可以建立若干个表。如果准备在 warehouse 数据库中创建名字为 bookList 的表，那么可以单击"数据库"下的 warehouse 数据库，在 bookDatabase 管理的"表"的选项上右击，选择"新建表"选项，将出现相应的新建表界面。

图 8.23 SQL Server 对象资源管理器

❸ SQL Server 的 JDBC 数据库连接器

可以登录 www.microsoft.com 下载 Microsoft JDBC Driver 6.0 for SQL Server，即下载 sqljdbc_6.0.8112.200_chs.tar.gz。在解压目录下的 sqljdbc_6.0\chs\jre8 子目录中可以找到 JDBC-SQLServer 连接器：sqljdbc42.jar。

将 SQLServer 数据库的 JDBC-SQLServer 数据库连接器 sqljdbc42.jar 保存到 Tomcat 安装目录下的 lib 文件夹中（例如 D:\apache-tomcat-9.0.26\lib），并重新启动 Tomcat 服务器。

应用程序加载 JDBC-SQLServer 数据库连接器的代码如下：

```
try {   Class.forName("com.microsoft.sqlserver.jdbc.SQLServerDriver");
}
catch(Exception e){
}
```

❹ 建立连接

假设 SQL Server 数据库服务器所驻留的计算机的 IP 地址是 192.168.100.1，SQL Server 数据库服务器占用的端口是 1433（默认端口）。应用程序要和 SQL Server 数据库服务器管理的数据库 bookDatabase 建立连接，而有权访问数据库 bookDatabase 的用户的 id 和密码分别是 sa、dog123456，那么建立连接的代码如下：

```
try{
    String url =
     "jdbc:sqlserver://192.168.100.1:1433;DatabaseName = warehouse";
```

```
            String user = "sa";
            String password = "dog123456";
            con = DriverManager.getConnection(url,user,password);
      }
      catch(SQLException e){
      }
```

注①：如果用户与要连接 SQL Server 2000 服务器驻留在同一计算机上，使用的 IP 地址可以是 127.0.0.1。

注②：对于本章的例子，只要将例子中加载针对 MySQL 的 JDBC 数据库连接器以及建立 MySQL 数据库连接的代码更换成加载针对 SQL Server 的 JDBC 数据库连接器和建立 SQL Server 数据库连接的代码，就可以访问 SQL Server 数据库。

▶ 8.9.2 连接 Microsoft Access 数据库

许多院校的实验环境都是 Microsoft 的 Windows 操作系统，在安装 Office 办公系统软件的同时就安装好了 Microsoft Access 数据库管理系统，例如 Microsoft Access 2010。这里不再介绍 Access 数据库本身的使用。如果喜欢用 Access 数据库，那么学习本节后，可以把前面的例子全部换成 Access 数据库，仅仅需要改变的就是数据库的连接方式而已。

用 Access 数据库管理系统建立一个名字是 bookDatabase.accdb 的数据库，并在数据库中建立了名字是 bookList 的表，并添加几条记录（bookDatabase.accdb 数据库中 bookList 的表的结构与 8.1.3 节的 MySQL 数据库的 bookList 的表的结构相同，仅仅是数据库不同而已）。数据库保存在 Tomcat 安装目录下的 bin 文件夹（bin 是 Tomcat 服务器运行时的当前目录）。

❶ 加载 JDBC-Access 连接器

登录 http://www.hxtt.com/access.zip，下载 JDBC-Access 连接器。解压下载 access.zip，在解压目录下\lib 子目录中的 Access_JDBC30.jar 就是 JDBC-Access 连接器，将 Access_JDBC30.jar 保存到 Tomcat 安装目录下的 lib 文件夹中（例如 D:\apache-tomcat-9.0.26\lib），并重新启动 Tomcat 服务器。

加载 Access 数据库连接器程序的代码是：

```
Class.forName("com.hxtt.sql.access.AccessDriver");
```

其中的 com.hxtt.sql.access 包是 Access_JDBC30.jar 提供的，该包中的 AccessDriver 类封装着驱动。

❷ 连接已有的数据库

将 bookDatabase.accdb 数据库保存在启动 Tomcat 服务器的当前目录下，即 Tomcat 安装目录的 bin 中。连接 bookDatabase.accdb 数据库的代码如下：

```
Stringpath = "./bookDatabase.accdb";
String loginName = "";
String password = "";
con =
DriverManager.getConnection("jdbc:Access://" + path,loginName,password);
```

例 8_9 和例 8_1 类似，仅仅是把 MySQL 数据库更换了成 Access 数据库。
例 8_9
example8_9.jsp（效果如图 8.24 所示）

```jsp
<%@ page contentType="text/html" %>
<%@ page pageEncoding="utf-8" %>
<%@ page import="java.sql.*" %>
<style>
    #tom{
        font-family:宋体;font-size:18;color:blue
    }
</style>
<HTML><body bgcolor=#EEDDFF>
<% Connection con=null;
    Statement sql;
    ResultSet rs;
    try{    //加载 JDBC-Access 连接器
        Class.forName("com.hxtt.sql.access.AccessDriver");
    }
    catch(Exception e){
        out.print("<h1>"+e);
    }
    String path="./bookDatabase.accdb";
    String loginName="";
    String password="";
    out.print("<table border=1>");
    out.print("<tr>");
    out.print("<th id=tom width=100>"+"ISBN");
    out.print("<th id=tom width=100>"+"名称");
    out.print("<th id=tom width=50>"+"价格");
    out.print("<th id=tom width=50>"+"日期");
    out.print("</tr>");
    try{
        con=DriverManager.getConnection("jdbc:Access://"+path,
                                        loginName,password);     //连接数据库
        sql=con.createStatement();
        String SQL="SELECT * FROM bookList";                      //SQL 语句
        rs=sql.executeQuery(SQL);                                 //查表
        while(rs.next()) {
            out.print("<tr>");
            out.print("<td id=tom>"+rs.getString(1)+"</td>");
            out.print("<td id=tom>"+rs.getString(2)+"</td>");
            out.print("<td id=tom>"+rs.getFloat(3)+"</td>");
            out.print("<td id=tom>"+rs.getDate(4)+"</td>");
            out.print("</tr>") ;
        }
        out.print("</table>");
        con.close();
    }
    catch(SQLException e) {
        out.print("<h1>"+e);
    }
%>
</body></HTML>
```

127.0.0.1:8080/ch8/example8_9.jsp

ISBN	名称	价格	日期
7302014655	高等数学	29.0	2021-06-20
7987302464259	Java2实用教程第5版（微课版）	59.9	2017-05-01
92306084659	数据库原理	56.8	2019-02-10

图 8.24 连接 Access 数据库

8.10 使用连接池

8.10.1 连接池简介

视频讲解

Web 应用程序必须首先和数据库建立连接，即得到 Connection 对象，才能进行后续的操作，例如查询、更新、添加记录等操作。和数据库建立连接属于比较耗时的操作，连接池的思想是，由一个管理者，例如 Tomcat 服务器，事先建立好若干个连接，即创建若干个 Connection 对象放在一起（存放在一个实现 DataSource 接口的对象中），称作一个连接池。当 Web 应用程序需要连接数据库时，只需到连接池中获得一个 Connection 对象即可。当 Web 应用程序不再需要 Connection 对象时，就让 Connection 对象调用 close()方法，这样就可以把这个 Connection 对象再返回到连接池中，以便其他 Web 应用程序使用这个 Connection 对象。需要注意的是，Tomcat 服务器提供的连接池中的 Connection 对象调用 close()方法不会关闭和数据库的 TCP 连接，其作用仅仅是把 Connection 对象返回到连接池（这一点和 Web 应用程序自己创建的 Connection 对象不同，其原理不必深究）。简单地说，连接池可以让 Web 应用程序方便地使用 Connection 对象（用完务必放回），再用再取，节省了创建 Connection 的时间，提高了 Web 应用程序访问数据库的效率。

8.10.2 建立连接池

视频讲解

❶ 连接池配置文件

为了让 Tomcat 服务器创建连接池，必须编写一个 XML 文件，Tomcat 服务器通过读取该文件创建连接池。XML 文件的名字必须是 context.xml。内容如下：

context.xml

```
<?xml version = "1.0" encoding = "utf-8" ?>
<Context>
  <Resource
    name = "gxy"
    type = "javax.sql.DataSource"
    driverClassName = "com.mysql.cj.jdbc.Driver"
    url = "jdbc:mysql://127.0.0.1:3306/bookDatabase?useSSL = false
             &amp;serverTimezone = CST&amp;characterEncoding = utf - 8"
    username = "root"
    password = ""
    maxActive = "15"
    maxIdle = "15"
```

```
        minIdle = "1"
        maxWait = "1000"
    />
</Context>
```

将连接池配置文件 context.xml 文件保存在 Web 服务目录的 META-INF 子目录中。例如,对于 Web 服务目录 ch8,需要在 ch8 下新建一个子目录 META-INF(字母大写),将连接池配置文件 context.xml 保存在\ch8\META-INF\目录路径下(不必重新启动 Tomcat 服务器)。Tomcat 服务器会立刻读取 context.xml,创建连接池。如果修改 context.xml 文件,重新保存,Tomcat 服务器会立刻再次读取 context.xml,创建连接池。

连接池配置文件 context.xml 中的 Resource 标记(元素)通知 Tomcat 服务器创建数据源(连接池),Resource 标记中各个属性的意义如下:

(1) name:设置连接池的名字(命名一个喜欢的名字即可),例如 gxy,该名字是连接池的 id,context.xml 文件中,如果有多个 Resource 标记(一个 Resource 标记对应一个连接池),必须保证这些 Resource 标记中的 name 互不相同,即 context.xml 文件可以给出多个不同连接池的配置信息。

(2) type:设置连接池的类型,这里必须是 javax.sql.DataSource,即 Tomcat 服务器把创建的 Connection 对象存放在实现 DataSource 接口的对象中,即连接池是一个 DataSource 型对象。Tomcat 服务器使用 org.apache.tomcat.dbcp.dbcp2 包中的 BasicDataSource 类创建连接池(包含该包的 jar 文档在 Tomcat 安装目录的\lib 子目录中)。

(3) driverClassName:设置数据库连接器,即数据库驱动的类。如果是为 MySQL 8.0 数据库建立连接池,driverClassName 的值就是 com.mysql.cj.jdbc.Driver(MySQL 5.6 是 com.mysql.jdbc.Driver),如果是 SQL Server 数据库,其值是 com.microsoft.sqlserver.jdbc.SQLServerDriver,如果是 Access 数据库,其值是 com.hxtt.sql.access.AccessDriver。不要忘记把相应的数据库连接器保存到 Tomcat 安装目录下的 lib 文件夹中,例如,把 JDBC-MySQL 数据库连接器 mysql-connector-java-8.0.18.jar 保存到 Tomcat 安装目录下的 lib 文件夹中,并重新启动了 Tomcat 服务器。

(4) url:设置连接数据库的 URL,注意 url 要小写。需要注意的是对于 url 中的 & 字符要写成"&",这是 XML 文件对特殊字符的一个特殊规定。

(5) username:给出可以访问数据库的用户名,例如 root。

(6) password:给出访问数据库的密码,例如,root 用户的默认密码是无密码。

(7) maxActive:设置连接池的大小,即连接池中处于活动状态 Connection 对象的数目 (Tomcat 服务器创建的 Connection 对象的最大数目)。maxActive 取值不可以超过数据库系统允许的最大连接数目(同一时刻允许的最多连接数目,细节参看有关数据库系统的说明)。在设计 Web 应用程序时,通过使用连接池,不仅可以提高访问数据库的效率,也可以防止数据库系统连接数目过多导致数据库系统性能下降。取值 0 表示不受限制。

(8) maxIdle:设置连接池中可处于空闲状态的 Connection 对象的最大数目,取值非正整数,例如 0,表示不限制,如果取正整数,例如 6,那么当空闲状态的 Connection 对象多余 6 个时,Tomcat 服务器会释放多余的 Connection 对象,即从连接池中删除某些 Connection 对象。

(9) minIdle:设置连接池中保证处于空闲状态的 Connection 对象的最小数目。例如,取值 1,当连接池中仅仅剩下两个空闲状态 Connection 对象时,此刻如果同时有两个用户需要使用连接池中的 Connection 对象,那么二者只能有一个获得 Connection 对象,另一个用户必须等待(因为 Tomcat 服务器必须保证连接池中有 1 个 Connection 对象空闲)。

(10) maxWait:设置连接池中没有空闲状态 Connection 对象可用时,用户请求获得连接池中 Connection 对象需要等待的最长时间(单位是毫秒),如果超出 maxWait 设置的时间,Tomcat 服务器将抛出一个 TimeoutException 给用户。取值负数,例如－1,表示用户可以无限时等待。

❷ 使用连接池

应用程序必须到 Tomcat 服务器中去获得连接池。步骤如下:

(1) Context 接口。

首先创建一个实现 Context(上下文)接口的对象:

```
Context context = new InitialContext();
```

然后让 context 去寻找 Tomcat 服务器曾绑定在运行环境中的另一个 Context 对象:

```
Context contextNeeded = (Context)context.lookup("java:comp/env");
```

其中的 java:comp/env 是 Tomcat 服务器绑定这个 Context 对象时使用的资源标志符(不可以写错)。

(2) 得到连接池。

Tomcat 服务器通过连接池配置文件 context.xml 将连接池绑定在 Context 对象 contextNeeded 中(见步骤(1))。绑定用的资源标志符是连接池配置文件 context.xml 中 name 给的值,例如 gxy。因此,从 contextNeeded 中获得连接池的代码是:

```
DataSource ds = (DataSource)contextNeeded.lookup("gxy");
```

(3) 从连接池中获得连接。

获得连接池之后,就可以从连接池中获得 Connection 对象。连接池中都是已经创建好的 Connection 对象,连接池调用 getConnection()方法,如果有空闲的 Connection 对象,该方法就返回一个 Connection 对象,如果没有空闲的 Connection 对象,该方法将使得用户线程处于 waiting 状态,即等待该方法返回 Connection 对象(见连接池配置文件 context.xml 中 maxWait 值的设置)。代码如下:

```
Connection con = ds.getConnection();
```

(4) 将连接返回连接池。

当 Web 应用程序不再需要 Connection 对象时,就让 Connection 对象调用 close()方法,这样就可以把这个 Connection 对象再放回连接池中,以便其他 Web 应用程序使用这个 Connection 对象。如果用户都忘记 Connection 对象再返回连接池中,将可能很快导致连接池中无 Connection 对象可用。

下面的例 8_10 和例 8_1 类似,不同的仅仅是例 8_10 使用了连接池。例 8_10 是一个简单的 JSP 页面,该页面中的 Java 程序片代码使用连接池中的 Connection 对象连接数据库 bookDatabase,查询 bookList 表中的全部记录(见 8.1.3 节曾建立的 bookDatabase 数据库)。

例 8_10

example8_10.jsp(效果如图 8.25 所示)

```jsp
<%@ page contentType="text/html" %>
<%@ page pageEncoding="utf-8" %>
<%@ page import="java.sql.*" %>
<%@ page import="javax.sql.DataSource" %>
<%@ page import="javax.naming.Context" %>
<%@ page import="javax.naming.InitialContext" %>
<style>
   #tom{
      font-family:宋体;font-size:18;color:blue
   }
</style>
<HTML><body bgcolor=#EEDDFF>
<%
   Context  context = new InitialContext();
   Context  contextNeeded = (Context)context.lookup("java:comp/env");
   DataSource  ds = (DataSource)contextNeeded.lookup("gxy");    //获得连接池
   out.print("连接池对象:" + ds.toString());
   Connection con = null;
   Statement sql;
   ResultSet rs;
   out.print("<table border=1>");
   out.print("<tr>");
   out.print("<th id=tom width=100>" + "ISBN");
   out.print("<th id=tom width=100>" + "名称");
   out.print("<th id=tom width=50>" + "价格");
   out.print("<th id=tom width=50>" + "日期");
   out.print("</tr>");
   try{
      con = ds.getConnection();               //使用连接池中的连接
      sql = con.createStatement();
      String SQL = "SELECT * FROM bookList";  //SQL 语句
      rs = sql.executeQuery(SQL);             //查表
      while(rs.next()) {
         out.print("<tr>");
         out.print("<td id=tom>" + rs.getString(1) + "</td>");
         out.print("<td id=tom>" + rs.getString(2) + "</td>");
         out.print("<td id=tom>" + rs.getFloat(3) + "</td>");
         out.print("<td id=tom>" + rs.getDate(4) + "</td>");
         out.print("</tr>");
      }
      out.print("</table>");
      con.close();                            //连接返回连接池
   }
   catch(SQLException e) {
      out.print("<h1>" + e);
   }
   finally{
      try{
```

```
            con.close();
        }
        catch(Exception ee){}
    }
%>
</body></HTML>
```

127.0.0.1:8080/ch8/example8_10.jsp

连接池对象:org.apache.tomcat.dbcp.dbcp2.BasicDataSource@6d4678d3

ISBN	名称	价格	日期
2306084657	春天	35.8	2020-03-20
5777564629	冬日	29.9	2019-12-23
6777564622	Java面向对象程序设计第3版（微课版）	59.9	2020-01-23
7302014655	高等数学	28.67	2021-06-20
7352014658	大学英语	58.5	2021-02-10
7987302464259	Java2实用教程第5版	56.0	2017-05-01
82306884758	计算机组成原理	66.7	2020-08-10
8302084658	月亮湾	56.0	2021-12-10
8306084656	四月天	39.2	2020-05-20
92306084659	数据库原理	56.9	2019-02-10
9787302198048	Java设计模式	29.0	1999-05-16
97873902488644	Java课程设计第3版	32.0	2018-01-10

图 8.25　使用连接池

8.11　标准化考试训练

视频讲解

我们很熟悉的标准化考试，就是在给出的选项中选出正确的答案。

8.11.1　功能概述

（1）Web应用程序——标准化考试训练有两个JSP页面example8_11_start.jsp和example8_11_exm.jsp。用户首先访问example8_11_start.jsp页面，在该页面输入训练题目的数量，单击提交键请求名字是givePaper的servlet。givePaper servlet将一套试题存放在名字是paperBean的session bean中，然后将用户定向到example8_11_exam.jsp页面。

（2）example8_11_exam.jsp页面显示paperBean中的数据，用户根据显示的内容开始答题。用户给出一道题目的回答，即给出用户的答案，单击“确认”提交键，请求名字是handleMess的servlet，handleMess servlet负责将用户给出的答案存储到paperBean中，并将用户再定向到example8_11_exam.jsp页面。用户单击“下一题”或“上一题”提交键，请求名字是handleMess的servlet，handleMess servlet负责将下一题或上一题的题目内容存储到paperBean中，并将用户再定向到example8_11_exam.jsp页面。用户单击该页面上“交卷”提交键，请求名字是giveScore的servlet，giveScore servlet负责计算用户的得分并显示得分。

8.11.2　数据库设计

❶ 数据库与表

采用的数据库是MySQL数据库，创建数据库的名称是Examination，在此数据库中建立的表的名称是 testQuesion，此表的列(字段)及意义如下：
- id(int)：存放题号；
- content(varchar)：存放试题内容；
- a(char)：存放试题提供的 a 选项；
- b(char)：存放试题提供的 b 选项；
- c(char)：存放试题提供的 c 选项；
- d(char)：存放试题提供的 d 选项；
- pic(varchar)：存放试题示意图的图像文件的名字；
- answer(char)：存放试题的答案。

testQuesion 表中的每一条记录(每一行)是一道试题的有关信息。

通过导入 sql 文件中的 SQL 语句，完成数据库、表的创建，以及向表中插入记录(试题)的操作。编写下列 sql 文件 createDatabase.sql，保存在 D:\myFile 目录中。使用命令行进入 MySQL 安装目录下的 bin 子目录，使用 mysql 命令行启动客户端：

```
mysql -u root -p
```

按要求输入密码后，执行操作 source D:/myFile/createDatabase.sql; 导入该 sql 文件中的 SQL 语句，完成数据库、表的创建。如果某个题目不需要图像，那么插入记录时让 pic 字段的值为''(只输入一对单引号)，即长度为 0 的字符序列，answer 字段的值用大写英文字母 A、B、C 或 D(有关创建数据库和表的细节见 8.1.3 节)。

createDatabase.sql

```sql
create database Examination;
use Examination;
drop table testQuesion;
create table testQuesion(id int not null,
            content varchar(200) character set gb2312,
            a char(10),
            b char(10),
            c char(10),
            d char(10),
            pic varchar(30),
            answer char(10),
            primary key(id));
insert into testQuesion values
(1,'前方路口信号灯亮表示什么意思?',
'A.准许通行','B.提醒注意','C.路口警示','D.禁止通行','p1.jpg','D'),
(2,'这个标志是何含义?',
'A.十字交叉路口','B.环形交叉路口','C.T型交叉路口','D.Y型交叉路口','p2.jpg','A'),
(3,'这个仪表是何含义?',
'A.发动机转速表','B.行驶速度表','C.区间里程表','D.百公里油耗表','p3.jpg','A'),
(4,'遇到这种情况怎样行驶?',
'A.靠左减速让行','B.靠右减速让行','C.靠左加速让行','D.保持原速','p1.jpg','B');
```

❷ 数据库连接池

标准化考试训练系统准备使用连接池连接数据库。打开 ch8\META-INF 下的连接池配置文件 context.xml，增加如下的内容(知识点见 8.11.2 节)。

context.xml

```xml
<?xml version = "1.0" encoding = "utf-8" ?>
<Context>
  <!-- 以下是新增加的内容(连接池)-->
  <Resource
    name = "examinationConn"
    type = "javax.sql.DataSource"
    driverClassName = "com.mysql.cj.jdbc.Driver"
    url = "jdbc:mysql://127.0.0.1:3306/Examination?useSSL = false
        &amp;serverTimezone = CST&amp;characterEncoding = utf-8"
    username = "root"
    password = ""
    maxActive = "20"
    maxIdle = "20"
    minIdle = "1"
    maxWait = "5000"
  />
</Context>
```

▶ 8.11.3　Web 应用设计

例 8_11　使用 MVC 模式(有关知识见第 7 章)设计标准化考试。

例 8_11

➢ **bean(模型)**

bean 的 id 是 paperBean,是 session bean,用于存储一套试题,paperBean 由 servlet 负责用 TestPaperBean 类创建。TestPaperBean 类需要组合的 Problem 类的实例。id 是 problem 的 bean,也是 session bean,用于存储一道题目,具体是哪一道题目由 paperBean 来确定。

用命令行进入 save\data 的父目录 classes,编译 Problem.java 和 TestPaperBean.java(见本章开始的约定):

```
javac save\data\ Problem.java
javac save\data\ TestPaperBean.java
```

Problem.java

```java
package save.data;
public class Problem {
    public String index;                                   //题目的编号
    public String content;                                 //存放题目内容
    public String imageName;                               //存放题目所带的图像的名字
    public String correctAnswer = "asWdweq*23456@34";      //存放题目的正确答案
    //用户回答的初始答案和 correctAnswer 不同,防止出题人忘记给正确答案
    public String userAnswer = ""  ;                       //初始值必须是不含任何字符的串
    public String choiceA = "",choiceB = "",choiceC = "",choiceD = "";      //存放选择
}
```

TestPaperBean.java

```java
package save.data;
import java.util.ArrayList;
```

```java
public class TestPaperBean {                            //试卷
    public boolean isGivenProblem ;                     //试卷上是否已经有题目
    public ArrayList<Problem> problemList = null;       //存放全部试题
    public int index = 0;
    public Problem currentProblem ;                     //当前需要显示的题目
    public double score = 0;                            //存放试卷的得分
    public TestPaperBean(){
        problemList = new ArrayList<Problem>();
    }
    public void add(Problem problem){
        problemList.add(problem);
    }
    public Problem getCurrentProblem() {
        return problemList.get(index);
    }
    public void nextProblem() {
        index++;
        if(index == problemList.size()) {
            index = problemList.size() - 1;             //到最后一个题目
        }
    }
    public void previousProblem() {
        index -- ;
        if(index < 0) {
            index = 0;                                  //到第一个题目
        }
    }
    public int getProlemAmount(){
        return problemList.size();
    }
}
```

> **JSP 页面（视图）**

视图部分由两个 JSP 页面构成，其中 example8_11_start.jsp 页面负责提供输入试题数量的视图，在该页面输入训练的试题数量，然后请求名字是 givePaper 的 servlet。servlet 负责查询数据库，并将结果存储到 id 为 paperBean 的 session bean 中，然后请求视图中的 example8_11_exam.jsp 页面显示 paperBean 中的数据。

example8_11_start.jsp（效果如图 8.26(a)所示）

```jsp
<%@ page contentType = "text/html" %>
<%@ page pageEncoding = "utf-8" %>
<style>
    #tom{
        font-family:宋体;font-size:26;color:blue
    }
</style>
<h1>标准化考试训练</h1>
<HTML><body bgcolor = #ffccff>
<form action = 'givePaper' id = 'tom' method = 'post'>
输入训练题目数量:<input type = 'text' id = 'tom' name = 'testAmount' value = 10 />
```

```
<br><input type='submit' id='tom' name='submit' value='提交'/>
</form>
</body></HTML>
```

example8_11_exam.jsp（效果如图8.26(b)所示）

```
<%@ page contentType="text/html" %>
<%@ page pageEncoding="utf-8" %>
<style>
    #tom{
        font-family:宋体;font-size:20;color:blue
    }
</style>
<jsp:useBean id="problem" class="save.data.Problem" scope="session"/>
<jsp:useBean id="paperBean"
             class="save.data.TestPaperBean" scope="session"/>
<HTML><body bgcolor=#DEEFF9>
<% request.setCharacterEncoding("utf-8");
   problem = paperBean.getCurrentProblem();
%>
<p id=tom><%=problem.index%>.
<%=problem.content%>
</p>
<% if(problem.imageName.length()>0) {
%>   <image src=image/<%=problem.imageName%> width=300 height=260>
     </image>
<% }
%>
<form action="handleMess" id=tom method=post>
    <input type="radio" id=tom name="R" value=A /><%=problem.choiceA%>
    <input type="radio" id=tom name="R" value=B /><%=problem.choiceB%>
<br><input type="radio" id=tom name="R" value=C /><%=problem.choiceC%>
    <input type="radio" id=tom name="R" value=D /><%=problem.choiceD%>
<br><input type="submit" id=tom value="确认(可反复确认)" name="submit" />
</form>
<form action="handleMess" method=post name=form>
  <input type="submit" id=tom value="上一题" name="submit" />
  <input type="submit" id=tom value="下一题" name="submit" />
</form>
<form action="giveScore" method=post>
    <input type="submit" id=tom value="交卷" name="submit" />
</form>
</body></HTML>
```

> **servlet**（控制器）

GiveTestPaper_Servlet 负责创建名字是 givePaper 的 servlet。givePape 查询数据库 testQuesion 表中的全部记录，将结果存放到 id 是 paperBean 的 session bean 中，然后请求 example8_11_exam.jsp 显示 paperBean 中的数据。

HandleMess_Servlet 负责创建名字是 handleMess 的 servlet。handleMess servlet 负责将用户给出的答案存储到 paperBean 中，并将用户再定向到 example8_11_exam.jsp 页面。用户单击"下一题"或"上一题"提交键，handleMess servlet 负责将下一题或上一题的题目内容存储

第 8 章　JSP中使用数据库

(a) 输入训练题目数量　　　　　　　(b) 答卷页面

(c) 看成绩页面

图 8.26　标准化考试训练

到 paperBean 中,并将用户再定向到 example8_11_exam.jsp 页面。

GiveScore_Servlet 负责创建名字是 giveScore 的 servlet。用户单击该页面上"交卷"提交键,giveScore servlet 负责计算用户的得分并显示得分。

用命令行进入 handle\data 的父目录 classes,编译 GiveTestPaper_Servlet.java、HandleMess_Servlet.java 和 GiveScore_Servlet.java(本章开始的约定):

```
javac -cp .;servlet-api.jar   handle/data/GiveTestPaper_Servlet.java
javac -cp .;servlet-api.jar   handle/data/HandleMess_Servlet.java
javac -cp .;servlet-api.jar   handle/data/GiveScore_Servlet.java
```

注意".;"和"servlet-api.jar"之间不要有空格,".;"的作用是保证 Java 源文件能使用 import 语句引入当前 classes 目录中其他自定义包中的类,比如 save.data 包中的 bean 类"(.;"是 javac 默认具有的功能,在使用-cp 参数时,尽量保留这一功能)。

GiveTestPaper_Servlet.java

```java
package handle.data;
import save.data.TestPaperBean;                  //引入 bean
import save.data.Problem;
import java.io.*;
import java.sql.*;
import javax.servlet.*;
import javax.servlet.http.*;
import java.util.*;
import javax.sql.DataSource;
import javax.naming.Context;
import javax.naming.InitialContext;
public class GiveTestPaper_Servlet extends HttpServlet{
```

```java
public void init(ServletConfig config) throws ServletException{
    super.init(config);
}
public void service(HttpServletRequest request,
                    HttpServletResponse response)
    throws ServletException, IOException{
    TestPaperBean paperBean = null;
    HttpSession session = request.getSession(true);
    paperBean = (TestPaperBean)session.getAttribute("paperBean");
    if(paperBean == null){
        paperBean = new TestPaperBean();                           //创建 JavaBean 对象
        session.setAttribute("paperBean",paperBean);
    }
    request.setCharacterEncoding("utf-8");
    String testAmountMess = request.getParameter("testAmount");
    if(testAmountMess.length() == 0) {
        response.setContentType("text/html;charset=uft-8");
        response.getWriter().print("必须给出题目数量");
        return;
    }
    int testAmount = Integer.parseInt(testAmountMess);             //考题数量
    if(paperBean.isGivenProblem == false){
        giveProblem(paperBean,testAmount,response);      }
    else {
        response.sendRedirect("example8_11_exam.jsp");
    }
}
void giveProblem(TestPaperBean paperBean, int testAmount,
                 HttpServletResponse response){
    paperBean.problemList.clear();
    Connection con = null;
    try{
        Context context = new InitialContext();;
        Context contextNeeded = (Context)context.lookup("java:comp/env");
        DataSource ds =
            (DataSource)contextNeeded.lookup("examinationConn");   //连接池
        con = ds.getConnection();                                  //使用连接池中的连接
        Statement sql =
        con.createStatement(ResultSet.TYPE_SCROLL_SENSITIVE,
                            ResultSet.CONCUR_UPDATABLE);
        ResultSet rs = sql.executeQuery("SELECT * FROM testQuesion");
        rs.last();
        int recordAmount = rs.getRow();                            //得到记录数
        int [] a = getRandomNumber(recordAmount);                  //将数字随机排列
        //提交的考题数量可能大于数据库表的记录
        testAmount = Math.min(a.length,testAmount);
        for(int i = 0;i < testAmount;i++){                         //随机抽取题目放入 paperBean
            rs.absolute(a[i]);                                     //游标移动到第 a[i]行
            Problem problem = new Problem();
            problem.index = "" + (i + 1);                          //题目编号
            problem.content = rs.getString(2);                     //题目内容
```

```java
                problem.choiceA = rs.getString(3);         //选择 A
                problem.choiceB = rs.getString(4);         //选择 A
                problem.choiceC = rs.getString(5);         //选择 A
                problem.choiceD = rs.getString(6);         //选择 A
                problem.imageName = rs.getString(7);       //题目所需图像名字
                problem.correctAnswer = rs.getString(8);   //题目的答案
                paperBean.add(problem);                    //paperBean 添加一道题目
            }
            paperBean.isGivenProblem = true;
            con.close();                                   //连接放回连接池
            response.sendRedirect("example8_11_exam.jsp");
        }
        catch(Exception e){
            try{
                con.close();                               //连接放回连接池
            }
            catch(SQLException exp){}
        }
    }
    public int [] getRandomNumber(int n){                  //随机排列 1 至 n
        ArrayList < Integer > list = new ArrayList < Integer >();
        for(int i = 1;i < = n;i++){
            list.add(i);
        }
        Collections.shuffle(list);                         //把 list 随机排列
        int a[ ] = new int[list.size()];
        for(int i = 0;i < a.length;i++){
            a[i] = list.get(i);
        }
        return a;
    }
}
```

HandleMess_Servlet.java

```java
package handle.data;
import save.data.TestPaperBean;                            //引入 bean
import java.io.*;
import java.sql.*;
import javax.servlet.*;
import javax.servlet.http.*;
import java.util.*;
public class HandleMess_Servlet extends HttpServlet{
    public void init(ServletConfig config) throws ServletException{
        super.init(config);
    }
    public void service(HttpServletRequest request,
        HttpServletResponse response) throws ServletException,IOException{
        TestPaperBean paperBean = null;
        HttpSession session = request.getSession(true);
        paperBean = (TestPaperBean)session.getAttribute("paperBean");
```

```java
         if(paperBean == null){
             response.sendRedirect("example8_11_start.jsp");
             return;
         }
         request.setCharacterEncoding("utf-8");
         String mess = request.getParameter("submit");
         String userAnswer = request.getParameter("R");        //用户给出的选择
         if(mess.length() == 0) {
             response.sendRedirect("example8_11_exam.jsp");    //定向到答题页面
             return;
         }
         if(mess.contains("确认")) {
             paperBean.getCurrentProblem().userAnswer = userAnswer;
         }
         else if(mess.contains("下一题")){
             paperBean.nextProblem();
         }
         else if(mess.contains("上一题")){
             paperBean.previousProblem();
         }
         response.sendRedirect("example8_11_exam.jsp");
    }
}
```

GiveScore_Servlet.java(效果如图 8.26(c)所示)

```java
package handle.data;
import save.data.TestPaperBean;                    //引入 bean
import save.data.Problem;
import java.util.ArrayList;
import java.io.*;
import java.sql.*;
import javax.servlet.*;
import javax.servlet.http.*;
import java.util.*;
public class GiveScore_Servlet extends HttpServlet{
    public void init(ServletConfig config) throws ServletException{
        super.init(config);
    }
    public void service(HttpServletRequest request,
       HttpServletResponse response) throws ServletException,IOException{
       TestPaperBean paperBean = null;
       HttpSession session = request.getSession(true);
       paperBean = (TestPaperBean)session.getAttribute("paperBean");
       if(paperBean == null){
           response.sendRedirect("example8_11_exam.jsp");    //定向到答题页面
           return;
       }
       request.setCharacterEncoding("utf-8");
       String mess = request.getParameter("submit");
       if(mess.length() == 0) {
```

```java
                response.sendRedirect("example8_11.exam.jsp");    //定向到答题页面
                return;
            }
            if(mess.contains("交卷")) {
                ArrayList<Problem> problemList = paperBean.problemList;
                for(int i = 0;i<problemList.size();i++){
                    Problem p = problemList.get(i);
                    boolean b = compare(p.userAnswer,p.correctAnswer);
                    if(b) {
                        paperBean.score++;
                    }
                }
                session.invalidate();                              //销毁用户的会话
                response.setContentType("text/html;charset = gb2312");
                response.getWriter().print("<h1>得分: " + paperBean.score);
                response.getWriter().print
                   ("<a href = 'example8_11_start.jsp'>重新输入题目数量</a>");
            }
        }
        private boolean compare(String s,String t){    //比较字符串包含的字符是否相同
            char a[] = s.toCharArray();
            char b[] = t.toCharArray();
            Arrays.sort(a);
            Arrays.sort(b);
            return Arrays.equals(a,b);
        }
}
```

➢ **web.xml**（部署文件）

向 ch8\WEB\INF\下的部署文件 web.xml 添加如下的 servlet 和 servlet-mapping 标记（知识点见 6.1.2 节），部署的 servlet 的名字分别是 givePaper、handleMess 和 giveScore，访问 servlet 的 url-pattern 分别是/givePaper、/handleMess 和/giveScore。

web.xml

```xml
<?xml version = "1.0" encoding = "utf - 8"?>
<web - app>
    <!-- 以下是 web.xml 文件新添加的内容 -->
    <servlet>
        <servlet - name>givePaper</servlet - name>
        <servlet - class>handle.data.GiveTestPaper_Servlet</servlet - class>
    </servlet>
    <servlet - mapping>
        <servlet - name>givePaper</servlet - name>
        <url - pattern>/givePaper</url - pattern>
    </servlet - mapping>
    <servlet>
        <servlet - name>handleMess</servlet - name>
        <servlet - class>handle.data.HandleMess_Servlet</servlet - class>
    </servlet>
    <servlet - mapping>
        <servlet - name>handleMess</servlet - name>
```

```
                <url-pattern>/handleMess</url-pattern>
            </servlet-mapping>
            <servlet>
                <servlet-name>giveScore</servlet-name>
                <servlet-class>handle.data.GiveScore_Servlet</servlet-class>
            </servlet>
            <servlet-mapping>
                <servlet-name>giveScore</servlet-name>
                <url-pattern>/giveScore</url-pattern>
            </servlet-mapping>
</web-app>
```

> **图像的存放位置**

Web 应用中的需要的图像文件保存在 Web 服务目录的 image 子目录中，图像文件的名字是数据库 Examination 中 testQuesion 表的 pic 字段的值，比如 p1.jpg、p2.jpg 等（见 8.11.2 节）。

8.12 上机实验

视频讲解

本书提供了详细的实验步骤要求，按步骤完成，可提升学习效果，积累经验，不断提高 Web 设计能力。

▶ **8.12.1 实验1 查询成绩**

❶ 实验目的

掌握创建数据库、建立表的方法，并使用 JSP 页面查询数据库中表的记录。

❷ 实验要求

（1）用管理员身份（右击 cmd.exe，选择以管理员身份运行 cmd）启动命令行窗口，然后进入 MySQL 安装目录的 bin 子目录下输入 mysqld 或 mysqld -nt，回车确认启动 MySQL 数据库服务器（如果启动失败，可能需要初始化，知识点见 8.1.2 节）。

（2）采用的数据库是 MySQL 数据库，创建数据库的名称是 Student，在此数据库中建立的表的名称是 scoreReport，此表的列（字段）及意义如下：

- id(char)：存放学号，主键；
- name(char)：存放姓名；
- mathScore(int)：存放数学成绩；
- englishScore(int)：存放英语成绩。

scoreReport 表中的每一条记录（每一行）是一名学生的有关信息。通过导入 sql 文件中的 SQL 语句，完成数据库、表的创建，以及向表中插入记录（学生成绩信息）的操作。编写下列 sql 文件 createStudent.sql，保存在 D:\myFile 目录中。使用命令行进入 MySQL 安装目录下的 bin 子目录，使用 mysql 命令行启动客户端：

```
mysql -u root -p
```

按要求输入密码后，执行操作 source D:/myFile/createSudent.sql；导入 createStudent.sql 中的 SQL 语句（createStudent.sql 按 ANSI 编码保存在 D:\myFile）完成数据库、表的创建（有关创建数据库和表的细节见 8.1.3 节）。

createStudent.sql

```
create database Student;
use Student;
drop table scoreReport;
create table scoreReport(id char(20) not null,
                        name char(30) character set gb2312,
                        mathScore int,
                        englishScore int,
                        primary key(id));
insert into scoreReport values
('A10001','张三',89,67),
('A10002','李四',70,97),
('A10003','刘二',82,85),
('A10004','孙大三',69,76),
('A10005','林小军',80,92),
('A10006','吴大进',65,71);
```

（3）编写两个 JSP 页面 inputMess.jsp 和 queryShow.jsp。inputMess.jsp 页面提供 1 个 form 表单，该表单提供用户输入要查询的学生的学号或姓名，然后单击名字是"按学号查询"或"按姓名查询"提交键，请求 queryShow.jsp 页面，queryShow.jsp 负责连接数据库，查询学生的成绩并显示查询结果，其中按姓名查询是模糊查询，按学号查询是精准查询。

（4）在 Tomcat 服务器的 webapps 目录下（例如 D:\apache-tomcat-9.0.26\webapps）新建一个名字是 ch8_practice_one 的 Web 服务目录。把 JSP 页面保存到 ch8_practice_one 中。

（5）将 MySQL 数据库的 JDBC-MySQL 数据库连接器 mysql-connector-java-8.0.18.jar 保存到 Tomcat 安装目录下的 lib 文件夹中（例如 D:\apache-tomcat-9.0.26\lib），并重新启动 Tomcat 服务器。

（6）用浏览器访问 JSP 页面 inputMess.jsp。

❸ 参考代码

参考代码运行效果如图 8.27 所示。

(a) 输入查询信息

(b) 显示查询结果

图 8.27 查询成绩

➢ 输入查询信息的页面

inputNumber.jsp（效果如图 8.27(a)所示）

```
<%@ page contentType="text/html" %>
<%@ page pageEncoding="utf-8" %>
<style>
    #tom{
        font-family:宋体;font-size:28;color:black
    }
```

```
</style>
<HTML><body bgcolor = #ffccff>
<form action = "queryShow.jsp" id = tom method = post>
输入学号或姓名查询成绩：<br>
   <input type = "text" id = tom name = "mess" /><br>
   <input type = "submit" id = tom name = "submit" value = "按姓名查询"/>
   <input type = "submit" id = tom name = "submit" value = "按学号查询"/>
</form>
</body></HTML>
```

> 查询效果页面

queryShow.jsp（效果如图 8.27(b)所示）

```
<%@ page contentType = "text/html" %>
<%@ page pageEncoding = "utf-8" %>
<%@ page import = "java.sql.*" %>
<style>
   #tom{
       font-family:宋体;font-size:18;color:blue
   }
</style>
<HTML><body bgcolor = #EEDDFF>
<% request.setCharacterEncoding("utf-8");
   String mess = request.getParameter("mess");
   mess = mess.trim();
   String submit = request.getParameter("submit");
   if(mess.length() == 0){
       response.sendRedirect("inputMess.jsp");
       return;
   }
   Connection con = null;
   Statement sql;
   ResultSet rs;
   try{    //加载 JDBC-MySQL 8.0 连接器
       Class.forName("com.mysql.cj.jdbc.Driver");
   }
   catch(Exception e){
       out.print("<h1>" + e);
   }
   String url = "jdbc:mysql://localhost:3306/Student?" +
   "useSSL = false&serverTimezone = CST&characterEncoding = utf-8";
   String user = "root";
   String password = "";
   out.print("<table border = 1>");
   out.print("<tr>");
   out.print("<th id = tom width = 100>" + "学号");
   out.print("<th id = tom width = 100>" + "姓名");
   out.print("<th id = tom width = 50>" + "数学成绩");
   out.print("<th id = tom width = 50>" + "英语成绩");
   out.print("</tr>");
   try{
```

```
            con = DriverManager.getConnection(url,user,password);      //连接数据库
            sql = con.createStatement();
            String SQL = null;
            if(submit.contains("姓名")){
                SQL = "SELECT * FROM scoreReport where name like '%" + mess + "%'";
            }
            else if(submit.contains("学号")){
                SQL = "SELECT * FROM scoreReport where id = '" + mess + "'";
            }
            rs = sql.executeQuery(SQL);                                //查表
            while(rs.next()) {
                out.print("<tr>");
                out.print("<td id=tom>" + rs.getString(1) + "</td>");
                out.print("<td id=tom>" + rs.getString(2) + "</td>");
                out.print("<td id=tom>" + rs.getInt(3) + "</td>");
                out.print("<td id=tom>" + rs.getInt(4) + "</td>");
                out.print("</tr>");
            }
            out.print("</table>");
            con.close();
        }
        catch(SQLException e) {
            out.print("<h1>" + e);
        }
    %>
    </body></HTML>
```

▶ 8.12.2 实验2 管理学生成绩

❶ 实验目的

学习怎样使用数据库连接池,怎样使用预处理语句对数据库的表进行更新,添加和删除操作。

❷ 实验要求

(1) 用管理员身份(右击 cmd.exe,选择以管理员身份运行 cmd)启动命令行窗口,然后进入 MySQL 安装目录的 bin 子目录下输入 mysqld 或 mysqld -nt,回车确认启动 MySQL 数据库服务器(如果启动失败,可能需要初始化,知识点见 8.1.2 节)。

(2) 使用的创建数据库是 8.12.1 节中的 Student 数据库。

(3) 编写一个 JSP 页面 adminStudent.jsp。该页面的 form 表单中用户可以输入某个学生的成绩信息:学号、姓名、数学成绩和英语成绩,单击"添加"提交键提交给本页面,本页面负责连接数据库,使用预处理语句向数据库的 scoreReport 表中插入一条记录。该页面的 form 表单中用户可以输入某个学生的学号,单击"更新"提交键提交给本页面,本页面负责连接数据库,使用预处理语句更新数据库的 scoreReport 表中的一条记录。该页面的 form 表单中,用户可以输入某个学生的学号,单击"删除"提交键提交给本页面,本页面负责连接数据库,使用预处理语句删除数据库的 scoreReport 表中的一条记录。

(4) 在 Tomcat 服务器的 webapps 目录下(例如 D:\apache-tomcat-9.0.26\webapps)新建名字是 ch8_practice_two 的 Web 服务目录。把 JSP 页面保存到 ch8_practice_two 目录中。

(5) 连接数据库采用数据库连接池。将 MySQL 数据库的 JDBC-MySQL 数据库连接器 mysql-connector-java-8.0.18.jar 保存到 Tomcat 安装目录下的 lib 文件夹中，并重新启动 Tomcat 服务器。

(6) 用浏览器访问 JSP 页面 adminStudent.jsp。

❸ 参考代码

参考代码运行效果如图 8.28 所示。

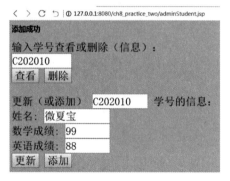

图 8.28 管理学生成绩

➤ 建立连接池

保存连接池配置文件 context.xml 到 ch8_practice_two\META-INF 下(知识点见 8.11.2 节)。

context.xml

```xml
<?xml version = "1.0" encoding = "utf - 8" ?>
<Context>
  <Resource
    name = "studentConn"
    type = "javax.sql.DataSource"
    driverClassName = "com.mysql.cj.jdbc.Driver"
    url = "jdbc:mysql://127.0.0.1:3306/Student?
        &amp;serverTimezone = CST&amp;characterEncoding = utf - 8"
    username = "root"
    password = ""
    maxActive = "5"
    maxIdle = "5"
    minIdle = "1"
    maxWait = "5000"
  />
</Context>
```

➤ **JSP 页面**

adminStudent.jsp(效果如图 8.28 所示)

```jsp
<%@ page contentType = "text/html" %>
<%@ page pageEncoding = "utf - 8" %>
<%@ page import = "java.sql.*" %>
<%@ page import = "javax.sql.DataSource" %>
<%@ page import = "javax.naming.Context" %>
<%@ page import = "javax.naming.InitialContext" %>
```

```jsp
<style>
  #tom{
      font-family:宋体;font-size:28;color:black
  }
</style>
<% request.setCharacterEncoding("utf-8");
   Connection con = null;
   PreparedStatement pre = null;                    //预处理语句
   ResultSet rs;
   Context  context = new InitialContext();
   Context  contextNeeded = (Context)context.lookup("java:comp/env");
   DataSource ds =
   (DataSource)contextNeeded.lookup("studentConn");  //连接池
   try{
      con = ds.getConnection();                    //使用连接池中的连接
   }
   catch(Exception exp){ }
   String updateSQL =
  "update scoreReport set name = ?,mathScore = ?,englishScore = ? where id = ?";
   String insertSQL = "insert into scoreReport values(?,?,?,?)";
   String deleteSQL = "delete from scoreReport where id = ?";
   String querySQL = "select * from scoreReport where id = ?";
   String mess = request.getParameter("submit");
   if(mess == null) mess = "";
   String id = request.getParameter("id");
   String name = request.getParameter("name");
   String math = request.getParameter("mathScore");
   String english = request.getParameter("englishScore");
   try{
     if(mess.contains("查看")){
        pre = con.prepareStatement(querySQL);
        pre.setString(1,id);
        rs = pre.executeQuery();
        if(rs.next()){
           id = rs.getString(1);
           name = rs.getString(2);
           math = rs.getString(3);
           english = rs.getString(4);
        }
     }
     else if(mess.contains("更新")){
        pre = con.prepareStatement(updateSQL);
        pre.setString(1,name);
        pre.setInt(2,Integer.parseInt(math));
        pre.setInt(3,Integer.parseInt(english));
        pre.setString(4,id);
        pre.executeUpdate();
        out.print("<h3>更新成功</h3>");
     }
     else if(mess.contains("添加")){
        pre = con.prepareStatement(insertSQL);
```

```
                pre.setString(1,id);
                pre.setString(2,name);
                pre.setInt(3,Integer.parseInt(math));
                pre.setInt(4,Integer.parseInt(english));
                pre.executeUpdate();
                out.print("<h3>添加成功</h3>");
            }
             else if(mess.contains("删除")){
                pre = con.prepareStatement(deleteSQL);
                pre.setString(1,id);
                pre.executeUpdate();
                out.print("<h3>删除成功</h3>");
            }
            con.close();                            //连接放回连接池
        }
        catch(SQLException e) {
            out.print("<h1>学号不能重复");
            try{
                con.close();                        //连接放回连接池
            }
            catch(SQLException exp){}
        }
%>
<HTML><body bgcolor=#ffccff>
<form action="" id=tom method=post>
输入学号查看或删除(信息):<br>
 <input type="text"    id=tom name="id" size=10 /><br>
 <input type="submit" id=tom name="submit" value="查看"/>
 <input type="submit" id=tom name="submit" value="删除"/>
</form>
<form action="" id=tom method=post>
更新(或添加)
<input type="text" id=tom name="id" value=<%=id%> size=9 />
学号的信息:
<br>姓名:
<input type="text" id=tom name="name" value='<%=name%>'size=11/>
<br>数学成绩:
<input type="text" id=tom name="mathScore" value='<%=math%>'size=7 />
<br>英语成绩:
<input type="text" id=tom name="englishScore" value='<%=english%>'size=7/>
<br><input type="submit" id=tom name="submit" value="更新"/>
    <input type="submit" id=tom name="submit" value="添加"/>
</form>
</body></HTML>
```

▶ 8.12.3 实验3 小星星广告网

❶ 实验目的

掌握借助数据库实现注册、登录的方法,让登录的用户可以上传图像(广告),未登录的用户只能浏览图像(广告)。本实验要求学生按照实验步骤完成各项操作,重点是阅读和调试代码(可以下载教材提供的源代码,下载信息也可参考耿祥义老师的教学辅助微信公众号javaviolin),以此掌握 Web 设计中的重要知识点和设计过程。

❷ 实验要求

(1)用管理员身份(右击 cmd.exe,选择以管理员身份运行 cmd)启动命令行窗口,然后进入 MySQL 安装目录的 bin 子目录下输入 mysqld 或 mysqld -nt ,回车确认启动 MySQL 数据库服务器(如果启动失败,可能需要初始化,知识点见 8.1.2 节)。

(2)打开一个新的命令行窗口(不必是管理员身份),进入 MySQL 安装目录下的 bin 子目录。执行 mysql.exe,即启动 mysql 命令行客户端(知识点见 8.1.3 节)。

(3)创建一个名字是 adertisement 的数据库,在该数据库中再创建名字是 user 的表。
user 表用于存储用户的注册信息,user 表的主键是 logname,各个字段值的要求如下:
- logname:存储注册的用户名(属性是字符型,主键)。
- password:存储密码(属性是字符型)。

(4)用文本编辑器编写下列用于创建数据库的 sql 文件 adver.sql,将该文件按 ANSI 编码保存到磁盘中,例如 D:\myFile 中。

adver.sql

```
create database adertisement;
use adertisement;
create table user
     (logname char(30) not null,
      password varchar(30) character set utf8,
      primary key(logname));
```

(5)在命令行客户端 mysql>状态下执行操作:

```
source D:/myFile/adver.sql;
```

导入该 adver.sql 文件中的 SQL 语句,完成数据库、表的创建(有关创建数据库和表的细节见 8.1.3 节)。

(6)在 Tomcat 服务器的 webapps 目录下(例如 D:\apache-tomcat-9.0.26\webapps)新建名字是 ch8_practice_three 的 Web 服务目录,后续的 JSP 页面按 UTF-8 编码保存在该服务目录中。ch8_practice_three 下再建立目录结构 ch8_practice_three\WEB-INF\classes(WEB-INF 字母大写),将 Tomcat 安装目录 lib 子目录中的 servlet-api.jar 文件复制到 \WEB-INF\classes 中(知识点见 6.1.1 节)。

(7)连接数据库采用数据库连接池。将 MySQL 数据库的 JDBC-MySQL 数据库连接器 mysql-connector-java-8.0.18.jar 保存到 Tomcat 安装目录下的 lib 文件夹中(例如 D:\apache-tomcat-9.0.26\lib),并重新启动 Tomcat 服务器。在 ch8_practice_three 下新建子目录 META-INF(字母大写),将连接池配置文件 context.xml 按 UTF-8 编码保存在 META-INF 子目录中(不必重新启动 Tomcat 服务器,知识点见 8.10.2 节)。context.xml 及内容如下。

context.xml

```
<?xml version = "1.0" encoding = "utf-8" ?>
<Context>
  <Resource
    name = "adverConn"
```

```
        type = "javax.sql.DataSource"
        driverClassName = "com.mysql.cj.jdbc.Driver"
        url = "jdbc:mysql://127.0.0.1:3306/adertisement?useSSL = false
            &amp;serverTimezone = CST&amp;characterEncoding = utf - 8"
        username = "root"
        password = ""
        maxActive = "15"
        maxIdle = "15"
        minIdle = "1"
        maxWait = "1000"
    />
</Context>
```

(8) 导航条与主页。导航条由注册、登录、发布公告和浏览广告组成。其他 JSP 页面通过使用 JSP 的<%@ include…%>标记将 head.txt(导航条文件)嵌入到自己的页面(知识点见 2.7 节)。head.txt 按 UTF-8 编码保存在 Web 服务目录 ch8_practice_three 中。主页，即名字是 index.jsp 的 JSP 页面由导航条、一条欢迎语构成。用户在浏览器的地址栏中输入 http://ch8_practice_three:8080/ 就可以访问该主页(不必输入主页的名字)。

head.txt

```
<style>
    #jerry{
        font - family:隶书;font - size:58;color:blue;
    }
    #tom{
        font - family:楷体;font - size:33;color:blue;
    }
</style>
<div align = "center">
<p id = jerry>小星星广告网</p>
<table width = "600" align = "center" border = "0">
    <tr valign = "bottom">
    <td id = tom><a href = "register.jsp">注册</a></td>
    <td id = tom><a href = "login.jsp">登录</a></td>
    <td id = tom><a href = "publishAdver.jsp">发布广告</a></td>
    <td id = tom><a href = "browseAdver.jsp">浏览广告</a></td>
    <td id = tom><a href = "index.jsp">主页</a></td>
</tr>
</table></div>
```

index.jsp(效果如图 8.29 所示)

```
<%@ page contentType = "text/html" %>
<%@ page pageEncoding = "utf - 8" %>
<title>小星星广告网</title>
<HEAD><%@ include file = "head.txt" %></HEAD>
<style>
    #ok{
```

```
            font-family:楷体;font-size:50;color:green
    }
</style>
<HTML><body bgcolor=pink>
<center id = ok>
欢迎注册,发布广告.
</center>
</body></HTML>
```

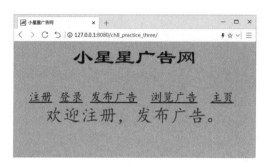

图 8.29　主页 index.jsp

（9）bean 与 servlet 管理。bean 的包名均为 save.data。在 classes 下建立目录结构\save\data，创建 bean 的类的 Java 源文件按 ANSI 编码保存在\WEB-INF\classes\save\data 中。servlet 的包名均为 handle.data。在 classes 下建立目录结构\handle\data，创建 servlet 的类的 Java 源文件按 ANSI 编码保存在\WEB-INF\classes\handle\data 中。

后续涉及编译 bean 和 servlet，均按下列要求操作。

用命令行进入 save\data 的父目录 classes，编译 bean 的 Java 源文件：

```
javac save\data\Java 源文件
```

用命令行进入 handle\data 的父目录 classes，编译 servlet 的 Java 源文件：

```
javac -cp .;servlet-api.jar  handle\data\Java 源文件
```

注意".;"和"servlet-api.jar"之间不要有空格，".;"的作用是保证 Java 源文件能使用 import 语句引入当前 classes 目录中其他自定义包中的类，例如 save.data 包中的 bean 类。

（10）web.xml（部署文件）。web.xml 文件按 UTF-8 编码保存在\WEB\INF\（知识点见 6.1.2 节）。web.xml 部署的 servlet 的名字分别是 registerServlet、loginServlet 和 upFile，访问 servlet 的 url-pattern 分别/registerServlet、/loginServlet 和/upFile。

web.xml

```
<?xml version = "1.0" encoding = "utf-8"?>
<web-app>
    <!-- 以下是 web.xml 文件新添加的内容 -->
  <servlet>
      <servlet-name>registerServlet</servlet-name>
      <servlet-class>handle.data.HandleRegister</servlet-class>
  </servlet>
  <servlet-mapping>
```

```
                <servlet-name>registerServlet</servlet-name>
                <url-pattern>/registerServlet</url-pattern>
        </servlet-mapping>
        <servlet>
                <servlet-name>loginServlet</servlet-name>
                <servlet-class>handle.data.HandleLogin</servlet-class>
        </servlet>
        <servlet-mapping>
                <servlet-name>loginServlet</servlet-name>
                <url-pattern>/loginServlet</url-pattern>
        </servlet-mapping>
        <servlet>
                <servlet-name>upFile</servlet-name>
                <servlet-class>handle.data.UpFile</servlet-class>
        </servlet>
        <servlet-mapping>
                <servlet-name>upFile</servlet-name>
                <url-pattern>/upFile</url-pattern>
        </servlet-mapping>
</web-app>
```

(11) 注册。采用 MVC 模式设计注册模块。

> **视图(JSP 页面)**

用户在 register.jsp 页面(见 head.txt 导航条)输入注册信息,请求名字是 registerServlet 的 servlet(见配置文件 web.xml),registerServlet 负责将注册信息写入数据库中的 user 表中。register.jsp 同时负责显示注册是否成功的信息。

register.jsp(效果如图 8.30 所示)

```
<%@ page contentType="text/html" %>
<%@ page pageEncoding="utf-8" %>
<jsp:useBean id="userBean" class="save.data.Register" scope="request"/>
<HEAD><%@ include file="head.txt" %></HEAD>
<title>注册页面</title>
<style>
    #ok{
        font-family:宋体;font-size:26;color:black;
    }
</style>
<HTML><body id=ok bgcolor = cyan>
<form action="registerServlet" id=ok method="post">
用户名由字母、数字、下画线构成, * 注释的项必须填写。<br>
*用户名:<input type=text id=ok name="logname" /><br>
*用户密码:<input type=password id=ok name="password"/><br>
*重复密码:<input type=password id=ok name="again_password"/><br>
<input type=submit    id=ok value="提交"><tr>
</form>
注册反馈:
<jsp:getProperty name="userBean" property="logname"/>
<jsp:getProperty name="userBean"    property="backNews" />
</body></HTML>
```

第 8 章　JSP中使用数据库

图 8.30　注册页面

> **模型（bean）**

该模块中 bean 的 id 是 userBean，是 request bean，由 servlet 控制器负责创建。

Register.java

```
package save.data;
public class Register{
    String logname = "",
           backNews = "请输入注册信息";
    public void setLogname(String logname){
        this.logname = logname;
    }
    public String getLogname(){
        return logname;
    }
    public void setBackNews(String backNews){
        this.backNews = backNews;
    }
    public String getBackNews(){
        return backNews;
    }
}
```

> **控制器（servlet）**

servlet 控制器是 HandleRegister 类的实例，名字是 registerServlet（见 web.xml），registerServlet 负责创建 id 是 userBean 的 request bean，将用户提交的信息写入到数据库的 user 表，并将注册反馈信息存储到 userBean 中，然后转发到 register.jsp 页面。另外，registerServlet 使用了 Encrypt 类的 static 方法将用户的密码加密后存放到数据库的 user 表中。

Encrypt.java

```
package handle.data;
public class Encrypt {
    static String encrypt(String sourceString,String password) {
        char [] p = password.toCharArray();
        int n = p.length;
        char [] c = sourceString.toCharArray();
        int m = c.length;
```

271

```java
        for(int k = 0;k < m;k++){
            int mima = c[k] + p[k % n];        //加密算法
            c[k] = (char)mima;
        }
        return new String(c);                  //返回密文
    }
}
```

HandleRegister.java

```java
package handle.data;
import save.data.Register;
import java.sql.*;
import java.io.*;
import javax.servlet.*;
import javax.servlet.http.*;
import javax.sql.DataSource;
import javax.naming.Context;
import javax.naming.InitialContext;
import javax.naming.NamingException;
public class HandleRegister extends HttpServlet {
    public void init(ServletConfig config) throws ServletException {
        super.init(config);
    }
    public void service(HttpServletRequest request,
                HttpServletResponse response)
                    throws ServletException, IOException {
        request.setCharacterEncoding("utf-8");
        Connection con = null;
        PreparedStatement sql = null;
        Register userBean = new Register();              //创建 bean
        request.setAttribute("userBean",userBean);
        String logname = request.getParameter("logname").trim();
        String password = request.getParameter("password").trim();
        String again_password = request.getParameter("again_password").trim();
        if(logname == null)
            logname = "";
        if(password == null)
            password = "";
        if(!password.equals(again_password)) {
            userBean.setBackNews("两次密码不同,注册失败,");
            RequestDispatcher dispatcher =
            request.getRequestDispatcher("register.jsp");
            dispatcher.forward(request, response);       //转发
            return;
        }
        boolean isLD = true;
        for(int i = 0;i < logname.length();i++){
            char c = logname.charAt(i);
            if(!(Character.isLetterOrDigit(c)||c == '_'))
                isLD = false;
        }
```

```java
            boolean boo = logname.length()> 0&&password.length()> 0&&isLD;
            String backNews = "";
            try{   Context context = new InitialContext();
                   Context contextNeeded =
                         (Context)context.lookup("java:comp/env");
                   DataSource ds =
                   (DataSource)contextNeeded.lookup("adverConn");       //连接池
                   con =  ds.getConnection();                           //使用连接池中的连接
                   String insertCondition = "INSERT INTO user VALUES (?,?)";
                   sql = con.prepareStatement(insertCondition);
                   if(boo){
                      sql.setString(1,logname);
                      password =
                      Encrypt.encrypt(password,"javajsp");              //给用户密码加密
                      sql.setString(2,password);
                      int m = sql.executeUpdate();
                      if(m!= 0){
                         backNews = "注册成功";
                         userBean.setBackNews(backNews);
                         userBean.setLogname(logname);
                      }
                   }
                   else {
                      backNews = "信息填写不完整或名字中有非法字符";
                      userBean.setBackNews(backNews);
                   }
                   con.close();                                         //连接返回连接池
            }
            catch(SQLException exp){
                   backNews = "该会员名已被使用,请您更换名字" + exp;
                   userBean.setBackNews(backNews);
            }
            catch(NamingException exp){
                   backNews = "没有设置连接池" + exp;
                   userBean.setBackNews(backNews);
            }
            finally{
               try{
                   con.close();
               }
               catch(Exception ee){}
            }
            RequestDispatcher dispatcher =
            request.getRequestDispatcher("register.jsp");
            dispatcher.forward(request, response);                      //转发
      }
}
```

(12) 登录。采用 MVC 模式设计登录模块。

> 视图（**JSP** 页面）

视图部分由一个 JSP 页面 login.jsp 构成（见 head.txt 导航条）。login.jsp 页面负责提供输入登录信息界面，并负责显示登录反馈信息，例如登录是否成功，是否已经登录等。

login.jsp（效果如图 8.31 所示）

```jsp
<%@ page contentType="text/html" %>
<%@ page pageEncoding="utf-8" %>
<jsp:useBean id="loginBean" class="save.data.Login" scope="session"/>
<HEAD><%@ include file="head.txt" %></HEAD>
<title>登录页面</title>
<style>
   #tom{
      font-family:宋体;font-size:30;color:black
   }
</style>
<HTML><body id=tom bgcolor=pink>
<form action="loginServlet" method="post">
登录用户:<input type=text id=tom name="logname" size=12><br>
输入密码:<input type=password id=tom name="password" size=12><br>
<input type=submit id=tom value="提交"><br>
登录反馈信息:<br>
登录名称:<br><jsp:getProperty name="loginBean" property="logname"/>
<jsp:getProperty name="loginBean" property="backNews"/>
</body></HTML>
```

图 8.31　登录页面

> **模型（bean）**

Login.java

该模块中 bean 的 id 是 loginBean，是 session bean，由 servlet 控制器负责创建或更新。

```java
package save.data;
import java.util.*;
public class Login {
    String logname = "",
           backNews = "未登录";
    public void setLogname(String logname){
        this.logname = logname;
    }
    public String getLogname(){
        return logname;
    }
    public void setBackNews(String s) {
        backNews = s;
    }
```

```
    public String getBackNews(){
        return backNews;
    }
}
```

> **控制器（servlet）**

servlet 是 HandleLogin 的实例，名字是 loginServlet（见 web.xml），负责连接数据库，查询 user 表，验证用户输入的会员名和密码是否在该表中，即验证是否是已注册的用户，如果用户是已注册的用户，就将用户设置成登录状态，即将用户的名称存放到 id 是 loginBean 的 session bean 中，并将用户转发到 login.jsp 页面查看登录反馈信息。如果用户不是注册用户，控制器将提示登录失败。另外，loginServlet 使用了 Encrypt 类的 static 方法将用户的密码加密后和数据库的 user 表中密码进行比对。

HandleLogin.java

```java
package handle.data;
import save.data.*;
import java.sql.*;
import java.io.*;
import javax.servlet.*;
import javax.servlet.http.*;
import javax.sql.DataSource;
import javax.naming.Context;
import javax.naming.InitialContext;
import javax.naming.NamingException;
public class HandleLogin extends HttpServlet{
    public void init(ServletConfig config) throws ServletException{
        super.init(config);
    }
    public void service(HttpServletRequest request,
                HttpServletResponse response)
                throws ServletException,IOException{
        request.setCharacterEncoding("utf-8");
        Connection con = null;
        Statement sql;
        String logname = request.getParameter("logname").trim(),
        password = request.getParameter("password").trim();
        password = Encrypt.encrypt(password,"javajsp");           //给用户密码加密
        boolean boo = (logname.length()>0)&&(password.length()>0);
        try{
            Context context = new InitialContext();
            Context contextNeeded =
                    (Context)context.lookup("java:comp/env");
            DataSource ds =
            (DataSource)contextNeeded.lookup("adverConn");        //连接池
             con = ds.getConnection();                            //使用连接池中的连接
            String condition = "select * from user where logname = '" +
                    logname + "' and password = '" + password + "'";
            sql = con.createStatement();
            if(boo){
                ResultSet rs = sql.executeQuery(condition);
```

```java
                    boolean m = rs.next();
                    if(m == true){
                        //调用登录成功的方法
                        success(request,response,logname,password);
                        RequestDispatcher dispatcher =
                        request.getRequestDispatcher("login.jsp");      //转发
                        dispatcher.forward(request,response);
                    }
                    else{
                        String backNews = "您输入的用户名不存在,或密码不般配";
                        //调用登录失败的方法
                        fail(request,response,logname,backNews);
                    }
                }
                else{
                    String backNews = "请输入用户名和密码";
                    fail(request,response,logname,backNews);
                }
                con.close();                                             //连接返回连接池
            }
            catch(SQLException exp){
                String backNews = "" + exp;
                fail(request,response,logname,backNews);
            }
            catch(NamingException exp){
                String backNews = "没有设置连接池" + exp;
                fail(request,response,logname,backNews);
            }
            finally{
              try{
                  con.close();
              }
              catch(Exception ee){}
            }
        }
        public void success(HttpServletRequest request,
                    HttpServletResponse response,
                    String logname,String password) {
            Login loginBean = null;
            HttpSession session = request.getSession(true);
            try{  loginBean = (Login)session.getAttribute("loginBean");
                    if(loginBean == null){
                        loginBean = new Login();                          //创建新的数据模型
                        session.setAttribute("loginBean",loginBean);
                        loginBean = (Login)session.getAttribute("loginBean");
                    }
                    String name = loginBean.getLogname();
                    if(name.equals(logname)) {
                        loginBean.setBackNews(logname + "已经登录了");
                        loginBean.setLogname(logname);
                    }
                    else {  //数据模型存储新的登录用户
```

```
                    loginBean.setBackNews(logname + "登录成功");
                    loginBean.setLogname(logname);
                }
            }
            catch(Exception ee){
                loginBean = new Login();
                session.setAttribute("loginBean",loginBean);
                loginBean.setBackNews("" + ee);
                loginBean.setLogname(logname);
            }
        }
        public void fail(HttpServletRequest request,
                        HttpServletResponse response,
                        String logname,String backNews) {
            response.setContentType("text/html;charset = utf - 8");
            try {
                PrintWriter out = response.getWriter();
                out.println("< html >< body >");
                out.println("< h2 >" + logname + "登录反馈结果< br >" + backNews + "</h2 >") ;
                out.println("返回登录页面或主页< br >");
                out.println("< a href = login.jsp >登录页面</a >");
                out.println("< br >< a href = index.jsp >主页</a >");
                out.println("</body ></html >");
            }
            catch(IOException exp){ }
        }
}
```

(13) 发布广告。采用 JSP＋servlet 设计发布广告模块。

➤ 视图(JSP 页面)

视图部分由一个 JSP 页面 publishAdver.jsp 构成(见 head.txt 导航条)。该页面负责选择、提交上传文件，并请求名字是 upFile 的 servlet。servlet 负责完成文件的上传，并显示反馈信息。用户可以使用该页面上传若干幅图像，代表自己的广告信息(上传文件的知识点可事先预习 9.4 节)。

publishAdver.jsp(效果如图 8.32 所示)

```
<% @ page contentType = "text/html" %>
<% @ page pageEncoding = "utf - 8" %>
< HEAD ><% @ include file = "head.txt" %></HEAD >
< title >发布广告页面</title >
< style >
    #tom{
        font - family:宋体;font - size:26;color:black;
    }
</style >
< HTML >< body id = tom bgcolor = #FFBBFF >
    选择要上传的文件:< br >
< form action = "upFile" method = "post" ENCTYPE = "multipart/form - data">
    < input type = FILE name = "file" id = tom size = "45">< br >
    < input type = "submit" id = tom name = "submit" value = "提交">
</form >
</body ></HTML >
```

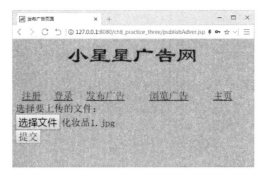

图 8.32　上传广告图片

➢ **servlet**

名字是 upFile 的 servlet 负责获取用户上传的全部信息，然后解析出上传文件的内容和文件名字。把文件的内容保存在 Web 服务目录的 image 子目录中，该文件名字是用户登录名尾加上用户上传的文件名（可预习 9.3 节）。

UpFile.java

```java
package handle.data;
import save.data.Login;
import java.io.*;
import javax.servlet.*;
import javax.servlet.http.*;
public class UpFile extends HttpServlet{
    public void init(ServletConfig config) throws ServletException{
        super.init(config);
    }
    public void service(HttpServletRequest request,
                        HttpServletResponse response)
                        throws ServletException,IOException{
        String backMess = "";
        request.setCharacterEncoding("utf-8");
        Login loginBean = null;
        HttpSession session = request.getSession(true);
        String fileName = null;
        try{
            loginBean = (Login)session.getAttribute("loginBean");
            if(loginBean == null){
                response.sendRedirect("login.jsp");          //重定向到登录页面
                return;
            }
            else {
                boolean b = loginBean.getLogname() == null||
                            loginBean.getLogname().length() == 0;
                if(b){
                    response.sendRedirect("login.jsp");      //重定向到登录页面
                    return;
                }
            }
```

```java
    catch(Exception exp){
        response.sendRedirect("login.jsp");              //重定向到登录页面
        return;
    }
    try{
        String tempFileName = (String)session.getId();
        String webDir = request.getContextPath();
        webDir = webDir.substring(1);
        File f = new File("");
        String path = f.getAbsolutePath();
        int index = path.indexOf("bin");
        String tomcatDir = path.substring(0,index);
        File dir = new File(tomcatDir + "/webapps/" + webDir + "/image");
        dir.mkdir();                                     //建立目录
        File fileTemp = new File(dir,tempFileName);
        RandomAccessFile randomWrite = new RandomAccessFile(fileTemp,"rw");
        InputStream in = request.getInputStream();
        byte b[] = new byte[10000];
        int n;
        while( (n = in.read(b))!= -1){
            randomWrite.write(b,0,n);
        }
        randomWrite.close();
        in.close();
        RandomAccessFile randomRead = new RandomAccessFile(fileTemp,"r");
        int second = 1;
        String secondLine = null;
        while(second <= 2) {
            secondLine = randomRead.readLine();
            second++;
        }
        int position = secondLine.lastIndexOf(" = ");
        fileName = secondLine.substring(position + 2,secondLine.length() - 1);
        randomRead.seek(0);
        long forthEndPosition = 0;
        int forth = 1;
        while((n = randomRead.readByte())!= -1&&(forth <= 4)){
            if(n == '\n'){
                forthEndPosition = randomRead.getFilePointer();
                forth++;
            }
        }
        byte cc[] = fileName.getBytes("iso - 8859 - 1");
        fileName = new String(cc,"utf - 8");
        fileName = (loginBean.getLogname()).concat(fileName);
        File fileUser =  new File(dir,fileName);
        randomWrite = new RandomAccessFile(fileUser,"rw");
        randomRead.seek(randomRead.length());
        long endPosition = randomRead.getFilePointer();
        long mark = endPosition;
        int j = 1;
```

```
            while((mark >= 0)&&(j <= 6)) {
                mark--;
                randomRead.seek(mark);
                n = randomRead.readByte();
                if(n == '\n'){
                    endPosition = randomRead.getFilePointer();
                    j++;
                }
            }
            randomRead.seek(forthEndPosition);
            long startPoint = randomRead.getFilePointer();
            while(startPoint < endPosition - 1){
                n = randomRead.readByte();
                randomWrite.write(n);
                startPoint = randomRead.getFilePointer();
            }
            randomWrite.close();
            randomRead.close();
            backMess = "上传成功";
            fileTemp.delete();
        }
        catch(Exception ee) {
            backMess = "没有选择文件或上传失败";
        }
        response.setContentType("text/html;charset=utf-8");
        try {
            PrintWriter out = response.getWriter();
            out.println("<html><body>");
            out.println("<h2>" + loginBean.getLogname() + ":" + backMess + "</h2>");
            out.println("<br>返回主页");
            out.println("<br><a href = index.jsp>主页</a>");
            out.println("</body></html>");
        }
        catch(IOException exp){}
    }
}
```

(14) 浏览广告。采用 JSP+bean 设计浏览广告模块。

➢ **视图(JSP 页面)**

视图部分由 browseAdver.jsp(见 head.txt 导航条)和 showAdver.jsp 两个 JSP 页面构成。browseAdver.jsp 页面负责查询数据库的 user 表,得到注册用户的名字,并为每个注册用户提供相应超链接,用户单击超链接将参数 logname 的值(注册用户的名字)传递给超链接请求的 showAdver.jsp 页面。用户在 showAdver.jsp 页面浏览注册用户发布的广告图片(另外,将 flower.jpg 图片保存到 image 目录中,对于没有上传广告图片的注册用户,showAdver.jsp 显示 flower.jpg)。

browseAdver.jsp(效果如图 8.33 所示)

```
<%@ page contentType = "text/html" %>
<%@ page pageEncoding = "utf-8" %>
```

```jsp
<%@ page import = "java.sql.*" %>
<%@ page import = "javax.sql.DataSource" %>
<%@ page import = "javax.naming.Context" %>
<%@ page import = "javax.naming.InitialContext" %>
<HEAD><%@ include file = "head.txt" %></HEAD>
<title>浏览广告页面</title>
<style>
    #tom{
        font-family:宋体;font-size:26;color:black;
    }
</style>
<HTML><body id = tom bgcolor = #EEDFF >
<%
    Context context = new InitialContext();
    Context contextNeeded = (Context)context.lookup("java:comp/env");
    DataSource ds = (DataSource)contextNeeded.lookup("adverConn");     //连接池
    Connection con  = null;
    Statement sql;
    ResultSet rs;
    try{
        con = ds.getConnection();                          //使用连接池中的连接
        sql = con.createStatement();
        String SQL = "SELECT logname FROM user";           //SQL 语句
        rs = sql.executeQuery(SQL);                        //查表
        while(rs.next()) {
            String logname = rs.getString(1);
            out.print("<br><a href = showAdver.jsp?logname = " + logname +
            ">浏览" + logname + "发布的广告</a>");
        }
        con.close() ;                                      //连接返回连接池
    }
    catch(SQLException e) {
        out.print("<h1>" + e);
    }
    finally{
        try{
            con.close();
        }
        catch(Exception ee){}
    }
%>
</body></HTML>
```

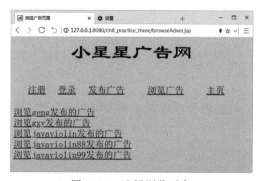

图 8.33　选择浏览对象

showAdver.jsp（效果如图 8.34 所示）

```jsp
<%@ page contentType="text/html" %>
<%@ page pageEncoding="utf-8" %>
<jsp:useBean id="play" class="save.data.Play" scope="session" />
<HEAD><%@ include file="head.txt" %></HEAD>
<title>浏览广告图</title>
<style>
    #textStyle{
        font-family:宋体;font-size:16;color:blue
    }
</style>
<% request.setCharacterEncoding("utf-8");
%>
<%
   String logname = request.getParameter("logname");        //获取用户名
   String webDir = request.getContextPath();                //获取当前Web服务目录的名称
   webDir = webDir.substring(1);                            //去掉名称前面的目录符号
%>
<jsp:setProperty name="play" property="logname" value="<%= logname %>"/>
<jsp:setProperty name="play" property="webDir" value="<%= webDir %>"/>
<jsp:setProperty name="play" property="index" param="index" />
<HTML><center><body bgcolor=pink><p id=textStyle>
</p><br><%= play.logname %>的广告：<br>
<image src=
image/<jsp:getProperty name="play"
       property="showImage"/> width=300 height=200></image><br>
<a href="?index=<%= play.getIndex()+1 %>&logname=<%= play.logname %>">
下一张</a>
<a href="?index=<%= play.getIndex()-1 %>&logname=<%= play.logname %>">
上一张</a>
</body><center></HTML>
```

图 8.34　浏览某注册用户的广告

➢ **bean**

名字是 play 的 session bean 负责存放广告图片的有关数据。

Play.java

```
Play.java
package save.data;
```

```java
import java.io.*;
import java.util.regex.Pattern;
import java.util.regex.Matcher;
public class Play {
    public String logname;
    String pictureName[];                      //存放 logname 用户广告图片文件名字的数组
    public String showImage;                   //存放当前要显示的图片
    public String webDir = "";                 //Web 服务目录的名字
    public String tomcatDir;                   //Tomcat 的安装目录，例如 apache-tomcat-9.0.26
    int index = 0;                             //存放图片文件的序号
    public Play() {
        File f = new File("");
        String path = f.getAbsolutePath();
        int index = path.indexOf("bin");       //bin 是 Tomcat 的安装目录下的子目录
        tomcatDir = path.substring(0,index);   //得到 Tomcat 的安装目录的名字
    }
    public void setLogname(String s){
        showImage = "";
        logname = s;
    }
    public void setWebDir(String s) {
        webDir = s;
        File dirImage = new File(tomcatDir + "/webapps/" + webDir + "/image");
        pictureName = dirImage.list(new FileStartName(logname));
    }
    public String getShowImage() {
        try {
            showImage = pictureName[index];
            return showImage;
        }
        catch(Exception exp){
            return "flower.jpg";
        }
    }
    public void setIndex(int i) {
        index = i;
        if(index >= pictureName.length)
            index = 0;
        if(index < 0)
            index = pictureName.length - 1;
    }
    public int getIndex() {
        return index ;
    }
}
class FileStartName implements FilenameFilter {
    String logname = null;
    Pattern pattern;
    Matcher matcher;
    FileStartName(String logname) {
        this.logname = logname;
```

```
            pattern = Pattern.compile(logname);
        }
        public boolean accept(File dir,String name) {
            matcher = pattern.matcher(name);
            if(matcher.find())
                return true;
            else
                return false;
        }
    }
```

❸ 参考代码

代码在实验步骤中。

8.13 小结

> JSP 使用 JDBC 提供的 API 和数据库进行交互信息。JDBC 技术在数据库开发中占有很重要的地位,JDBC 操作不同的数据库仅仅是连接方式上的差异而已,使用 JDBC 的应用程序一旦和数据库建立连接,就可以使用 JDBC 提供的 API 操作数据库。
> 当查询 ResultSet 对象中的数据时,不可以关闭和数据库的连接。
> 使用 PreparedStatement 对象可以提高操作数据库的效率。
> 使用数据库连接池可以提高连接数据库的效率。

习题 8

1. 启动 MySQL 数据库服务器的命令是哪个?
2. 启动 MySQL 数据库服务器命令行客户端的命令是哪个?
3. 操作数据库之前可以不和数据库建立连接吗?
4. 查询表中记录的 SQL 语句的基本语句格式是怎样的?
5. 更新表中记录的 SQL 语句的基本语句格式是怎样的?
6. 删除表中记录的 SQL 语句的基本语句格式是怎样的?
7. 向表中插入(添加)记录的 SQL 语句的基本语句格式是怎样的?
8. 加载 MySQL 的 JDBC-数据库连接器的代码是什么?
9. 加载 SQL Server 的 JDBC-数据库连接器的代码是什么?
10. 使用预处理语句的好处是什么?
11. 建立连接池的配置文件的名字是什么? 应该保存在哪个目录中?
12. 使用连接池修改例 8_2。
13. 参照例 8_2,编写一个查询 Access 数据库的 JSP 页面。
14. 使用 MVC 结构,设计一个用户注册的 Web 应用程序。

第 9 章 JSP中的文件操作

本章导读

　　主要内容
　　　❖ File 类
　　　❖ RandomAccessFile 类
　　　❖ 文件上传
　　　❖ 文件下载
　　难点
　　　❖ 文件上传
　　关键实践
　　　❖ 听学《新概念英语》

　　有时服务器需要将用户提交的信息保存到文件,或根据用户的要求将服务器上的文件的内容显示到用户端。JSP 通过 Java 的输入输出流来实现文件的读写操作。

　　为了更好地体现一个 Web 应用将数据的处理和显示相分离,除个别例子用于说明基本知识外,本章大部分例子都采用 MVC 模式(有关 MVC 的知识请参见第 7 章)。

　　本章在 Tomcat 安装目录的 webapps 目录下建立 Web 服务目录 ch9。另外,需要在 Web 服务目录 ch9 下建立目录结构 ch9\WEB-INF\classes(WEB-INF 字母大写)。

　　本章使用的 bean 的包名(除非特别说明)均为 save.data。在 classes 下建立目录结构 \save\data,创建 bean 的类的字节码文件保存在\WEB-INF\classes\save\data 中(知识点见 5.1.2 节)。本章的 servlet 的包名(除非特别说明)均为 handle.data。在 classes 下建立目录结构\handle\data,创建 servlet 的类的字节码文件保存在\WEB-INF\classes\handle\data 中(知识点见 6.1.1 节)。

　　本章使用的 javax.servlet 和 javax.servlet.http 包中的类不在 JDK 提供的核心类库中,为了方便编译 Java 源文件,请事先将 Tomcat 安装目录 lib 子目录中的 servlet-api.jar 文件复制(不要剪贴)到\ch9\WEB-INF\classes\中(知识点见 6.1.1 节)。另外,保存 Java 源文件时,"保存类型"选择为"所有文件","将"编码"选择为"ANSI"。保存 JSP 文件和部署文件 web.xml 时,"保存类型"选择为"所有文件","将"编码"选择为"UTF-8"。

视频讲解

9.1　File 类

　　File 对象用来获取文件本身的一些信息,例如文件所在的目录、文件的长度、文件读写权限等,不涉及对文件的读写操作。创建一个 File 对象的构造方法有 3 个:

```
File(String filename);
File(String directoryPath,String filename);
```

```
File(File f, String filename);
```

对于第一个构造方法，filename 是文件名字或文件的绝对路径，如 filename="Hello.txt" 或 filename="c:/mybook/A.txt"。对于第二个构造方法 directoryPath 是文件的路径，filename 是文件名字，如 directoryPath="c:/mybook/"，filename="A.txt"。对于第三个构造方法，参数 f 是要指定成一个目录的文件，filename 是文件名字，如 f=new File("c:/mybook")，filename="A.txt"。

需要注意的是，当使用第一个构造方法 File(String filename) 创建文件时，filename 是文件名字，那么该文件会被认为是与当前应用程序在同一目录中。由于 Tomcat 服务器是在 bin 下启动执行的，所以该文件被认为在 Tomcat 安装目录的\bin 中。

在下面的例 9_1 中，使用 File 类的一些方法获取用户访问的当前 JSP 页面的一些信息，如图 9.1 所示。该例子中让内置对象 request 调用 getContextPath() 方法获取当前 Web 服务目录的名称（request 的常用方法见 4.1.3 节）。

例 9_1

example9_1.jsp

```jsp
<%@ page contentType="text/html" %>
<%@ page pageEncoding="utf-8" %>
<%@ page import="java.io.*" %>
<style>
    #tom{
        font-family:宋体;font-size:28;color:blue
    }
</style>
<HTML><body id=tom bgcolor=#FFBBFF>
<% String jspPage = request.getServletPath();    //请求的页面的名称
    out.print(jspPage+"<br>");
    String webDir = request.getContextPath();    //当前 Web 目录的名称
    out.print(webDir+"<br>");
    jspPage = jspPage.substring(1);              //去掉名称前面的目录符号"/"
    webDir = webDir.substring(1);                //去掉名称前面的目录符号"/"
    out.print(jspPage+"<br>");
     out.print(webDir+"<br>");
    File f = new File("");                       //文件 f 被认为在 Tomcat 安装目录的\bin 中
    String path = f.getAbsolutePath();
    out.print(path+"<br>");
    int index = path.indexOf("bin");
    String tomcatDir = path.substring(0,index);  //Tomcat 的安装目录
    File file = new File(tomcatDir+"/webapps/"+webDir,jspPage);
%>
<b>
文件<%=file.getName()%>是可读的吗?<%=file.canRead()%>
<br>文件<%=file.getName()%>的长度：<%=file.length()%>字节.
<br><%=file.getName()%>的父目录(即所在目录)
是:<br><%=file.getParent()%>
<br><%=file.getName()%>的绝对路径是
<br><%=file.getAbsolutePath()%>
</b></body></HTML>
```

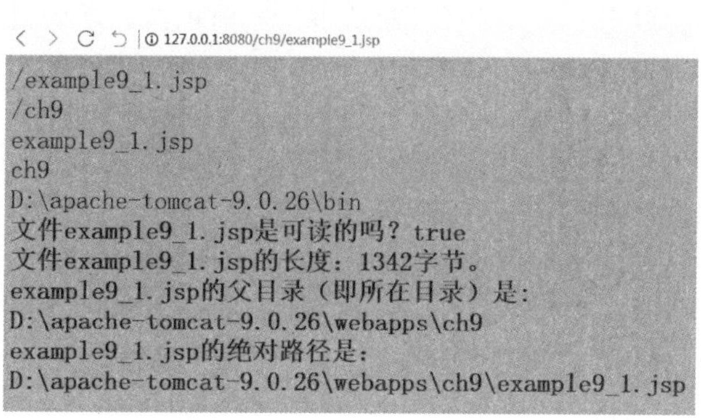

图 9.1 File 类获取文件的信息

9.2　RandomAccessFile 类

视频讲解

需要对一个文件进行读写操作时，可以创建一个 RandomAccessFile 对象，RandomAccessFile 对象可以读取文件的数据，也可以向文件写入数据。

RandomAccessFile 类的两个构造方法：

- RandomAccessFile(String name,String mode)：参数 name 用来确定一个文件名，参数 mode 取 "r"（只读）或 "rw"（可读写），决定对文件的访问权利。
- RandomAccessFile(File file,String mode)：参数 file 是一个 File 对象，参数 mode 取 "r"（只读）或 "rw"（可读写），决定对文件的访问权利。

RandomAccessFile 流对文件的读写方式很灵活。例如 RandomAccessFile 流的 seek(long a) 方法移动在文件中的读写位置，其中 seek(long a) 方法的参数 a 确定读写位置，即距离文件开头的字节数目。另外，RandomAccessFile 流还可以调用 getFilePointer() 方法获取当前流在文件中的读写位置。在后面的 9.3 节实现文件上传时，就借助了 RandomAccessFile 流的强大功能。

RandomAccessFile 类的常用方法有：

- getFilePointer() 获取当前流在文件中的读写的位置。
- length() 获取文件的长度。
- readByte() 从文件中读取一个字节。
- readDouble() 从文件中读取一个双精度浮点值（8 个字节）。
- readInt() 从文件中读取一个 int 值（4 个字节）。
- readLine() 从文件中读取一个文本行。
- readUTF() 从文件中读取一个 UTF 字符串。
- seek(long a) 定位当前流在文件中的读写的位置。
- write(byte b[]) 写 b.length 个字节到文件。
- writeDouble(double v) 向文件写入一个双精度浮点值。
- writeInt(int v) 向文件写入一个 int 值。
- writeUTF(String s) 写入一个 UTF 字符串。

需要注意的是，RondomAccessFile 流的 readLine() 方法在读取含有非 ASCII 字符的文件时（例如含有汉字的文件）会出现"乱码"现象，因此，需要把 readLine() 读取的字符串用"ISO-

8859-1"编码重新编码存放到 byte 数组中,然后再用当前机器的默认编码或指定编码将该数组转化为字符串,操作如下:

(1) 读取 String str＝in.readLine();。

(2) 用"ISO-8859-1"重新编码 byte b[]＝str.getBytes("ISO-8859-1");。

(3) 使用默认编码将字节数组转化为字符串 new String(b);,如果机器默认编码是以 GB 2312 编码,那么 new String(b);等同于 new String(b, "GB 2312");。

例 9_2 的 Web 应用程序使用 JSP＋servlet(知识点见第 6 章)。在例子中有一个 JSP 页面 example9_2 和一个 servlet。用户在 JSP 页面 example9_2 选择一个文件以及文件的编码类型,提交给 servlet,该 servlet 负责读取和显示文件的内容。

例 9_2

> **JSP 页面(视图)**

视图部分由 example9_2.jsp 构成。用户在 example9_2_choiceFile.jsp 页面输入文件的路径和名字,选择文件的编码类型,请求名字为 handleFile 的 servlet 对象。handleFile 负责读取文件并显示文件的内容。

example9_2.jsp(效果如图 9.2(a)(b)所示)

```
<%@ page contentType="text/html" %>
<%@ page pageEncoding="utf-8" %>
<%@ page import="java.io.*" %>
<style>
   #tom{
      font-family:宋体;font-size:28;color:black
   }
</style>
<HTML><body id=tom bgcolor=#FFBBFF>
<form action="handleFile" method="post">
输入文件的路径(如:d:/2000):<br>
<input type="text" id=tom name="filePath" size=12>
<br>输入文件的名字(如:Hello.java):<br>
<input type="text" id=tom name="fileName" size=18><br>
文件的编码类型:<br>
```

(a) 输入目录和文件名　　　　　　　(b) 显示文件内容

图 9.2　读取文件

```
    <input type = "radio" name = "R" value = "yes"/>是 UTF-8
    <input type = "radio" name = "R" value = "no" checked = "default"/>不是 UTF-8
<br><input type = "submit" id = tom value = "读取" name = "submit">
</form>
</body></HTML>
```

> **servlet**

Example9_2_Servlet 负责创建名字是 handleFile 的 servlet。handleFile 负责读取文件的内容,并显示文件的内容。

用命令行进入 handle\data 的父目录 classes,编译 Example9_2_Servlet.java(见本章开始的约定):

```
javac -cp servlet-api.jar   handle/data/Example9_2_Servlet.java
```

Example9_2_Servlet.java

```java
package handle.data;
import java.io.*;
import javax.servlet.*;
import javax.servlet.http.*;
public class Example9_2_Servlet extends HttpServlet{
    public void init(ServletConfig config) throws ServletException{
        super.init(config);
    }
    public void service(HttpServletRequest request,
        HttpServletResponse response) throws ServletException,IOException{
        String fileContent = "";
        request.setCharacterEncoding("utf-8");
        String filePath = request.getParameter("filePath");
        String fileName = request.getParameter("fileName");
        String isUTF = request.getParameter("R");
        if(filePath == null||fileName == null)
            return;
        if(fileName.length() == 0||fileName.length() == 0)
            return;
        try{   File f = new File(filePath,fileName);
            RandomAccessFile randomAccess = new RandomAccessFile(f,"r");
            String s = null;
            StringBuffer stringbuffer = new StringBuffer();
            while ((s = randomAccess.readLine())!= null){
                byte b[] = s.getBytes("iso-8859-1");
                if(isUTF.contains("yes"))
                    stringbuffer.append("\n" + new String(b,"utf-8"));
                else
                    stringbuffer.append("\n" + new String(b));
            }
            fileContent = new String(stringbuffer);
        }
        catch(Exception exp){
            fileContent = "读取失败" + exp.toString();
        }
```

```
        response.setContentType("text/plain;charset = utf - 8");
        PrintWriter out = response.getWriter();
        out.print(fileContent);
    }
}
```

9.3 文件上传

视频讲解

用户通过一个 JSP 页面上传文件给服务器时,form 表单必须将 ENCTYPE 的属性值设成 multipart/form-data,并含有 File 类型的 GUI 组件。含有 File 类型 GUI 组件的 form 表单如下所示:

```
< form action = "JSP 页面或 servlet" method = "post" ENCTYPE = "multipart/form - data"
    < input type = "File" name = "picture" >
    < input type = "submit" value = "提交">
</form>
```

Tomcat 服务器可以让内置对象 request 调用方法 getInputStream()获得一个输入流,通过这个输入流读入用户上传的全部信息,包括文件的内容以及表单域的信息。

在下面的例 9_3 中,用户通过 example9_3.jsp 页面上传如下的文本文件 A.txt。

A.txt

request 获得一个输入流读取用户上传的全部信息,包括表单的头信息以及上传文件的内容。以后将讨论如何去掉表单的信息,获取文件的内容。

在 example9_3_accept.jsp 页面,内置对象 request 调用方法 getInputStream()获得一个输入流 inputStream,用 RandomAccessFile 创建 randomAccess 对象。输入流 inputStream 读取用户上传的信息,输出流 randomAccess 将读取的信息写入文件 B.txt。用户上传的全部信息,包括文件 A.txt 的内容以及表单域的信息存放于服务器的 D:/2000 目录中的 B.txt 文件中。文件 B.txt 的前 4 行(包括一个空行)以及倒数 5 行(包括一个空行)是表单域的内容,中间部分是上传文件 A.txt 的内容。B.txt 的内容如图 9.3 所示。

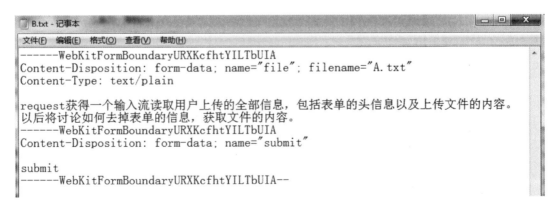

图 9.3 上传的全部信息

第 9 章　JSP中的文件操作

例 9_3
example9_3.jsp

```
<%@ page contentType = "text/html" %>
<%@ page pageEncoding = "utf-8" %>
<HTML><body>
选择要上传的文件:<br>
<form action =
  "example9_3_accept.jsp" method = "post" ENCTYPE = "multipart/form-data">
  <input type = FILE  name = "file" size = "38" />
  <br><input type = "submit" name = "submit" value = "submit" />
</form>
</body></HTML>
```

example9_3_accept.jsp

```
<%@ page contentType = "text/html" %>
<%@ page pageEncoding = "utf-8" %>
<%@ page import = "java.io.*" %>
<HTML><body>
<% try{
      InputStream inputStream = request.getInputStream();
      File dir = new File("D:/2000");
      dir.mkdir();
      File f = new File(dir,"B.txt");
      RandomAccessFile randomAccess = new RandomAccessFile(f,"rw");
      byte b[] = new byte[1000];
      int n;
      while((n = inputStream.read(b))!= -1){
         randomAccess.write(b,0,n);
      }
      randomAccess.close();
      inputStream.close();
      out.print("文件已上传");              //这个 out 是内置对象
   }
   catch(IOException ee){
      out.print("上传失败" + ee);            //这个 out 是内置对象
   }
%>
</body></HTML>>
```

通过上面的讨论知道，根据 HTTP 协议，文件表单提交的信息中，前面 4 行和后面的 5 行是表单本身的信息，中间部分才是用户提交的文件的内容。

在下面的例 9_4 中，通过输入、输出流技术获取文件的内容，即去掉表单的信息。

根据不同用户的 session 对象互不相同这一特点，将用户提交的全部信息首先保存成一个临时文件，该临时文件的名字是用户的 session 对象的 id，然后读取该临时文件的第 2 行，因为这一行中含有用户上传的文件的名字，再获取第 4 行结束的位置，以及倒数第 6 行结束的位置，因为这两个位置之间的内容是上传文件的内容，然后将这部分内容存入文件，该文件的名字和用户上传的文件的名字保持一致，最后删除临时文件。

例 9_4 的 Web 应用程序使用 MVC 模式（MVC 的知识见第 7 章）。

例 9_4
➤ bean（模型）

Example9_4_Bean.java 负责创建 id 是 fileBean 的 request bean。request bean 负责存放文件的相关内容，如文件的名字等。用命令行进入 save\data 的父目录 classes，编译 Example9_2_Bean.java（见本章开始的约定）：

```
javac save\data\Example9_4_Bean.java
```

Example9_2_Bean.java（负责创建 request bean）

```java
package save.data;
public class Example9_4_Bean {
    String fileName = "";
    String mess = "";
    public void setFileName(String str){
        fileName = str;
    }
    public String getFileName(){
        return fileName;
    }
    public void setMess(String str){
        mess = str;
    }
    public String getMess(){
        return mess;
    }
}
```

➤ JSP 页面（视图）

example9_4.jsp 页面负责选择、提交上传文件，并请求名字是 upFile 的 servlet。example9_4.jsp 页面同时负责显示 fileBean 中的数据。如果 fileBean 中存放的文件不是图像文件就显示文件的名字，如果是图像文件，不仅显示文件的名字，同时也显示当前图像。

example9_4.jsp（效果如图 9.4 所示）

```jsp
<%@ page contentType="text/html" %>
<%@ page pageEncoding="utf-8" %>
<jsp:useBean id="fileBean" class="save.data.Example9_4_Bean" scope="request"/>
<style>
    #tom{
        font-family:宋体;font-size:28;color:blue
    }
</style>
<HTML><body id=tom bgcolor=#FFBBFF>
    选择要上传的文件：<br>
<form action="upFile" method="post" ENCTYPE="multipart/form-data">
    <input type=FILE name="file" id=tom size="45"><br>
    <input type="submit" id=tom name="submit" value="提交">
</form>
<br>上传的文件名字：
    <jsp:getProperty name="fileBean" property="fileName"/><br>
```

第9章 JSP中的文件操作

```
上传反馈：
  <jsp:getProperty name = "fileBean" property = "mess"/>
<%
   String name = fileBean.getFileName();
   boolean boo = name.endsWith(".jpg");
   boo = boo||name.endsWith(".gif");
   if(boo) {
%>    <image src = "image/<% = name %>" width = 380 height = 300>
      <% = name %></image>
<% }
   else {
%>    <% = name %>
<% }
%>
</body></HTML>
```

图 9.4 上传文件

➤ **servlet**（控制器）

Example9_4_Servlet 负责创建名字是 upFile 的 servlet。upFile 负责获取用户上传的全部信息，然后解析出上传文件的内容和文件名字。文件的内容保存在 Web 服务目录的 image 子目录中，将文件名字存放到 id 是 fileBean 的 request bean 中，然后用转发的方法请求 example9_4.jsp 显示 fileBean 中的数据。

用命令行进入 handle\data 的父目录 classes，编译 Example9_4_Servlet.java（见本章开始的约定）：

```
javac -cp .;servlet-api.jar  handle/data/Example9_2_Servlet.java
```

注意"．；"和"servlet-api.jar"之间不要有空格，"．；"的作用是保证 Java 源文件能使用 import 语句引入当前 classes 目录中其他自定义包中的类，例如 save.data 包中的 bean 类（"．；"是 javac 默认具有的功能，在使用-cp 参数时，尽量保留这一功能）。

Example9_4_Servlet.java

```
package handle.data;
```

```java
import save.data.Example9_4_Bean;
import java.io.*;
import javax.servlet.*;
import javax.servlet.http.*;
public class Example9_4_Servlet extends HttpServlet{
    public void init(ServletConfig config) throws ServletException{
        super.init(config);
    }
    public void service(HttpServletRequest request,
                        HttpServletResponse response)
                        throws ServletException,IOException{
        request.setCharacterEncoding("utf-8");
        Example9_4_Bean fileBean = new Example9_4_Bean();        //创建 JavaBean 对象
        request.setAttribute("fileBean",fileBean);
        String fileName = null;
        HttpSession session = request.getSession(true);
        try{
            //用客户的 session 对象的 id 建立一个临时文件
            String tempFileName = (String)session.getId();
            String webDir = request.getContextPath();           //获取当前 Web 服务目录的名称
            webDir = webDir.substring(1);                       //去掉名称前面的目录符号"/"
            File f = new File("");                              //文件 f 被认为在 Tomcat 安装目录的\bin 中
            String path = f.getAbsolutePath();
            int index = path.indexOf("bin");
            String tomcatDir = path.substring(0,index);         //Tomcat 的安装目录
            //文件上传到 image 文件夹中
            File dir = new File(tomcatDir + "/webapps/" + webDir + "/image");
            dir.mkdir();                                        //建立目录
            File fileTemp = new File(dir,tempFileName);         //建立临时文件 fileTemp
            RandomAccessFile randomWrite = new RandomAccessFile(fileTemp,"rw");
            //将客户上传的全部信息存入 fileTemp
            InputStream in = request.getInputStream();
            byte b[] = new byte[10000];
            int n;
            while( (n = in.read(b))!= -1){
                randomWrite.write(b,0,n);
            }
            randomWrite.close();
            in.close();
            //读取临时文件 fileTemp,从中获取上传文件的名字和上传文件的内容
            RandomAccessFile randomRead = new RandomAccessFile(fileTemp,"r");
            //读出 fileTemp 的第 2 行,析取出上传文件的名字
            int second = 1;
            String secondLine = null;
            while(second <= 2) {
                secondLine = randomRead.readLine();
                second++;
            }
            //获取 fileTemp 中第 2 行中"filename"之后 "=" 出现的位置
            int position = secondLine.lastIndexOf(" = ");
            //客户上传的文件的名字是
            fileName = secondLine.substring(position + 2,secondLine.length() - 1);
```

```
                randomRead.seek(0);                    //再定位到文件 fileTemp 的开头
                //获取第 4 行回车符号的位置
                long    forthEndPosition = 0;
                int forth = 1;
                while((n = randomRead.readByte())!= -1&&(forth <= 4)){
                    if(n == '\n'){
                        forthEndPosition = randomRead.getFilePointer();
                        forth++;
                    }
                }
                //根据客户上传文件的名字,将该文件存入磁盘
                byte   cc[] = fileName.getBytes("iso-8859-1");
                fileName = new String(cc,"utf-8");
                File fileUser = new File(dir,fileName);
                randomWrite = new RandomAccessFile(fileUser,"rw");
                //确定出文件 fileTemp 中包含客户上传的文件的内容的最后位置,即倒数第 6 行
                randomRead.seek(randomRead.length());
                long endPosition = randomRead.getFilePointer();
                long mark = endPosition;
                int j = 1;
                while((mark >= 0)&&(j <= 6)) {
                    mark--;
                    randomRead.seek(mark);
                    n = randomRead.readByte();
                    if(n == '\n'){
                        endPosition = randomRead.getFilePointer();
                        j++;
                    }
                }
                //将 randomRead 流指向文件 fileTemp 的第 4 行结束的位置
                randomRead.seek(forthEndPosition);
                long startPoint = randomRead.getFilePointer();
                //从 fileTemp 读出客户上传的文件存入 fileUser
                //读取第 4 行结束位置和倒数第 6 行之间的内容
                while(startPoint < endPosition - 1){
                    n = randomRead.readByte();
                    randomWrite.write(n);
                    startPoint = randomRead.getFilePointer();
                }
                randomWrite.close();
                randomRead.close();
                fileBean.setMess("上传成功");
                fileBean.setFileName(fileName);
                fileTemp.delete();                        //删除临时文件
            }
            catch(Exception ee) {
                fileBean.setMess("没有选择文件或上传失败");
            }
        RequestDispatcher dispatcher =
        request.getRequestDispatcher("example9_4.jsp");
        dispatcher.forward(request, response);
    }
}
```

➤ **web.xml**（部署文件）

向 ch9\WEB\INF\下的部署文件 web.xml 添加如下的 servlet 和 servlet-mapping 标记（知识点见 6.1.2 节），部署的 servlet 的名字是 upFile，访问 servlet 的 url-pattern 是/upFile。

web.xml

```xml
<?xml version = "1.0" encoding = "utf-8"?>
<web-app>
    <!-- 以下是 web.xml 文件新添加的内容 -->
    <servlet>
        <servlet-name>upFile</servlet-name>
        <servlet-class>handle.data.Example9_3_Servlet</servlet-class>
    </servlet>
    <servlet-mapping>
        <servlet-name>upFile</servlet-name>
        <url-pattern>/upFile</url-pattern>
    </servlet-mapping>
</web-app>
```

9.4 文件下载

视频讲解

JSP 内置对象 response 调用方法 getOutputStream()可以获取一个指向用户的输出流，服务器将文件写入这个流，用户就可以下载这个文件了。当提供下载功能时，应当使用 response 对象向用户发送 HTTP 头信息，这样用户的浏览器就会调用相应的外部程序打开下载的文件。response 调用 setHeader 方法添加下载头的格式如下：

```
response.setHeader("Content-disposition","attachment;filename = "文件名");
```

例 9_5 的 Web 应用程序使用 JSP + servlet（知识点见第 6 章）。在例 9_5 中，用户在 example9_5.jsp 页面选择一个要下载的文件，将该文件的名字提交给名字是 loadFile 的 servlet，servlet 将用户选择的文件发送给用户，即提供下载。

例 9_5

➤ **JSP 页面**

example9_5.jsp（效果如图 9.5(a)所示）

```jsp
<%@ page contentType = "text/html" %>
<%@ page pageEncoding = "utf-8" %>
<style>
    #tom{
        font-family:宋体;font-size:22;color:blue
    }
</style>
<HTML><body id = tom bgcolor = #FFBBFE>
<form action = "loadFile" method = post>
选择要下载的文件:<br>
  <select name = "filePath" id = tom size = 1>
    <Option Selected value = "d:/2000/E.java">E.java
```

```
            <Option value = "d:/2000/first.jsp">first.jsp
            <Option value = "d:/2000/book.zip">book.zip
            <Option value = "d:/2000/A.txt">A.txt
        </select>
<br><input type = "submit" id = tom value = "提交" />
</form>
</body></HTML>
```

➢ **Servlet 类**

用命令行进入 handle\data 的父目录 classes,编译 Example9_5_Servlet.java(见本章开始的约定):

```
javac -cp servlet-api.jarhandle\data\Example9_5_Servlet.java
```

Example9_5_Servlet.java(效果如图 9.5(b)所示)

```java
package handle.data;
import java.io.*;
import javax.servlet.*;
import javax.servlet.http.*;
public class Example9_5_Servlet extends HttpServlet{
    public void init(ServletConfig config) throws ServletException{
        super.init(config);
    }
    public void service(HttpServletRequest request,
        HttpServletResponse response) throws ServletException,IOException{
        request.setCharacterEncoding("utf-8");
        String filePath = request.getParameter("filePath");
        String fileName = filePath.substring(filePath.lastIndexOf("/")+1);
        response.setHeader
        ("Content-disposition","attachment;filename = "+fileName);
        try{    //读取文件,并发送给用户下载
            File f = new File(filePath);
            FileInputStream in = new FileInputStream(f);
            OutputStream out = response.getOutputStream();
            int n = 0;
            byte b[] = new byte[1204];
            while((n = in.read(b))!= -1)
                out.write(b,0,n);
            out.close();
            in.close();
        }
        catch(Exception exp){}
    }
}
```

➢ **web.xml 文件**

向 ch9\WEB\INF\下的部署文件 web.xml 添加如下的 servlet 和 servlet-mapping 标记(知识点见 6.1.2 节),部署的 servlet 的名字是 loadFile,访问 servlet 的 url-pattern 是 loadFile。

(a) 选择下文件   (b) 下载对话框

图 9.5   文件下载

web.xml

```xml
<?xml version = "1.0" encoding = "utf-8"?>
<web-app>
    <!-- 以下是web.xml文件新添加的内容 -->
    <servlet>
        <servlet-name>loadFile</servlet-name>
        <servlet-class>handle.data.Example9_5_Servlet</servlet-class>
    </servlet>
    <servlet-mapping>
        <servlet-name>loadFile</servlet-name>
        <url-pattern>/loadFile</url-pattern>
    </servlet-mapping>
</web-app>
```

9.5 上机实验

视频讲解

本书提供了详细的实验步骤要求,按步骤完成,可提升学习效果,积累经验,不断提高Web设计能力。

▶ 9.5.1 实验1 查看JSP源文件

❶ 实验目的

掌握使用输入流读取服务器端的文件的方法。

❷ 实验要求

(1) 编写 readJSPFile.jsp,该页面提供一个 form 表单,用户单击 submit 提交键请求 readJSPFile.jsp,readJSPFile.jsp 将显示源文件的内容。

(2) 在 Tomcat 服务器的 webapps 目录下(例如 D:\apache-tomcat-9.0.26\webapps)新建一个名字是 ch9_practice_one 的服务目录。把 JSP 页面都保存到 ch9_practice_one 目录中。

(3) 用浏览器访问 JSP 页面 inputVertex.jsp。

❸ 参考代码

参考代码运行效果如图 9.6 所示。

第 9 章 JSP 中的文件操作

```
127.0.0.1:8080/ch9_practice_one/readJSPFile.jsp

<style>
    #tom{
        font-family:宋体;font-size:28;color:blue
    }
</style>
<HTML><body id =tom bgcolor=#FFBBFF>
JSP文件源代码：<br>

<%@ page contentType="text/html" %>
<%@ page pageEncoding = "utf-8" %>
<%@ page import="java.io.*"%>
<style>
    #tom{
        font-family:宋体;font-size:28;color:blue
    }
</style>
<HTML><body id =tom bgcolor=#FFBBFF>
<% StringBuffer sourceCode=new StringBuffer();//存放JSP文件的源代码。
```

图 9.6 查看 JSP 源文件

readJSPFile.jsp

```jsp
<%@ page contentType = "text/html" %>
<%@ page pageEncoding = "utf-8" %>
<%@ page import = "java.io.*" %>
<style>
    #tom{
        font-family:宋体;font-size:28;color:blue
    }
</style>
<HTML><body id = tom bgcolor = #FFBBFF>
<% StringBuffer sourceCode = new StringBuffer();          //存放 JSP 文件的源代码
    request.setCharacterEncoding("utf-8");
    String mess = request.getParameter("submit");
    if(mess == null)
        mess = "";
    if(mess.contains("源码")){
        response.setContentType("text/plain");
        String jspPage = request.getServletPath();
        String webDir = request.getContextPath();
        jspPage = jspPage.substring(1);
        webDir = webDir.substring(1);
        File f = new File("");
        String path = f.getAbsolutePath();
        int index = path.indexOf("bin");
        String tomcatDir = path.substring(0,index);
        File jspFile = new File(tomcatDir + "/webapps/" + webDir,jspPage);
        try{
            RandomAccessFile randomAccess =
            new RandomAccessFile(jspFile,"r");
            String s = null;
            StringBuffer stringbuffer = new StringBuffer();
            while ((s = randomAccess.readLine())!= null){
                byte b[] = s.getBytes("iso-8859-1");
                sourceCode.append("\n" + new String(b,"utf-8"));
            }
```

```
        }
        catch(Exception exp){
            out.println(exp);
        }
    }
%>
JSP 文件源代码：< br >
< % = sourceCode % >
< br >
< form action = "" method = post >
< input type = "submit" id = tom name = "submit" value = "看本页面的源码" />
</form>
</body></HTML>
```

9.5.2 实验 2 听学《新概念英语》

❶ 实验目的

掌握使用 MVC 模式读取文件的方法。

❷ 实验要求

(1) 编写一个创建 bean 的类，该类创建的 request 可以存储一个文本文件的相关信息，例如文件名、内容以及和该文本文件相关的音频文件的名字等信息。

(2) 编写一个 JSP 页面 studyEnglish.jsp。该页面提供一个 form 表单，该表单提供用户在下拉列表里选择文件名字(该文件的内容是《新概念英语》的某篇课文内容)。用户选择文件后单击"学习"提交键请求名字是 readFile 的 servlet。

(3) 编写创建 servlet 的 Servlet 类，该类创建的 servlet 根据用户提交的文件名字，读取该文件的内容，并将文件的名字、内容，以及相关的音频文件的名字保存到 request bean 中。servlet 请求 studyEnglish.jsp 页面显示 request bean 中的数据。

(4) 在 Tomcat 服务器的 webapps 目录下(例如 D:\apache-tomcat-9.0.26\webapps)新建名字是 ch9_practice_two 的 Web 服务目录。把 JSP 页面保存到 ch9_practice_two 目录中。在 ch9_practice_two 目录下建立子目录 WEB-INF(字母大写)，然后在 WEB-INF 下再建立子目录 classes，将创建 servlet 和 bean 的类的 Java 源文件按照包名保存在 classes 的相应子目录中。

(5) 向 ch9_practice_two\WEB\INF\下的部署文件 web.xml 添加 servlet 和 servlet-mapping 标记(知识点见 6.1.2 节)，部署 servlet 的名字和访问 servlet 的 url-pattern。

(6) 准备若干个.txt 文本文件，这些.txt 文件的编码均为 ANSI 编码，每个文本文件的内容是《新概念英语》的一篇课文，例如 lession1.txt、lession2.txt…。为每个.txt 文件准备一个相关的音频文件，格式可以是 mp3 等浏览器支持的格式，音频文件的名字和相应的文本文件的名字相同，例如 lession1.mp3、lession2.mp3…。文本文件保存在 ch9_practice_two 的\englishText 子目录中，mp3 文件保存在 ch9_practice_two 的\englishAudio 子目录中。

(7) 用浏览器访问 JSP 页面 studyEnglish.jsp。

❸ 参考代码

参考代码运行效果如图 9.7 所示。

➤ **bean**(模型)

用命令行进入 save\data 的父目录 classes，编译 EnglishBean.java(参考本章开始的约定)：

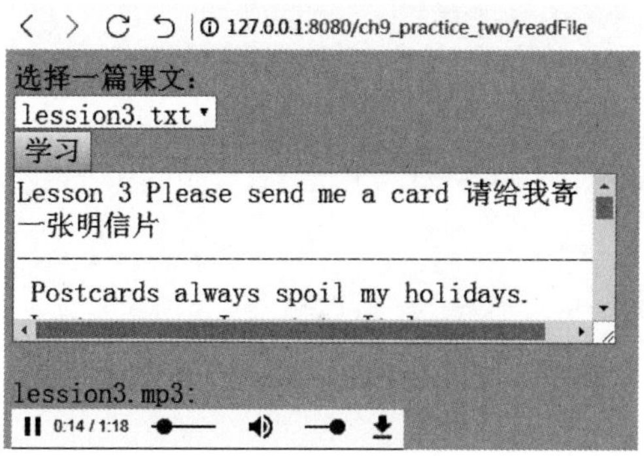

第9章　JSP中的文件操作

图 9.7　听学《新概念英语》

```
javac save\data\EnglishBean.java
```

EnglishBean.java

```java
package save.data;
public class EnglishBean {
    public String fileName = "";
    public String fileContent = "";
    public String fileAudio;                //附加的音频文件名
    public void setFileName(String str){
        fileName = str;
    }
    public String getFileName(){
        return fileName;
    }
    public void setFileContent(String str){
        fileContent = str;
    }
    public String getFileContent(){
        return fileContent;
    }
    public void setFileAudio(String str){
        fileAudio = str;
    }
    public String getFileAudio(){
        return fileAudio;
    }
}
```

➤ **JSP 页面（视图）**

studyEnglish.jsp 页面提供一个 form 表单，该表单提供用户在下拉列表里选择文件名字（该文件的内容是《新概念英语》的某篇课文内容）。用户选择文件后单击"学习"提交键请求名字是 readFile 的 servlet。

studyEnglish.jsp（效果如图 9.7 所示）

```jsp
<%@ page import = "java.io.*" %>
<%@ page contentType = "text/html" %>
```

```
<%@ page pageEncoding = "utf-8" %>
<jsp:useBean id = "english" class = "save.data.EnglishBean" scope = "request"/>
<style>
    #tom{
        font-family:宋体;font-size:22;color:blue
    }
</style>
<HTML><body id = tom bgcolor = #ffccff>
<form action = "readFile" id = tom method = post>
选择一篇课文:<br>
<% File f = new File(".");                    //文件 f 被认为在 Tomcat 安装目录的\bin 中
   String jspPage = request.getServletPath();
   String webDir = request.getContextPath();  //获取当前 Web 服务目录的名称
   jspPage = jspPage.substring(1);            //去掉名称前面的目录符号"/"
   webDir = webDir.substring(1);              //去掉名称前面的目录符号"/"
   String path = f.getAbsolutePath();
   int index = path.indexOf("bin");
   String tomcatDir = path.substring(0,index);
   String filePath = tomcatDir + "/webapps/" + webDir + "/englishText";
   File fileDir = new File(filePath);
   String name [] = fileDir.list();           //返回 englishText 中全部文本文件的名字
%>   <select id = tom name = "fileName" size = 1>
<%   for(int i = 0;i< name.length;i++){
%>      <option selected value = '<% = name[i] %>'/><% = name[i] %>
<%   }
%> </select>
     <input type = hidden name = 'filePath' value = '<% = filePath %>' />
<br><input type = submit id = tom value = "学习" /><br>
<textArea id = tom rows = 5 cols = 40 ><% = english.getFileContent() %>
</textArea><br>
<% if(english.getFileName().length()>0) {
%><br>
  <% = english.getFileAudio() %>:<br>
  <embed src = 'englishAudio/<% = english.getFileAudio() %>'
         height = 56 autostart = 'false'>
  </embed>
<% }
%>
</form>
</body></HTML>
```

> **servlet**(控制器)

ReadFile_Servlet 类负责创建名字是 readFile 的 servlet。readFile 负责创建 id 是 english 的 request bean,并读取文件,将有关信息存储到 request bean 中,然后请求 studyEnglish.jsp 页面显示 request bean 中的数据。

用命令行进入 handle\data 的父目录 classes,编译 ReadFile_Servlet.java(参考本章开始的约定):

```
javac -cp .;servlet-api.jar  handle/data/ReadFile_Servlet.java
```

事先将 Tomcat 安装目录 lib 子目录中的 servlet-api.jar 文件复制(不要剪贴)到\ch7_

practice_one\WEB-INF\classes 中（知识点见 6.1.1 节）。

注意".;"和"servlet-api.jar"之间不要有空格，".;"的作用是保证 Java 源文件能使用 import 语句引入当前 classes 目录中其他自定义包中的类，例如 save.data 包中的 bean 类。

ReadFile_Servlet.java

```java
package handle.data;
import save.data.EnglishBean;
import java.io.*;
import javax.servlet.*;
import javax.servlet.http.*;
public class ReadFile_Servlet extends HttpServlet{
    public void init(ServletConfig config) throws ServletException{
        super.init(config);
    }
    public void service(HttpServletRequest request,HttpServletResponse
                response)throws ServletException,IOException{
        EnglishBean fileBean = new EnglishBean();           //创建 bean 对象
        request.setAttribute("english",fileBean);
        String fileContent = "";
        request.setCharacterEncoding("utf-8");
        String filePath = request.getParameter("filePath");
        String fileName = request.getParameter("fileName");
        if(filePath == null||fileName == null)
            return;
        if(fileName.length() == 0||fileName.length() == 0)
            return;
        fileBean.setFileName(fileName);
        String audioFileName =
        fileName.substring(0,fileName.lastIndexOf("."))+".mp3";
        fileBean.setFileAudio(audioFileName);
        try{   File f = new File(filePath,fileName);
            RandomAccessFile randomAccess = new RandomAccessFile(f,"r");
            String s = null;
            StringBuffer stringbuffer = new StringBuffer();
            while ((s = randomAccess.readLine())!= null){
                byte b[] = s.getBytes("iso-8859-1");
                stringbuffer.append("\n" + new String(b));
            }
            fileContent = new String(stringbuffer);
            fileBean.setFileContent(fileContent);
        }
        catch(Exception exp){
            fileContent = "读取失败" + exp.toString();
        }
        RequestDispatcher dispatcher =
        request.getRequestDispatcher("studyEnglish.jsp");
        dispatcher.forward(request, response);
    }
}
```

> **web.xml（部署文件）**

向 ch9_practice_two\WEB\INF\下的 web.xml 添加如下的 servlet 和 servlet-mapping 标记（知识点见 6.1.2 节），部署的 servlet 的名字是 readFile，访问 servlet 的 url-pattern 是 readFile。

web.xml

```xml
<?xml version = "1.0" encoding = "utf - 8"?>
<web - app>
    <!-- 以下是 web.xml 文件新添加的内容 -->
    <servlet>
        <servlet - name>readFile</servlet - name>
        <servlet - class>handle.data.ReadFile_Servlet</servlet - class>
    </servlet>
    <servlet - mapping>
        <servlet - name>readFile</servlet - name>
        <url - pattern>/readFile</url - pattern>
    </servlet - mapping>
</web - app>
```

9.6 小结

> RandomAccessFile 对象在读写文件时可以调用 seek 方法改变读写位置。
> Tomcat 服务器可以让内置对象 request 调用方法 getInputStream() 获得一个输入流，通过这个输入流读入用户上传的全部信息，包括文件的内容以及表单域的信息。
> JSP 内置对象 response 调用方法 getOutputStream() 可以获取一个指向用户的输出流，服务器将文件写入这个流，用户就可以下载文件。

习题 9

1. File 对象能读写文件吗？
2. File 对象怎样获取文件的长度？
3. RandomAccessFile 类创建的流在读写文件时有什么特点？
4. 参考 9.6.2 节的代码编写一个播放视频的 Web 应用程序。用户通过一个下拉列表选择视频的名字，单击提交键可以看到视频的文本介绍，以及播放视频的 GUI 控件。

第 10 章　手机销售网

这一章讲述如何用 JSP 技术建立一个简单的手机销售网,其目的是掌握一般 Web 应用中常用基本模块的开发方法。系统采用 MVC 模式实现各个模块,数据库使用的是 MySQL 数据库。

视频讲解

10.1　系统模块构成

系统主要模块如图 10.1 所示。

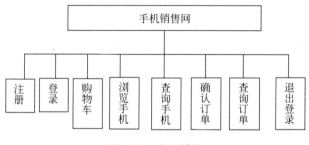

图 10.1　主要模块

10.2　Web 目录结构

在 Tomcat 安装目录的 webapps 目录下建立 Web 服务目录(即网站)ch10。另外,需要在 Web 服务目录 ch10 下建立目录结构 ch10\WEB-INF\classes(WEB-INF 字母大写)。

本章使用的 bean 的包名(除非特别说明)均为 save.data。在 classes 下建立目录\save\data,创建 bean 的类的 Java 源文件按 ANSI 编码保存在\WEB-INF\classes\save\data 中(知识点见 5.1.2 节)。本章的 servlet 的包名(除非特别说明)均为 handle.data。在 classes 下建立目录\handle\data,创建 servlet 的类的 Java 源文件按 ANSI 编码保存在\WEB-INF\classes\handle\data 中(知识点见 6.1.1 节)。

本章使用的 javax.servlet 和 javax.servlet.http 包中的类不在 JDK 提供的核心类库中,为了方便编译 Java 源文件,请事先将 Tomcat 安装目录 lib 子目录中的 servlet-api.jar 文件复制(不要剪贴)到\ch10\WEB-INF\classes 中(知识点见 6.1.1 节)。另外,保存 Java 源文件时,"保存类型"选择为"所有文件",将"编码"选择为"ANSI"。保存 JSP 文件和部署文件 web.xml 时,"保存类型"选择为"所有文件",将"编码"选择为"UTF-8"。

将 MySQL 数据库的 JDBC-MySQL 数据库连接器,例如 mysql-connector-java-8.0.18.jar 保存到 Tomcat 安装目录下的 lib 文件夹中(例如 D:\apache-tomcat-9.0.26\lib),并重新启动 Tomcat 服务器。

10.3 数据库设计与连接

当一个 Web 应用涉及和数据库交互数据时，数据库的设计（主要是数据库中各个表的设计）就显得尤为重要（要认真学习好数据库原理这门课程）。在 Web 应用程序的其他模块开始之前，首先要进行数据库设计，数据库设计是 Web 应用开发中一个非常重要的环节，数据库设计好之后才能进入 Web 应用程序的设计阶段。

▶ 10.3.1 数据库设计

使用 MySQL 建立一个数据库 mobileDatabase，该库共有 5 个表，以下是这些表的名称、结构和用途（有关建立数据库和表的操作细节见 8.1 节）。

❶ user 表的结构

user 表用于存储用户的注册信息，user 表的主键是 logname，各个字段值的说明如下：
- logname：存储注册的用户名（属性是字符型，主键）。
- password：存储密码（属性是字符型）。
- phone：存储电话（属性是字符型）。
- address：存储地址（属性是字符型）。
- realname：存储姓名（属性是字符型）。

❷ mobileClassify 表的结构

mobileClassify 表存储手机的类别（例如可以按操作系统分类为 iOS 手机、Android 手机等）。各个字段值的说明如下：
- id：手机的分类号（属性是 int 型，号码自动增加，主键）。
- name：手机的分类名称（属性是字符型）。

❸ mobileForm 表的结构

mobileForm 表存储手机的基本信息，mobileForm 表的主键是 mobile_version，各个字段值的说明如下：
- mobile_version：手机的产品标识号（属性是字符型，主键）。
- mobile_name：手机的名称（属性是字符型）。
- mobile_made：手机的制造商（属性是字符型）。
- mobile_price：手机的价格（属性是单精度浮点型）。
- mobile _mess：手机产品介绍（属性是字符型）
- mobile_pic：存储和手机相关的一幅图像文件的名字（属性是字符型）。
- id：作为 mobileClassify 表中 id 的外键。

❹ shoppingForm 表的结构

用户选择商品添加到购物车之后，可能不会马上购买该商品。因此不使用 session 担当购物车的角色，理由是，session 一旦销毁，购物车就消失了。因此，使用数据库中的表存放用户选择的商品，即担当购物车的角色。

shoppingForm 表存储用户的购物车信息，表的主键是 logname，字段值的说明如下：
- logname：存储注册的用户名（属性是字符型，作为 user 表中 logname 的外键）。
- goodsId：商品 id（属性是字符型）。

- goodsName：商品名称（属性是字符型）。
- goodsPrice：商品的单价（属性是单精度浮点型）。
- goodsAmount：商品的数量（属性是 int 型）。

❺ orderForm 表的结构

orderForm 表存储订单信息，orderForm 表的主键是 orderNumber，字段值的说明如下：

- orderNumber：存储订单序号（属性是 int 型，自动增加）。
- logname：存储注册的用户名（属性是字符型）。
- mess：订单信息（属性是字符型）。

❻ 创建数据库与表

通过导入 sql 文件中的 SQL 语句，完成 mobileDatabase 数据库以及表的创建，同时完成向 mobileClassify 表和 mobileForm 表中插入记录的操作。首先编写下列 sql 文件 createMobileshop. sql，保存在 D:\myFile 目录中（参见教材提供的源代码，在 ch10 目录下）。使用命令行进入 MySQL 安装目录下的 bin 子目录，使用 mysql 命令行启动客户端：

```
mysql  -u root -p
```

按要求输入密码后，在命令行客户端"mysql>"状态下执行操作：

```
source D:/myFile/createMobile.sql;
```

导入该 sql 文件中的 SQL 语句，完成数据库、表的创建（有关创建数据库和表的细节见 8.1.3 节）。

createMobile. sql

```
create database mobileDatabase;
use mobileDatabase;
create table user(logname char(30) not null,
            password varchar(30) character set utf8,
            phone char(20),
            addess char(50) character set gb2312,
            realname char(60) character set gb2312,
            primary key(logname));

create table mobileClassify(id int not null auto_increment,
            name varchar(50) character set gb2312,
            primary key(id));

create table mobileForm(mobile_version char(30) not null,
            mobile_name varchar(50) character set gb2312,
            mobile_made varchar(20) character set gb2312,
            mobile_price float,
            mobile_mess varchar(600) character set gb2312,
            mobile_pic varchar(20) character set gb2312,
            id int not null auto_increment,
            primary key(mobile_version),
            foreign key(id) references mobileClassify(id));

create table shoppingForm(goodsId char(30) not null,
```

```
              logname char(30) not null ,
              goodsName varchar(50) character set gb2312,
              goodsPrice float,
              goodsAmount int,
              foreign key(logname) references user(logname));
create table orderForm(orderNumber int not null auto_increment,
              logname char(30) character set gb2312,
              mess varchar(5000) character set gb2312,
              primary key(orderNumber));
```

然后再编写下列 sql 文件 insertRecord.sql,保存在 D:\myFile 目录中,在命令行客户端"mysql>"状态下执行操作:

```
source D:/myFile/insertRecord.sql;
```

导入该 sql 文件中的 SQL 语句,完成向 mobileClassify 表和 mobileForm 表中插入记录的操作,操作效果如图 10.2 所示。

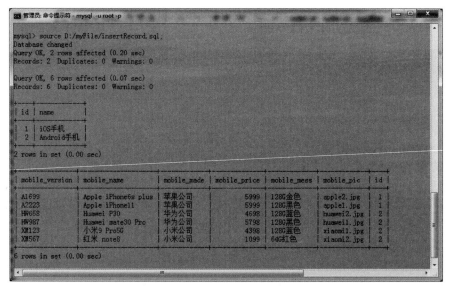

图 10.2 向数据库中的表插入记录

insertRecord.sql

```
use mobileDatabase;
insert into mobileClassify values
    (1,'iOS 手机'),
    (2,'Android 手机');
insert into mobileForm values
    ('A2223','Apple iPhone11','苹果公司',5999,'128G 黑色','apple1.jpg',1),
    ('A1699','Apple iPhone6s plus','苹果公司',5999,'128G 金色','apple2.jpg',1),
    ('HW987','Huawei mate30 Pro','华为公司',5798,'128G 黑色','huawei1.jpg',2),
    ('HW658','Huawei P30','华为公司',4698,'128G 蓝色','huawei2.jpg',2),
    ('XM123','小米 9 Pro5G','小米公司',4398,'128G 蓝色','xiaomi1.jpg',2),
    ('XM567','红米 note8','小米公司',1099,'64G 红色','xiaomi2.jpg',2);
select * from mobileClassify;
select * from mobileForm;
```

> 注：设计者可以参见教材的 8.1 节，选择自己喜欢的方式，例如，用某种 MySQL GUI 客户端管理工具创建数据库和相关的表。

▶ 10.3.2 数据库连接

手机销售网使用连接池连接数据库。将连接池配置文件 context.xml 保存在 Web 服务目录 ch10 的 META-INF 子目录中（不必重新启动 Tomcat 服务器，知识点见 8.10.2 节）。context.xml 及内容如下（根据实际项目，可增加连接池的大小）。

context.xml

```xml
<?xml version = "1.0" encoding = "utf-8" ?>
<Context>
  <Resource
    name = "mobileConn"
    type = "javax.sql.DataSource"
    driverClassName = "com.mysql.cj.jdbc.Driver"
    url = "jdbc:mysql://127.0.0.1:3306/mobileDatabase?
        &amp;serverTimezone = CST&amp;characterEncoding = utf-8"
    username = "root"
    password = ""
    maxActive = "15"
    maxIdle = "15"
    minIdle = "1"
    maxWait = "1000"
  />
</Context>
```

10.4　Web 应用模块管理

手机销售网采用 MVC 模式设计，由分视图、数据模型和控制器三个模块组成，即 JSP 页面、bean 和 servlet 三个模块。

▶ 10.4.1 页面管理

JSP 页面保存在 Web 服务目录 ch10 中。所有的页面包含一个导航条，该导航条由注册、登录、浏览手机、查看订单等功能组成。为了便于维护，其他 JSP 页面通过使用 JSP 的<%@ include …%>标记将 head.txt（导航条文件）嵌入到自己的页面（知识点见 2.7 节）。head.txt 按 UTF-8 编码保存在 Web 服务目录 ch10 中。head.txt 的内容如下：

head.txt

```
<style>
  #jerry{
     font-family:隶书;font-size:58;color:blue;
  }
  #tom{
     font-family:楷体;font-size:33;color:blue;
```

```
        }
    </style>
    <div align="center">
    <p id=jerry>小蜜蜂手机销售网</p>
    <table   cellSpacing="1" cellPadding="1" width="900" align="center" border="0">
       <tr valign="bottom">
         <td id=tom><a href="inputRegisterMess.jsp">注册</a></td>
         <td id=tom><a href="login.jsp">登录</a></td>
         <td id=tom><a href="lookMobile.jsp">浏览手机</a></td>
         <td id=tom><a href="searchMobile.jsp">查询手机</a></td>
         <td id=tom><a href="lookShoppingCar.jsp">查看购物车</a></td>
         <td id=tom><a href="lookOrderForm.jsp">查看订单</a></td>
         <td id=tom><a href="exitServlet">退出</font></td>
         <td id=tom><a href="index.jsp">主页</a></td>
    </tr>
    </table></div>
```

主页 index.jsp 由导航条、一行欢迎语和一幅背景图片 backgroud.jpg 构成。backgroud.jpg 保存在 ch10 目录的 image 子目录中(网站用的图像文件均存在 image 目录中,见稍后的 10.4.4 节)。用户可以通过在浏览器的地址栏中输入"http://服务器 IP:8080/index.jsp"或"http://服务器 IP:8080/"访问该主页,主页运行效果如图 10.3 所示。

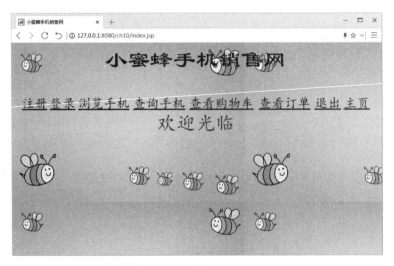

图 10.3 主页 index.jsp

index.jsp(效果如图 10.3 所示)

```
<%@ page contentType="text/html" %>
<%@ page pageEncoding="utf-8" %>
<title>小蜜蜂手机销售网</title>
<HEAD><%@ include file="head.txt" %></HEAD>
<style>
    #ok{
        font-family:楷体;font-size:50;color:green
    }
</style>
```

```
<HTML><body background=image/background.jpg>
<center id=ok>
欢迎光临
</center>
</body></HTML>
```

▶ 10.4.2　bean 与 servlet 管理

bean 类的包名均为 save.data，servlet 类的包名均为 handle.data（本章开始的约定）。
用命令行进入 save\data 的父目录 classes，编译 bean 的 Java 源文件：

```
javac save\data\Java 源文件
```

用命令行进入 handle\data 的父目录 classes，编译 servlet 的 Java 源文件：

```
javac -cp .;servlet-api.jar  handle\data\Java 源文件
```

注意"．；"和"servlet-api.jar"之间不要有空格，"．；"的作用是保证 Java 源文件能使用 import 语句引入当前 classes 目录中其他自定义包中的类，例如 save.data 包中的 bean 类。

▶ 10.4.3　web.xml（部署文件）

向 ch10\WEB\INF\下的 web.xml 添加如下的 servlet 和 servlet-mapping 标记（知识点见 6.1.2 节），部署 servlet 的名字和访问 servlet 的 url-pattern。

web.xml

```
<?xml version="1.0" encoding="utf-8"?>
<web-app>
<servlet>
    <servlet-name>registerServlet</servlet-name>
    <servlet-class>handle.data.HandleRegister</servlet-class>
</servlet>
<servlet-mapping>
    <servlet-name>registerServlet</servlet-name>
    <url-pattern>/registerServlet</url-pattern>
</servlet-mapping>
<servlet>
    <servlet-name>loginServlet</servlet-name>
    <servlet-class>handle.data.HandleLogin</servlet-class>
</servlet>
<servlet-mapping>
    <servlet-name>loginServlet</servlet-name>
    <url-pattern>/loginServlet</url-pattern>
</servlet-mapping>
<servlet>
    <servlet-name>deleteServlet</servlet-name>
    <servlet-class>handle.data.HandleDelete</servlet-class>
</servlet>
<servlet-mapping>
    <servlet-name>deleteServlet</servlet-name>
```

```xml
        <url-pattern>/deleteServlet</url-pattern>
    </servlet-mapping>
    <servlet>
        <servlet-name>updateServlet</servlet-name>
        <servlet-class>handle.data.HandleUpdate</servlet-class>
    </servlet>
    <servlet-mapping>
        <servlet-name>updateServlet</servlet-name>
        <url-pattern>/updateServlet</url-pattern>
    </servlet-mapping>
    <servlet>
        <servlet-name>buyServlet</servlet-name>
        <servlet-class>handle.data.HandleBuyGoods</servlet-class>
    </servlet>
    <servlet-mapping>
        <servlet-name>buyServlet</servlet-name>
        <url-pattern>/buyServlet</url-pattern>
    </servlet-mapping>
    <servlet>
        <servlet-name>queryServlet</servlet-name>
        <servlet-class>handle.data.QueryAllRecord</servlet-class>
    </servlet>
    <servlet-mapping>
        <servlet-name>queryServlet</servlet-name>
        <url-pattern>/queryServlet</url-pattern>
    </servlet-mapping>
    <servlet>
        <servlet-name>putGoodsServlet</servlet-name>
        <servlet-class>handle.data.PutGoodsToCar</servlet-class>
    </servlet>
    <servlet-mapping>
        <servlet-name>putGoodsServlet</servlet-name>
        <url-pattern>/putGoodsServlet</url-pattern>
    </servlet-mapping>
    <servlet>
        <servlet-name>searchByConditionServlet</servlet-name>
        <servlet-class>handle.data.SearchByCondition</servlet-class>
    </servlet>
    <servlet-mapping>
        <servlet-name>searchByConditionServlet</servlet-name>
        <url-pattern>/searchByConditionServlet</url-pattern>
    </servlet-mapping>
    <servlet>
        <servlet-name>exitServlet</servlet-name>
        <servlet-class>handle.data.HandleExit</servlet-class>
    </servlet>
    <servlet-mapping>
        <servlet-name>exitServlet</servlet-name>
        <url-pattern>/exitServlet</url-pattern>
    </servlet-mapping>
</web-app>
```

10.4.4 图像管理

本系统用到的图片所对应的图像文件,例如手机图片用的图像文件等,均保存在 Web 服务目录 ch10 的子目录 image 中。

10.5 会员注册

当新会员注册时,该模块要求用户必须输入会员名、密码信息,否则不允许注册。用户的注册信息存入数据库的 user 表中。

10.5.1 视图(JSP 页面)

该模块视图部分由一个 JSP 页面构成,用户在 input RegisterMess.jsp 页面输入注册信息,请求名字是 registerServlet 的 servlet(见 10.4.3 节的 web.xml),registerServlet 负责将注册信息写入数据库中的 user 表中。inputRegisterMess.jsp 同时负责显示注册是否成功的信息。

inputRegisterMess.jsp(效果如图 10.4 所示)

```
<%@ page contentType = "text/html" %>
<%@ page pageEncoding = "utf-8" %>
<jsp:useBean id = "userBean" class = "save.data.Register" scope = "request"/>
<HEAD><%@ include file = "head.txt" %></HEAD>
<title>注册页面</title>
<style>
    #ok{
        font-family:宋体;font-size:26;color:black;
    }
    #yes{
        font-family:黑体;font-size:18;color:black;
    }
</style>
<HTML><body id = ok background = image/back.jpg>
<div align = "center">
<form action = "registerServlet" method = "post">
<table id = ok>
    用户名由字母、数字、下画线构成,*注释的项必须填写
    <tr><td>*用户名称:</td>
        <td><input type = text id = ok name = "logname" /></td>
        <td>*用户密码:</td><td><input type = password id = ok name = "password"/>
        </td></tr>
    <tr><td>*重复密码:</td><td>
        <input type = password id = ok name = "again_password"/></td>
        <td>联系电话:</td><td><input type = text id = ok name = "phone"/></td></tr>
    <tr><td>邮寄地址:</td><td><input type = text id = ok name = "address"/></td>
        <td>真实姓名:</td><td><input type = text id = ok name = "realname"/></td>
        <td><input type = submit  id = ok value = "提交"></td> </tr>
</table>
</form>
```

```
</div>
<div align = "center">
注册反馈:
<jsp:getProperty name = "userBean" property = "backNews" />
<table id = yes border = 3>
    <tr><td>会员名称:</td>
    <td><jsp:getProperty name = "userBean" property = "logname"/></td>
    </tr>
    <tr><td>姓名:</td>
    <td><jsp:getProperty name = "userBean" property = "realname"/></td>
    </tr>
    <tr><td>地址:</td>
    <td><jsp:getProperty name = "userBean" property = "address"/></td>
    </tr>
    <tr><td>电话:</td>
    <td><jsp:getProperty name = "userBean" property = "phone"/></td>
    </tr>
</table></div>
</body></HTML>
```

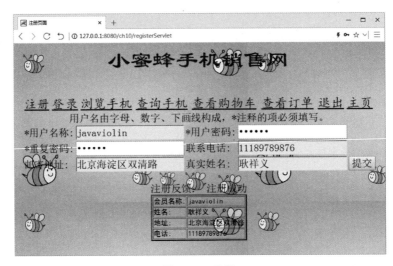

图 10.4 注册

▶ 10.5.2 模型(bean)

bean 的管理见 10.4.2 节。下列 bean 用来存储用户注册的信息。在该模块中 bean 的 id 是 userBean,生命周期是 request。该 bean 由 servlet 控制器负责创建或更新。

Register.java

```
package save.data;
public class Register{
    String  logname = "" , phone = "",
            address = "",realname = "",
            backNews = "请输入信息";
    public void setLogname(String logname){
```

```
            this.logname = logname;
        }
        public String getLogname(){
            return logname;
        }
        public void setPhone(String phone){
            this.phone = phone;
        }
        public String getPhone(){
            return phone;
        }
        public void setAddress(String address){
            this.address = address;
        }
        public String getAddress(){
            return address;
        }
        public void setRealname(String realname){
            this.realname = realname;
        }
        public String getRealname(){
            return realname;
        }
        public void setBackNews(String backNews){
            this.backNews = backNews;
        }
        public String getBackNews(){
            return backNews;
        }
    }
```

10.5.3 控制器(servlet)

servlet 的管理见 10.4.2 节。servlet 控制器是 HandleRegister 类的实例,名字是 registerServlet(见 10.4.3 节的 web.xml),registerServlet 负责创建 id 是 userBean 的 request bean,将用户提交的信息写入到数据库的 user 表中,并将注册反馈信息存储到 userBean 中,然后转发到 inputRegisterMess.jsp 页面。另外,registerServlet 使用了 Encrypt 类的 static 方法,将用户的密码加密后存放到数据库的 user 表中。

Encrypt.java

```
package handle.data;
public class Encrypt {
    static String encrypt(String sourceString,String password) {
        char [] p = password.toCharArray();
        int n = p.length;
        char [] c = sourceString.toCharArray();
        int m = c.length;
        for(int k = 0;k < m;k++){
            int mima = c[k] + p[k % n];          //加密算法
            c[k] = (char)mima;
```

```java
        }
        return new String(c);                        //返回密文
    }
}
```

HandleRegister.java

```java
package handle.data;
import save.data.Register;
import java.sql.*;
import java.io.*;
import javax.servlet.*;
import javax.servlet.http.*;
import javax.sql.DataSource;
import javax.naming.Context;
import javax.naming.InitialContext;
import javax.naming.NamingException;
public class HandleRegister extends HttpServlet {
    public void init(ServletConfig config) throws ServletException {
        super.init(config);
    }
    public   void   service(HttpServletRequest request,
                    HttpServletResponse response)
                    throws ServletException, IOException {
        request.setCharacterEncoding("utf-8");
        Connection con = null;
        PreparedStatement sql = null;
        Register userBean = new Register();                      //创建 bean
        request.setAttribute("userBean",userBean);
        String logname = request.getParameter("logname").trim();
        String password = request.getParameter("password").trim();
        String again_password = request.getParameter("again_password").trim();
        String phone = request.getParameter("phone").trim();
        String address = request.getParameter("address").trim();
        String realname = request.getParameter("realname").trim();
        if(logname == null)
            logname = "";
        if(password == null)
            password = "";
        if(!password.equals(again_password)) {
            userBean.setBackNews("两次密码不同,注册失败");
            RequestDispatcher dispatcher =
            request.getRequestDispatcher("inputRegisterMess.jsp");
            dispatcher.forward(request, response);                //转发
            return;
        }
        boolean isLD = true;
        for(int i = 0; i < logname.length(); i++){
            char c = logname.charAt(i);
            if(!(Character.isLetterOrDigit(c)||c == '_'))
                isLD = false;
        }
```

```java
        boolean boo = logname.length()>0&&password.length()>0&&isLD;
        String backNews = "";
        try{ Context   context = new InitialContext();
             Context   contextNeeded =
             (Context)context.lookup("java:comp/env");
             DataSource ds =
             (DataSource)contextNeeded.lookup("mobileConn");      //获得连接池
             con = ds.getConnection();                            //使用连接池中的连接
             String insertCondition = "INSERT INTO user VALUES (?,?,?,?,?)";
             sql = con.prepareStatement(insertCondition);
             if(boo){
                sql.setString(1,logname);
                password =
                Encrypt.encrypt(password,"javajsp");              //给用户密码加密
                sql.setString(2,password);
                sql.setString(3,phone);
                sql.setString(4,address);
                sql.setString(5,realname);
                int m = sql.executeUpdate();
                if(m!= 0){
                   backNews = "注册成功";
                   userBean.setBackNews(backNews);
                   userBean.setLogname(logname);
                   userBean.setPhone(phone);
                   userBean.setAddress(address);
                   userBean.setRealname(realname);
                }
             }
             else {
                 backNews = "信息填写不完整或名字中有非法字符";
                 userBean.setBackNews(backNews);
             }
             con.close();                                         //连接返回连接池
        }
        catch(SQLException exp){
             backNews = "该会员名已被使用,请您更换名字" + exp;
             userBean.setBackNews(backNews);
        }
        catch(NamingException exp){
             backNews = "没有设置连接池" + exp;
             userBean.setBackNews(backNews);
        }
        finally{
           try{
               con.close();
           }
           catch(Exception ee){}
        }
        RequestDispatcher dispatcher =
        request.getRequestDispatcher("inputRegisterMess.jsp");
        dispatcher.forward(request, response);                    //转发
    }
}
```

10.6 会员登录

用户可在该模块输入自己的会员名和密码,系统将对会员名和密码进行验证,如果输入的用户名或密码有错误,将提示"用户名或密码不正确"。

该模块视图部分由一个 JSP 页面 longin.jsp 构成,用户在该 JSP 页面输入登录信息,请求名字是 loginServlet 的 servlet。loginServlet 负责验证会员名和密码是否正确,将登录是否成功的信息存储到 id 是 loginBean 的 bean 中。loginServlet 负责让视图显示 bean 中的数据。

▶ 10.6.1 视图(JSP 页面)

视图部分由一个 JSP 页面 login.jsp 构成。login.jsp 页面负责提供输入登录信息界面,并负责显示登录反馈信息,例如登录是否成功、是否已经登录等。效果如图 10.5 所示。

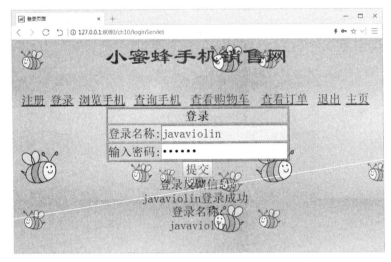

图 10.5 登录

login.jsp(效果如图 10.5 所示)

```
<%@ page contentType = "text/html" %>
<%@ page pageEncoding = "utf-8" %>
<jsp:useBean id = "loginBean" class = "save.data.Login" scope = "session"/>
<HEAD><%@ include file = "head.txt" %></HEAD>
<title>登录页面</title>
<style>
    #tom{
        font-family:宋体;font-size:30;color:black
    }
</style>
<HTML><body id = tom background = image/back.jpg>
<div align = "center">
<table id = tom border = 1>
<tr><th>登录</th></tr>
<form action = "loginServlet" method = "post">
```

```
<tr><td>登录名称:< input type = text id = tom name = "logname"></td></tr>
<tr><td>输入密码:< input type = password id = tom name = "password"></td></tr>
</table>
< input type = submit id = tom value = "提交">
</form></div >
< div align = "center" >
登录反馈信息:< br >
< jsp:getProperty name = "loginBean" property = "backNews"/>
< br >登录名称:< br >< jsp:getProperty name = "loginBean" property = "logname"/>
< div ></body></HTML >
```

▶ 10.6.2 模型(bean)

bean 的管理见 10.4.2 节。该模块中 bean 的 id 是 loginBean,是 session bean,由控制器负责创建或更新。

Login.java

```
package save.data;
import java.util.*;
public class Login {
    String logname = "",
           backNews = "未登录";
    public void setLogname(String logname){
        this.logname = logname;
    }
    public String getLogname(){
        return logname;
    }
    public void setBackNews(String s) {
        backNews = s;
    }
    public String getBackNews(){
        return backNews;
    }
}
```

▶ 10.6.3 控制器(servlet)

servlet 的管理见 10.4.2 节。servlet 控制器是 HandleLogin 的实例,名字是 loginServlet (见 10.4.3 节的 web.xml),负责连接数据库,查询 user 表,验证用户输入的会员名和密码是否在该表中,即验证是否是已注册的用户,如果用户是已注册的用户,就将用户设置成登录状态,即将用户的名称存放到 id 是 loginBean 的 session bean 中(见 10.5.2 节的 bean 模型),并将用户转发到 login.jsp 页面查看登录反馈信息。如果用户不是注册用户,控制器将提示登录失败。另外,loginServlet 使用了 Encrypt 类(见 10.5.3 节)的 static 方法,将用户的密码加密后和数据库的 user 表中密码进行比对。

HandleLogin.java

```
package handle.data;
```

```java
import save.data.*;
import java.sql.*;
import java.io.*;
import javax.servlet.*;
import javax.servlet.http.*;
import javax.sql.DataSource;
import javax.naming.Context;
import javax.naming.InitialContext;
import javax.naming.NamingException;
public class HandleLogin extends HttpServlet{
    public void init(ServletConfig config) throws ServletException{
        super.init(config);
    }
    public void service(HttpServletRequest request,
                        HttpServletResponse response)
                        throws ServletException,IOException{
        request.setCharacterEncoding("utf-8");
        Connection con = null;
        Statement sql;
        String logname = request.getParameter("logname").trim(),
        password = request.getParameter("password").trim();
        password = Encrypt.encrypt(password,"javajsp");          //给用户密码加密
        boolean boo = (logname.length()>0)&&(password.length()>0);
        try{
            Context   context = new InitialContext();
            Context   contextNeeded =
            (Context)context.lookup("java:comp/env");
            DataSource ds =
            (DataSource)contextNeeded.lookup("mobileConn");      //获得连接池
             con = ds.getConnection();                           //使用连接池中的连接
            String condition = "select * from user where logname = '" +
            logname + "' and password = '" + password + "'";
            sql = con.createStatement();
            if(boo){
                ResultSet rs = sql.executeQuery(condition);
                boolean m = rs.next();
                if(m == true){
                    //调用登录成功的方法
                    success(request,response,logname,password);
                    RequestDispatcher dispatcher =
                    request.getRequestDispatcher("login.jsp");   //转发
                    dispatcher.forward(request,response);
                }
                else{
                    String backNews = "您输入的用户名不存在,或密码不匹配";
                    //调用登录失败的方法
                    fail(request,response,logname,backNews);
                }
            }
            else{
                String backNews = "请输入用户名和密码";
```

```java
                    fail(request,response,logname,backNews);
                }
                con.close();                          //连接返回连接池
        }
        catch(SQLException exp){
            String backNews = "" + exp;
            fail(request,response,logname,backNews);
        }
        catch(NamingException exp){
            String backNews = "没有设置连接池" + exp;
            fail(request,response,logname,backNews);
        }
        finally{
          try{
              con.close();
          }
          catch(Exception ee){}
        }
    }
    public void success(HttpServletRequest request,
                    HttpServletResponse response,
                    String logname,String password) {
        Login loginBean = null;
        HttpSession session = request.getSession(true);
        try{  loginBean = (Login)session.getAttribute("loginBean");
              if(loginBean == null){
                  loginBean = new Login();              //创建新的数据模型
                  session.setAttribute("loginBean",loginBean);
                  loginBean = (Login)session.getAttribute("loginBean");
              }
              String name = loginBean.getLogname();
              if(name.equals(logname)) {
                  loginBean.setBackNews(logname + "已经登录了");
                  loginBean.setLogname(logname);
              }
              else {                                    //数据模型存储新的登录用户
                  loginBean.setBackNews(logname + "登录成功");
                  loginBean.setLogname(logname);
              }
        }
        catch(Exception ee){
            loginBean = new Login();
            session.setAttribute("loginBean",loginBean);
            loginBean.setBackNews(ee.toString());
            loginBean.setLogname(logname);
        }
    }
    public void fail(HttpServletRequest request,
                    HttpServletResponse response,
```

```
                    String logname,String backNews) {
         response.setContentType("text/html;charset = utf - 8");
         try {
              PrintWriter out = response.getWriter();
              out.println("< html >< body >");
              out.println("< h2 >" + logname + "登录反馈结果< br >" + backNews + "</h2 >") ;
              out.println("返回登录页面或主页< br >");
              out.println("< a href = login.jsp >登录页面</a >");
              out.println("< br >< a href = index.jsp >主页</a >");
              out.println("</body ></html >");
         }
         catch(IOException exp){ }
     }
}
```

10.7 浏览手机

用户选择手机分类后(见10.3.1节给出的mobileclassfies表),该模块可以用分页的方式显示mobileform表中的记录。

10.7.1 视图(JSP页面)

视图部分由3个JSP页面lookMobile.jsp、byPageShow.jsp和showDetail.jsp构成。在lookMobile.jsp页面选择某个分类(效果如图10.6所示),例如Android手机,然后提交给servlet控制器queryServlet(见10.4.3节的web.xml),该控制器将查询结果放到id是dataBean的session bean中,然后将显示dataBean中的数据的任务交给byPageShow.jsp页面(效果如图10.7所示)。byPageShow.jsp用分页方式显示dataBean中的数据。用户在byPageShow.jsp页面看到商品后,可以选择"查看细节"链接到showDetail.jsp页面(效果如图10.8所示),查看商品的细节。

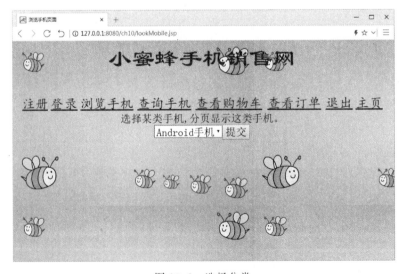

图10.6 选择分类

第10章 手机销售网

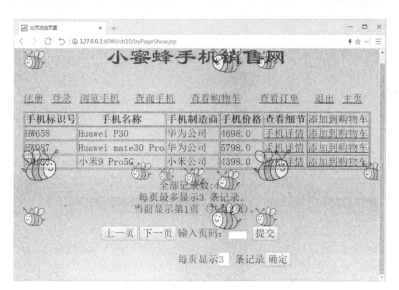

图10.7 分页显示商品

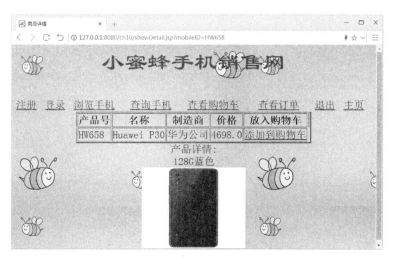

图10.8 查看商品详情

lookMobile.jsp（效果如图10.6所示）

```
<%@ page contentType = "text/html" %>
<%@ page pageEncoding = "utf-8" %>
<%@ page import = "java.sql.*" %>
<%@ page import = "javax.sql.DataSource" %>
<%@ page import = "javax.naming.Context" %>
<%@ page import = "javax.naming.InitialContext" %>
<HEAD><%@ include file = "head.txt" %></HEAD>
<title>浏览手机页面</title>
<style>
   #ok{
      font-family:宋体;font-size:26;color:black;
   }
</style>
```

```jsp
<HTML><body id = ok background = image/back.jpg>
<div align = "center">
选择某类手机,分页显示这类手机。
<% Connection con = null;
   Statement sql;
   ResultSet rs;
   Context context = new InitialContext();
   Context contextNeeded = (Context)context.lookup("java:comp/env");
   DataSource ds = (DataSource)contextNeeded.lookup("mobileConn");     //连接池
   try {
      con = ds.getConnection();                                         //使用连接池中的连接
      sql = con.createStatement();
      //读取 mobileClassify 表,获得分类
      rs = sql.executeQuery("SELECT * FROM mobileClassify");
      out.print("<form action = 'queryServlet' id = ok method = 'post'>");
      out.print("<select id = ok name = 'fenleiNumber'>");
      while(rs.next()){
          int id = rs.getInt(1);
          String mobileCategory = rs.getString(2);
          out.print("<option value = " + id + ">" + mobileCategory + "</option>");
      }
      out.print("</select>");
      out.print("<input type = 'submit' id = ok value = '提交'>");
      out.print("</form>");
      rs.close();
      con.close();                                                      //连接返回连接池
   }
   catch(SQLException e){
      out.print(e);
   }
   finally{
      try{
         con.close();
      }
      catch(Exception ee){}
   }
%>
</div></body></HTML>
```

byPageShow.jsp(效果如图 10.7 所示)

```jsp
<%@ page contentType = "text/html" %>
<%@ page pageEncoding = "utf-8" %>
<HEAD><%@ include file = "head.txt" %></HEAD>
<title>分页浏览页面</title>
<style>
   #tom{
      font - family:宋体;font - size:26;color:black
   }
</style>
<jsp:useBean id = "dataBean" class = "save.data.Record_Bean"
```

```jsp
            scope = "session"/>
<HTML><body background = image/back.jpg><center>
<jsp:setProperty name = "dataBean" property = "pageSize" param = "pageSize"/>
<jsp:setProperty name = "dataBean"
    property = "currentPage" param = "currentPage"/>
</p>
<table id = tom border = 1>
<% String[][] table = dataBean.getTableRecord();
    if(table == null) {
        out.print("没有记录");
        return;
    }
%>  <tr>
        <th>手机标识号</th>
        <th>手机名称</th>
        <th>手机制造商</th>
        <th>手机价格</th>
        <th>查看细节</th>
        <td>添加到购物车</td>
    </tr>
<%  int totalRecord = table.length;
    int pageSize = dataBean.getPageSize();              //每页显示的记录数
    int totalPages = dataBean.getTotalPages();
    if(totalRecord % pageSize == 0)
        totalPages = totalRecord/pageSize;              //总页数
    else
        totalPages = totalRecord/pageSize + 1;
    dataBean.setPageSize(pageSize);
    dataBean.setTotalPages(totalPages);
    if(totalPages >= 1) {
        if(dataBean.getCurrentPage()< 1)
            dataBean.setCurrentPage(dataBean.getTotalPages());
        if(dataBean.getCurrentPage()> dataBean.getTotalPages())
            dataBean.setCurrentPage(1);
        int index = (dataBean.getCurrentPage() - 1) * pageSize;
        int start = index;                              //table 的 currentPage 页起始位置
        for(int i = index;i < pageSize + index;i++) {
            if(i == totalRecord)
                break;
            out.print("<tr>");
            for(int j = 0;j < table[0].length;j++) {
                out.print("<td>" + table[i][j] + "</td>");
            }
            String detail =
             "<a href = 'showDetail.jsp?mobileID = " + table[i][0] + "'>
             手机详情</a>";
            out.print("<td>" + detail + "</td>");
            String shopping =
             "<a href = 'putGoodsServlet?mobileID = " + table[i][0] + "'>
             添加到购物车</a>";
            out.print("<td>" + shopping + "</td>");
```

```
            out.print("</tr>");
        }
    }
%>
</table>
<p id = tom>全部记录数:<jsp:getProperty name = "dataBean"
                        property = "totalRecords"/>
<br>每页最多显示<jsp:getProperty name = "dataBean" property = "pageSize"/>
条记录
<br>当前显示第<jsp:getProperty name = "dataBean" property = "currentPage"/>页
(共有<jsp:getProperty name = "dataBean" property = "totalPages"/>页).</p>
<table id = tom>
<tr>
  <td><form action = "" method = post>
    <input type = hidden name = "currentPage"
              value = "<% = dataBean.getCurrentPage() - 1 %>"/>
    <input type = submit id = tom value = "上一页"/></form>
  </td>
  <td><form action = "" method = post>
    <input type = hidden name = "currentPage"
              value = "<% = dataBean.getCurrentPage() + 1 %>"/>
    <input type = submit id = tom  value = "下一页"></form>
  </td>
   <td><form action = "" id = tom method = post>
    输入页码:<input type = textid = tom  name = "currentPage" size = 2>
    <input type = submit id = tom value = "提交"></form>
   </td>
</tr>
<tr><td></td><td></td>
  <td><form action = "" id = tom method = post>
   每页显示<input type = text id = tom
   name = "pageSize" value = <% = dataBean.getPageSize() %> size = 1>
   条记录<input type = submit id = tom value = "确定"></form></td>
</tr>
</table>
</center></body></HTML>
```

showDetail.jsp(效果如图 10.8 所示)

```
<%@ page contentType = "text/html" %>
<%@ page pageEncoding = "utf - 8" %>
<%@ page import = "save.data.Login" %>
<%@ page import = "javax.sql.DataSource" %>
<%@ page import = "javax.naming.Context" %>
<%@ page import = "javax.naming.InitialContext" %>
<%@ page import = "java.sql.*" %>
<jsp:useBean id = "loginBean" class = "save.data.Login" scope = "session"/>
<HEAD><%@ include file = "head.txt" %></HEAD>
<title>商品详情</title>
<style>
    #tom{
```

```
        font-family:宋体;font-size:26;color:black
    }
</style>
<HTML><body background = image/back.jpg id = tom><center>
<%    try{
          loginBean = (Login)session.getAttribute("loginBean");
          if(loginBean == null){
            response.sendRedirect("login.jsp");              //重定向到登录页面
            return;
          }
          else {
            boolean b = loginBean.getLogname() == null||
                   loginBean.getLogname().length() == 0;
            if(b){
              response.sendRedirect("login.jsp");            //重定向到登录页面
              return;
            }
          }
      }
      catch(Exception exp){
          response.sendRedirect("login.jsp");                //重定向到登录页面
          return;
      }
   String mobileID = request.getParameter("mobileID");
   if(mobileID == null) {
       out.print("没有产品号,无法查看细节");
       return;
   }
   Context context = new InitialContext();
   Context contextNeeded = (Context)context.lookup("java:comp/env");
   DataSource ds = (DataSource)contextNeeded.lookup("mobileConn");      //连接池
   Connection con = null;
   Statement sql;
   ResultSet rs;
   try{
     con =  ds.getConnection();                              //使用连接池中的连接
     sql = con.createStatement();
     String query =
     "SELECT * FROM mobileForm where mobile_version = '" + mobileID + "'";
     rs = sql.executeQuery(query);
     out.print("<table id = tom border = 2 >");
     out.print("<tr>");
     out.print("<th>产品号");
     out.print("<th>名称");
     out.print("<th>制造商");
     out.print("<th>价格");
     out.print("<th>放入购物车<th>");
     out.print("</tr>");
     String picture = "background.jpg";
     String detailMess = "";
     while(rs.next()){
```

```
            mobileID = rs.getString(1);
            String name = rs.getString(2);
            String maker = rs.getString(3);
            String price = rs.getString(4);
            detailMess = rs.getString(5);
            picture = rs.getString(6);
            out.print("<tr>");
            out.print("<td>" + mobileID + "</td>");
            out.print("<td>" + name + "</td>");
            out.print("<td>" + maker + "</td>");
            out.print("<td>" + price + "</td>");
            String shopping =
               "<a href = 'putGoodsServlet?mobileID = " + mobileID + "'>添加到购物车</a>";
            out.print("<td>" + shopping + "</td>");
            out.print("</tr>");
        }
        out.print("</table>");
        out.print("产品详情:<br>");
        out.println("<div align = center>" + detailMess + "<div>");
        String pic = "<img src = 'image/" + picture + "' width = 260 height = 200 ></img>";
        out.print(pic);                              //产品图片
        con.close();                                 //连接返回连接池
    }
    catch(SQLException exp){}
    finally{
        try{
            con.close();
        }
        catch(Exception ee){}
    }
%>
</center></body></HTML>
```

10.7.2 模型(bean)

本模块的 bean 的 id 是 dataBean(Record_Bean 类的实例),生命周期是 session,用于存储数据库中的记录。servlet 控制器 queryServlet(见 10.4.3 节的 web.xml)把从数据库查询到的记录存到 dataBean 中。

Record_Bean.java

```
package save.data;
public class Record_Bean{
    String [][] tableRecord = null;              //存放查询到的记录
    int pageSize = 3;                            //每页显示的记录数
    int totalPages = 1;                          //分页后的总页数
    int currentPage = 1 ;                        //当前显示页
    int totalRecords ;                           //全部记录
    public void setTableRecord(String [][] s){
        tableRecord = s;
    }
```

```java
    public String [][] getTableRecord(){
        return tableRecord;
    }
    public void setPageSize(int size){
        pageSize = size;
    }
    public int getPageSize(){
        return pageSize;
    }
    public int getTotalPages(){
        return totalPages;
    }
    public void setTotalPages(int n){
        totalPages = n;
    }
    public void setCurrentPage(int n){
        currentPage = n;
    }
    public int getCurrentPage(){
        return currentPage ;
    }
    public int getTotalRecords(){
        totalRecords = tableRecord.length;
        return totalRecords ;
    }
}
```

10.7.3 控制器(servlet)

本模块有两个控制器 queryServlet(QueryAllRecord 类的实例)、putGoodsServlet(PutGoodsToCar 类的实例),见 10.4.3 节的 web.xml。servlet 控制器 queryServlet 把从数据库 mobileForm 表中查询到的记录存到 dataBean 中,然后将用户重定向到 byPageShow.jsp 页面。当用户在 byPageShow.jsp 页面或 showDetail.jsp 看到商品时,每个商品都后缀了一个"添加到购物车"超链接,用户单击该超链接后,putGoodsServlet 控制器将该产品放入用户的购物车,即向数据库中的 shoppingForm 表插入记录。

QueryAllRecord.java

```java
package handle.data;
import save.data.Record_Bean;
import java.sql.*;
import java.io.*;
import javax.servlet.*;
import javax.servlet.http.*;
import javax.sql.DataSource;
import javax.naming.Context;
import javax.naming.InitialContext;
public class QueryAllRecord extends HttpServlet{
    public void init(ServletConfig config) throws ServletException{
        super.init(config);
```

```java
    }
    public void service(HttpServletRequest request,
            HttpServletResponse response)
            throws ServletException,IOException{
        request.setCharacterEncoding("utf-8");
        String idNumber = request.getParameter("fenleiNumber");
        if(idNumber == null)
            idNumber = "1";
        int id = Integer.parseInt(idNumber);
        HttpSession session = request.getSession(true);
        Connection con = null;
        Record_Bean dataBean = null;
        try{
            dataBean = (Record_Bean)session.getAttribute("dataBean");
            if(dataBean == null){
                dataBean = new Record_Bean();                    //创建 bean
                session.setAttribute("dataBean",dataBean);       //是 session bean
            }
        }
        catch(Exception exp){}
        try{
            Context   context = new InitialContext();
            Context   contextNeeded = (Context)context.lookup("java:comp/env");
            DataSource ds =
            (DataSource)contextNeeded.lookup("mobileConn");      //获得连接池
             con = ds.getConnection();                           //使用连接池中的连接
            Statement sql = con.createStatement(ResultSet.TYPE_SCROLL_SENSITIVE,
                            ResultSet.CONCUR_READ_ONLY);
            String query =
            "SELECT mobile_version,mobile_name,mobile_made,mobile_price " +
            "FROM mobileForm where id = " + id;
            ResultSet rs = sql.executeQuery(query);
            ResultSetMetaData metaData = rs.getMetaData();
            int columnCount = metaData.getColumnCount();         //得到结果集的列数
            rs.last();
            int rows = rs.getRow();                              //得到记录数
            String [][] tableRecord = dataBean.getTableRecord();
            tableRecord = new String[rows][columnCount];
            rs.beforeFirst();
            int i = 0;
            while(rs.next()){
               for(int k = 0;k < columnCount;k++)
                   tableRecord[i][k] = rs.getString(k + 1);
                   i++;
            }
            dataBean.setTableRecord(tableRecord);                //更新 bean
            con.close();                                         //连接返回连接池
            response.sendRedirect("byPageShow.jsp");             //重定向
        }
        catch(Exception e){
            response.getWriter().print("" + e);
```

```java
        }
        finally{
           try{
               con.close();
           }
           catch(Exception ee){}
        }
    }
}
```

PutGoodsToCar.java

```java
package handle.data;
import save.data.Login;
import javax.sql.DataSource;
import javax.naming.Context;
import javax.naming.InitialContext;
import javax.naming.NamingException;
import java.io.*;
import java.sql.*;
import javax.servlet.*;
import javax.servlet.http.*;
public class PutGoodsToCar extends HttpServlet {
    public void init(ServletConfig config) throws ServletException {
        super.init(config);
    }
    public  void  service(HttpServletRequest request,
                       HttpServletResponse response)
                       throws ServletException,IOException {
      request.setCharacterEncoding("utf-8");
      Connection con = null;
      PreparedStatement pre = null;              //预处理语句
      ResultSet rs;
      String mobileID = request.getParameter("mobileID");
      Login loginBean = null;
      HttpSession session = request.getSession(true);
      try{
         loginBean = (Login)session.getAttribute("loginBean");
         if(loginBean == null){
           response.sendRedirect("login.jsp");         //重定向到登录页面
           return;
         }
         else {
            boolean b  = loginBean.getLogname() == null||
                    loginBean.getLogname().length() == 0;
           if(b){
             response.sendRedirect("login.jsp");        //重定向到登录页面
             return;
           }
         }
```

```java
catch(Exception exp){
      response.sendRedirect("login.jsp");                //重定向到登录页面
      return;
}
try {
   Context context = new InitialContext();
   Context contextNeeded = (Context)context.lookup("java:comp/env");
   DataSource ds =
   (DataSource)contextNeeded.lookup("mobileConn");       //获得连接池
   con = ds.getConnection();                             //使用连接池中的连接
   String queryMobileForm =
   "select * from mobileForm where mobile_version = ?";  //查询商品表
   String queryShoppingForm =
   "select goodsAmount from shoppingForm where goodsId = ?";  //购物车表
    String updateSQL =
   "update shoppingForm set goodsAmount = ? where goodsId = ?"; //更新
   String insertSQL =
   "insert into shoppingForm values(?,?,?,?,?)";         //添加到购物车
   pre = con.prepareStatement(queryShoppingForm);
   pre.setString(1,mobileID);
   rs = pre.executeQuery();
   if(rs.next()){                                        //该货物已经在购物车中
       int amount = rs.getInt(1);
       amount++;
       pre = con.prepareStatement(updateSQL);
       pre.setInt(1,amount);
       pre.setString(2,mobileID);
       pre.executeUpdate();                              //更新购物车中该货物的数量
   }
      else {                                             //向购物车添加商品
       pre = con.prepareStatement(queryMobileForm);
       pre.setString(1,mobileID);
       rs = pre.executeQuery();
       if(rs.next()){
           pre = con.prepareStatement(insertSQL);
           pre.setString(1,rs.getString("mobile_version"));
           pre.setString(2,loginBean.getLogname());
           pre.setString(3,rs.getString("mobile_name"));
           pre.setFloat(4,rs.getFloat("mobile_price"));
           pre.setInt(5,1);
           pre.executeUpdate();                          //向购物车中添加该货物
       }
   }
   con.close();
   response.sendRedirect("lookShoppingCar.jsp");         //查看购物车
}
catch(SQLException exp){
    response.getWriter().print("" + exp);
}
catch(NamingException exp){}
finally{
```

```
            try{
                con.close();
            }
            catch(Exception ee){}
        }
    }
}
```

10.8 查看购物车

登录的用户可以通过该模块视图部分 lookShoppingCart.jsp 查看购物车中的商品,并选择是否删除某个商品或更新某个商品的数量。该模块有 updateServlet、deleteServlet 和 buyServlet 三个 servlet 控制器。updateServlet 负责更新商品的数量,deleteServlet 负责删除购物车中的商品,buyServlet 负责将用户购物车中的商品信息存放到数据库的 oderform 表中,即生成订单。

▶ 10.8.1 视图(JSP 页面)

视图部分由一个 JSP 页面 lookShoppingCar.jsp 构成,负责显示购物车中的商品,即显示 carBean 中的数据,并提供修改购物车功能和生成订单的 form 表单。

lookShoppingCar.jsp(效果如图 10.9 所示)

```
<%@ page import = "save.data.Login" %>
<%@ page import = "java.sql.*" %>
<%@ page import = "javax.sql.DataSource" %>
<%@ page import = "javax.naming.Context" %>
<%@ page import = "javax.naming.InitialContext" %>
<%@ page contentType = "text/html" %>
<%@ page pageEncoding = "utf-8" %>
<jsp:useBean id = "loginBean" class = "save.data.Login" scope = "session"/>
<HEAD><%@ include file = "head.txt" %></HEAD>
<title>查看购物车</title>
<style>
    #tom{
        font-family:宋体;font-size:26;color:blue
    }
</style>
<HTML><body background = image/back.jpg id = tom >
<div align = "center">
<%    if(loginBean == null){
        response.sendRedirect("login.jsp");          //重定向到登录页面
        return;
    }
    else {
        boolean b = loginBean.getLogname() == null||
                    loginBean.getLogname().length() == 0;
        if(b){
            response.sendRedirect("login.jsp");      //重定向到登录页面
```

```java
            return;
        }
    }
    Context context = new InitialContext();
    Context contextNeeded = (Context)context.lookup("java:comp/env");
    DataSource ds = (DataSource)contextNeeded.lookup("mobileConn");          //连接池
    Connection con = null;
    Statement sql;
    ResultSet rs;
    out.print("<table border=1>");
    out.print("<tr>");
    out.print("<th id=tom width=120>" + "手机标识(id)");
    out.print("<th id=tom width=120>" + "手机名称");
    out.print("<th id=tom width=120>" + "手机价格");
    out.print("<th id=tom width=120>" + "购买数量");
    out.print("<th id=tom width=50>" + "修改数量");
    out.print("<th id=tom width=50>" + "删除商品");
    out.print("</tr>");
    try{
        con = ds.getConnection();                                             //使用连接池中的连接
        sql = con.createStatement();
        String SQL =
        "SELECT goodsId,goodsName,goodsPrice,goodsAmount FROM shoppingForm" +
        " where logname = '" + loginBean.getLogname() + "'";
        rs = sql.executeQuery(SQL);                                           //查 shoppingForm 表
        String goodsId = "";
        String name = "";
        float price = 0;
        int amount = 0;
        String orderForm = null;                                              //订单
        while(rs.next()) {
            goodsId = rs.getString(1);
            name = rs.getString(2);
            price = rs.getFloat(3);
            amount = rs.getInt(4);
            out.print("<tr>");
            out.print("<td id=tom>" + goodsId + "</td>");
            out.print("<td id=tom>" + name + "</td>");
            out.print("<td id=tom>" + price + "</td>");
            out.print("<td id=tom>" + amount + "</td>");
            String update = "<form   action = 'updateServlet' method = 'post'>" +
            "<input type = 'text' id = tom name = 'update' size = 3 value = "
            + amount + " />" +
            "<input type = 'hidden' name = 'goodsId' value = " + goodsId + " />" +
            "<input type = 'submit' id = tom value = '更新数量'   ></form>";
            String del = "<form   action = 'deleteServlet' method = 'post'>" +
            "<input type = 'hidden' name = 'goodsId' value = " + goodsId + " />" +
            "<input type = 'submit' id = tom value = '删除该商品' /></form>";
            out.print("<td id=tom>" + update + "</td>");
            out.print("<td id=tom>" + del + "</td>");
            out.print("</tr>") ;
        }
```

```
            out.print("</table>");
            orderForm = "< form action = 'buyServlet' method = 'post'>" +
            "< input type = 'hidden' name = 'logname'
                   value = '" + loginBean.getLogname() + "'/>" +
            "< input type = 'submit' id = tom value = '生成订单(同时清空购物车)'></form>";
            out.print(orderForm);
            con.close(); //把连接返回连接池
        }
        catch(SQLException e) {
            out.print("< h1 >" + e + "</h1 >");
        }
        finally{
            try{
                con.close();
            }
            catch(Exception ee){}
        }
    %>
</div></body></HTML>
```

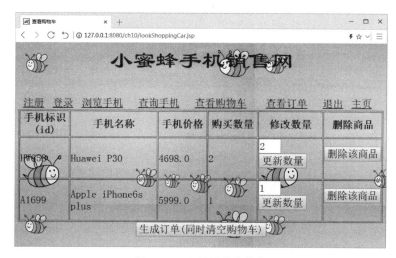

图 10.9　查看管理购物车

10.8.2　模型(bean)

本模块的 bean 的 id 为 loginBean,用于验证是否是登录用户(见 10.6.2 节)。

10.8.3　控制器(servlet)

本模块有 3 个 servlet 控制器,updateServlet、deleteServlet 和 buyServlet(见 10.4.3 节的 web.xml)。当用户查看购物车时不仅可以看到购物车中的物品,还可以单击 form 表单的"更新、删除"提交键。单击 form 表单的"更新数量"提交键,updateServlet 负责更新商品的数量;单击 form 表单的"删除该商品"提交键,deleteServlet 负责从购物车中删除该商品;单击 form 表单的"生成订单"提交键,buyServlet 负责将用户购物车中的商品信息存放到数据库的 oderform 表中、生成订单,同时删除用户购物车中的商品。

HandleUpdate.java

```java
package handle.data;
import save.data.Login;
import javax.sql.*;
import javax.sql.DataSource;
import javax.naming.Context;
import javax.naming.InitialContext;
import javax.naming.NamingException;
import java.io.*;
import java.sql.*;
import javax.servlet.*;
import javax.servlet.http.*;
public class HandleUpdate extends HttpServlet {
    public void init(ServletConfig config) throws ServletException {
        super.init(config);
    }
    public void service(HttpServletRequest request,
                        HttpServletResponse response)
                        throws ServletException, IOException {
        request.setCharacterEncoding("utf-8");
        String amount = request.getParameter("update");
        String goodsId = request.getParameter("goodsId");
        if(amount == null)
            amount = "1";
        int newAmount = 0;
        try{
            newAmount = Integer.parseInt(amount);
            if(newAmount < 0){
                newAmount = 1;
            }
        }
        catch(NumberFormatException exp){
            newAmount = 1;
        }
        Connection con = null;
        PreparedStatement pre = null;                    //预处理语句
        Login loginBean = null;
        HttpSession session = request.getSession(true);
        try{
            loginBean = (Login)session.getAttribute("loginBean");
            if(loginBean == null){
                response.sendRedirect("login.jsp");      //重定向到登录页面
                return;
            }
            else {
                boolean b = loginBean.getLogname() == null ||
                        loginBean.getLogname().length() == 0;
                if(b){
                    response.sendRedirect("login.jsp");  //重定向到登录页面
                    return;
                }
```

```java
            }
        }
        catch(Exception exp){
            response.sendRedirect("login.jsp");              //重定向到登录页面
            return;
        }
        Context contextNeeded = null;
        try {
          Context context = new InitialContext();
            contextNeeded = (Context)context.lookup("java:comp/env");
          DataSource ds =
          (DataSource)contextNeeded.lookup("mobileConn");    //获得连接池
          con = ds.getConnection();                          //使用连接池中的连接
          String updateSQL =
          "update shoppingForm set goodsAmount = ? where goodsId = ?";  //购物车
            pre = con.prepareStatement(updateSQL);
            pre.setInt(1,newAmount);
            pre.setString(2,goodsId);
            pre.executeUpdate();
            con.close();                                     //连接放回连接池
            response.sendRedirect("lookShoppingCar.jsp");    //查看购物车
        }
        catch(SQLException e) {
            response.getWriter().print("" + e);
        }
        catch(NamingException exp){
            response.getWriter().print("" + exp);
        }
        finally{
            try{
                con.close();
            }
            catch(Exception ee){}
        }
    }
}
```

HandleDelete.java

```java
package handle.data;
import save.data.Login;
import javax.sql.DataSource;
import javax.naming.Context;
import javax.naming.InitialContext;
import javax.naming.NamingException;
import java.io.*;
import java.sql.*;
import javax.servlet.*;
import javax.servlet.http.*;
public class HandleDelete extends HttpServlet {
    public void init(ServletConfig config) throws ServletException {
```

```java
        super.init(config);
    }
    public void service(HttpServletRequest request,
                        HttpServletResponse response)
                        throws ServletException,IOException {
        request.setCharacterEncoding("utf-8");
        String goodsId = request.getParameter("goodsId");
        Connection con = null;
        PreparedStatement pre = null;                    //预处理语句
        Login loginBean = null;
        HttpSession session = request.getSession(true);
        try{
            loginBean = (Login)session.getAttribute("loginBean");
            if(loginBean == null){
                response.sendRedirect("login.jsp");      //重定向到登录页面
                return;
            }
            else {
                boolean b = loginBean.getLogname() == null||
                        loginBean.getLogname().length() == 0;
                if(b){
                    response.sendRedirect("login.jsp");  //重定向到登录页面
                    return;
                }
            }
        }
        catch(Exception exp){
            response.sendRedirect("login.jsp");          //重定向到登录页面
            return;
        }
        try {
          Context context = new InitialContext();
          Context contextNeeded = (Context)context.lookup("java:comp/env");
          DataSource ds =
          (DataSource)contextNeeded.lookup("mobileConn");//获得连接池
          con = ds.getConnection();                      //使用连接池中的连接
          String deleteSQL =
          "delete from shoppingForm where goodsId = ?";  //从购物车中删除货物
           pre = con.prepareStatement(deleteSQL);
           pre.setString(1,goodsId);
           pre.executeUpdate();
           con.close();                                  //连接放回连接池
           response.sendRedirect("lookShoppingCar.jsp"); //查看购物车
        }
        catch(SQLException e) {
            response.getWriter().print("" + e);
        }
        catch(NamingException exp){
            response.getWriter().print("" + exp);
        }
        finally{
```

```java
            try{
                con.close();
            }
            catch(Exception ee){}
        }
    }
}
```

HandleBuyGoods.java

```java
package handle.data;
import save.data.Login;
import javax.sql.DataSource;
import javax.naming.Context;
import javax.naming.InitialContext;
import java.io.*;
import java.sql.*;
import javax.servlet.*;
import javax.servlet.http.*;
public class HandleBuyGoods extends HttpServlet {
    public void init(ServletConfig config) throws ServletException {
        super.init(config);
    }
    public void service(HttpServletRequest request,
                   HttpServletResponse response)
                   throws ServletException,IOException {
        request.setCharacterEncoding("utf-8");
        String logname = request.getParameter("logname");
        Connection con = null;
        PreparedStatement pre = null;            //预处理语句
        ResultSet rs;
        Login loginBean = null;
        HttpSession session = request.getSession(true);
        try{
            loginBean = (Login)session.getAttribute("loginBean");
            if(loginBean == null){
                response.sendRedirect("login.jsp");    //重定向到登录页面
                return;
            }
            else {
                boolean b = loginBean.getLogname() == null||
                       loginBean.getLogname().length() == 0;
                if(b){
                    response.sendRedirect("login.jsp");  //重定向到登录页面
                    return;
                }
            }
        }
        catch(Exception exp){
            response.sendRedirect("login.jsp");     //重定向到登录页面
            return;
```

```java
        }
        try {
            Context context = new InitialContext();
            Context contextNeeded = (Context)context.lookup("java:comp/env");
            DataSource ds =
            (DataSource)contextNeeded.lookup("mobileConn");              //获得连接池
            con = ds.getConnection();                                     //使用连接池中的连接
            String querySQL =
            "select * from shoppingForm where logname = ?";              //购物车
            String insertSQL = "insert into orderForm values(?,?,?)";    //订单表
            String deleteSQL = "delete from shoppingForm where logname = ?";
            pre = con.prepareStatement(querySQL);
            pre.setString(1,logname);
            rs = pre.executeQuery();                                      //查询购物车
            StringBuffer buffer = new StringBuffer();
            float sum = 0;
            boolean canCreateForm = false;                                //是否可以产生订单
            while(rs.next()){
                canCreateForm = true;
                String goodsId = rs.getString(1);
                logname = rs.getString(2);
                String goodsName = rs.getString(3);
                float price = rs.getFloat(4);
                int amount = rs.getInt(5);
                sum += price * amount;
                buffer.append("<br>商品 id:" + goodsId + ",名称:" + goodsName +
                "单价" + price + "数量" + amount);
            }
            if(canCreateForm == false){
                response.setContentType("text/html;charset = utf - 8");
                PrintWriter out = response.getWriter();
                out.println("<html><body>");
                out.println("<h2>" + logname + "请先选择商品添加到购物车");
                out.println("<br><a href = index.jsp>主页</a></h2>");
                out.println("</body></html>");
                return;
            }
            String strSum = String.format(" % 10.2f",sum);
            buffer.append("<br>" + logname + "<br>购物车的商品总价:" + strSum);
            pre = con.prepareStatement(insertSQL);
            pre.setInt(1,0);                                              //订单号会自动增加
            pre.setString(2,logname);
            pre.setString(3,new String(buffer));
            pre.executeUpdate();                                          //添加到订单表
            pre = con.prepareStatement(deleteSQL);
            pre.setString(1,logname);
            pre.executeUpdate();                                          //删除 logname 的购物车中货物
            con.close();                                                  //连接放回连接池
            response.sendRedirect("lookOrderForm.jsp");                   //查看订单
        }
        catch(Exception e) {
```

```
            response.getWriter().print("" + e);
        }
        finally{
          try{
              con.close();
          }
          catch(Exception ee){}
        }
    }
}
```

10.9 查询手机

本模块用到了 10.7 节给出的模型 dataBean 和视图 byPageShow.jsp。在视图部分 searchMobile.jsp 输入查询条件给 servlet 控制器 searchByConditionServlet（SearchByCondition 类的实例），searchByConditionServlet 控制器负责查询数据库，并将查询结果存放到数据模型 dataBean 中，然后将用户重定向到 byPageShow.jsp 页面查看 dataBean 中的数据。

▶ 10.9.1 视图（JSP 页面）

视图部分由两个 JSP 页面 searchMobile.jsp 和 byPageShow.jsp 构成，其中 byPageShow.jsp 见 10.7 节的视图部分。用户在 searchMobile.jsp 页面输入查询信息，提交给 searchByConditionServlet 控制器，该控制器将查询结果存放到 dataBean 中。byPageShow.jsp 页面负责显示 dataBean 中的数据。

searMobile.jsp（效果如图 10.10 及图 10.11 所示）

```
<%@ page contentType = "text/html" %>
<%@ page pageEncoding = "utf-8" %>
<HEAD><%@ include file = "head.txt" %></HEAD>
<title>查询页面</title>
<style>
   #tom{
      font-family:宋体;font-size:26;color:black;
   }
</style>
<HTML><body bgcolor = pink id = tom>
<div align = "center">
<p id = tom>查询时可以输入手机的版本号或手机名称及价格.<br>
手机名称支持模糊查询
<br>输入价格是在两个值之间的价格,格式是：价格1-价格2<br>
例如：897.98-10000
</p>
<form action = "searchByConditionServlet" id = tom method = "post">
    <br>输入查询信息:<input type = text id = tom name = "searchMess"><br>
    <input type = radio name = "radio" id = tom value = "mobile_version"/>
      手机版本号
    <input type = radio name = "radio" id = tom value = "mobile_name">
      手机名称
```

```
        < input type = radio name = "radio" value = "mobile_price">
          手机价格
        < br >< input type = submit id = tom value = "提交">
    </form>
</div ></body ></HTML>
```

图 10.10　输入查询信息车

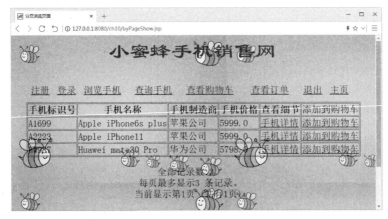

图 10.11　显示查询结果

10.9.2　模型(bean)

本模块的 bean 是 id 为 dataBean 的 session bean(见 10.7.2 节)。

10.9.3　控制器(servlet)

servlet 控制器 searchByConditionServlet(SearchByCondition 类的实例)(见 10.4.3 节的 web.xml)把从数据库 mobileForm 表中查询到的记录存到 dataBean 中,然后将用户重定向到 byPageShow.jsp 页面。

SearchByCondition.java

```
package handle.data;
import save.data.Record_Bean;
```

```java
import java.sql.*;
import java.io.*;
import javax.servlet.*;
import javax.servlet.http.*;
import javax.sql.DataSource;
import javax.naming.Context;
import javax.naming.InitialContext;
public class SearchByCondition extends HttpServlet{
    public void init(ServletConfig config) throws ServletException{
        super.init(config);
    }
    public void service(HttpServletRequest request,
                        HttpServletResponse response)
                        throws ServletException,IOException{
        request.setCharacterEncoding("utf-8");
        HttpSession session = request.getSession(true);
        String searchMess = request.getParameter("searchMess");
        String radioMess = request.getParameter("radio");
        if(searchMess == null||searchMess.length() == 0) {
            response.getWriter().print("没有查询信息,无法查询");
            return;
        }
        Connection con = null;
        String queryCondition = "";
        float max = 0,min = 0;
        if(radioMess.contains("mobile_version")){
           queryCondition =
           "SELECT mobile_version,mobile_name,mobile_made,mobile_price " +
           "FROM mobileForm where mobile_version = '" + searchMess + "'";
        }
        else if(radioMess.contains("mobile_name")) {
           queryCondition =
           "SELECT mobile_version,mobile_name,mobile_made,mobile_price " +
           "FROM mobileForm where mobile_name like '%" + searchMess + "%'";
        }
        else if(radioMess.contains("mobile_price")) {
            String priceMess[] = searchMess.split("[-]+");
            min = Float.parseFloat(priceMess[0]);
            max = Float.parseFloat(priceMess[1]);
            queryCondition =
           "SELECT mobile_version,mobile_name,mobile_made,mobile_price " +
           "FROM mobileForm where mobile_price <= " + max +
                           " and mobile_price >= " + min;
        }
        Record_Bean dataBean = null;
        try{
            dataBean = (Record_Bean)session.getAttribute("dataBean");
            if(dataBean == null){
                dataBean = new Record_Bean();                   //创建 bean
                session.setAttribute("dataBean",dataBean);      //是 session bean
            }
        }
```

```
        catch(Exception exp){}
        try{
            Context context = new InitialContext();
            Context contextNeeded = (Context)context.lookup("java:comp/env");
            DataSource ds =
            (DataSource)contextNeeded.lookup("mobileConn");        //获得连接池
             con = ds.getConnection();                              //使用连接池中的连接
            Statement sql = con.createStatement();
            ResultSet rs = sql.executeQuery(queryCondition);
            ResultSetMetaData metaData = rs.getMetaData();
            int columnCount = metaData.getColumnCount();          //得到结果集的列数
            rs.last();
            int rows = rs.getRow();                                //得到记录数
            String [][] tableRecord = dataBean.getTableRecord();
            tableRecord = new String[rows][columnCount];
            rs.beforeFirst();
            int i = 0;
            while(rs.next()){
               for(int k = 0;k < columnCount;k++)
                   tableRecord[i][k] = rs.getString(k + 1);
                   i++;
            }
            dataBean.setTableRecord(tableRecord);                  //更新 bean
            con.close();                                           //连接返回连接池
            response.sendRedirect("byPageShow.jsp");               //重定向
        }
        catch(Exception e){
            response.getWriter().print("" + e);
        }
        finally{
           try{
              con.close();
           }
           catch(Exception ee){}
        }
    }
}
```

10.10 查询订单

▶ 10.10.1 视图(JSP 页面)

视图部分由一个 JSP 页面 lookOrderForm.jsp 构成,负责显示用户的订单信息。
lookOrderForm.jsp(效果如图 10.12 所示)

```
<%@ page contentType = "text/html" %>
<%@ page pageEncoding = "utf - 8" %>
<%@ page import = "javax.sql.DataSource" %>
<%@ page import = "javax.naming.Context" %>
```

```jsp
<%@ page import = "javax.naming.InitialContext" %>
<%@ page import = "java.sql.*" %>
<jsp:useBean id = "loginBean" class = "save.data.Login" scope = "session"/>
<HEAD><%@ include file = "head.txt" %></HEAD>
<title>查看订单</title>
<style>
    #tom{
        font-family:宋体;font-size:26;color:black
    }
</style>
<HTML><body bgcolor = cyan id = tom>
<div align = "center">
<%   if(loginBean == null){
        response.sendRedirect("login.jsp");              //重定向到登录页面
        return;
     }
     else {
        boolean b = loginBean.getLogname() == null||
            loginBean.getLogname().length() == 0;
        if(b){
            response.sendRedirect("login.jsp");          //重定向到登录页面
            return;
        }
     }
     Context context = new InitialContext();
     Context contextNeeded = (Context)context.lookup("java:comp/env");
     DataSource ds = (DataSource)contextNeeded.lookup("mobileConn");        //连接池
     Connection con = null;
     Statement sql;
     ResultSet rs;
     out.print("<table border = 1>");
     out.print("<tr>");
     out.print("<th id = tom width = 100>" + "订单序号");
     out.print("<th id = tom width = 100>" + "用户名称");
     out.print("<th id = tom width = 200>" + "订单信息");
     out.print("</tr>");
     try{
        con = ds.getConnection();                        //使用连接池中的连接
        sql = con.createStatement();
        String SQL =
       "SELECT * FROM orderForm where logname = '" + loginBean.getLogname() + "'";
        rs = sql.executeQuery(SQL);                      //查表
        while(rs.next()) {
            out.print("<tr>");
            out.print("<td id = tom>" + rs.getString(1) + "</td>");
            out.print("<td id = tom>" + rs.getString(2) + "</td>");
            out.print("<td id = tom>" + rs.getString(3) + "</td>");
            out.print("</tr>") ;
        }
        out.print("</table>");
        con.close() ;                                    //连接返回连接池
```

```
        }
        catch(SQLException e) {
            out.print("<h1>" + e + "</h1>");
        }
        finally{
            try{
                con.close();
            }
            catch(Exception ee){}
        }
    %>
</div></body></HTML>
```

图 10.12　显示订单

▶ 10.10.2　模型(bean)

本模块用的 bean 的 id 为 loginBean,用于验证是否是登录用户(见 10.6.2 节)。

▶ 10.10.3　控制器(servlet)

本模块无控制器。

10.11　退出登录

该模块只有一个名字为 exitServlet 的 servlet 控制器(见 10.4.3 节的 web.xml)。用户单击导航条上的"退出"超链接,请求 exitServlet,exitServlet 负责销毁用户的 session 对象,导致登录失效。

HandleExit.java

```
package handle.data;
import javax.servlet.*;
import javax.servlet.http.*;
import java.io.*;
public class HandleExit extends HttpServlet {
```

```java
    public void init(ServletConfig config) throws ServletException{
        super.init(config);
    }
    public void service(HttpServletRequest request,
                        HttpServletResponse response)
                        throws ServletException, IOException {
        HttpSession session = request.getSession(true);
        session.invalidate();                      //销毁用户的 session 对象
        response.sendRedirect("index.jsp");        //返回主页
    }
}
```

图书资源支持

感谢您一直以来对清华版图书的支持和爱护。为了配合本书的使用,本书提供配套的资源,有需求的读者请扫描下方的"书圈"微信公众号二维码,在图书专区下载,也可以拨打电话或发送电子邮件咨询。

如果您在使用本书的过程中遇到了什么问题,或者有相关图书出版计划,也请您发邮件告诉我们,以便我们更好地为您服务。

我们的联系方式:

地　　址：北京市海淀区双清路学研大厦 A 座 701

邮　　编：100084

电　　话：010-83470236　010-83470237

资源下载：http://www.tup.com.cn

客服邮箱：2301891038@qq.com

QQ：2301891038（请写明您的单位和姓名）

书　圈

扫一扫，获取最新目录

课　程　直　播

用微信扫一扫右边的二维码，即可关注清华大学出版社公众号"书圈"。